History of Meteorology

Mladjen Ćurić • Vlado Spiridonov

History of Meteorology

Springer

Mladjen Ćurić
Department of Meteorology
University of Belgrade
Belgrade, Serbia

Vlado Spiridonov
Institute of Physics, Faculty of Natural Sciences and Mathematics
Ss. Cyril and Methodius University in Skopje
Skopje, North Macedonia

ISBN 978-3-031-45031-0   ISBN 978-3-031-45032-7 (eBook)
https://doi.org/10.1007/978-3-031-45032-7

This Springer imprint is published by the registered company Springer Nature Switzerland AG
The registered company address is: Gewerbestrasse 11, 6330 Cham, Switzerland

Paper in this product is recyclable.

Preface

Welcome to this book, which is divided into three main parts. The first part explores the fascinating journey of civilization's development, while the second part delves into the progress of science. Finally, the third part provides a comprehensive account of the evolution of meteorology as a scientific discipline. Our intention is to offer readers a glimpse into the unifying spirit of science, breaking down barriers between narrow specialties.

We believe that this book offers a captivating subject, particularly given the growing interest in climate change and extreme weather phenomena. It covers a long pre-modern period, shedding light on the lesser-known history of meteorology. Our aim was to provide a clear and detailed layout, offering informative content that encompasses the history of meteorology up until approximately a century ago, culminating with the influential Bergen School. Moreover, we draw connections between this period and modern developments in theoretical meteorology, including topics such as cyclones as baroclinic instability, the quasi-geostrophic framework, Q vectors, coupled ocean-atmosphere dynamics, and numerical weather prediction. These connections are especially relevant in relation to climate variability and change.

The book begins by exploring the earliest stages of planet Earth and its atmosphere. Subsequently, it delves into the early development of meteorology, including the earliest weather records, the teachings of the ancient Greeks, Aristotle's contributions to meteorology, and the state of meteorology during the Dark Ages. A significant portion of the content is dedicated to the origins of quantitative meteorology, the invention of basic meteorological instruments, and the establishment of meteorological measurements and measurement networks. This section also encompasses modern meteorological instruments that utilize radar, satellites, and remote sensing technologies.

In our exposition, we aimed not only to present major events but also to capture the entire spectrum of development. Rather than simply listing historical facts, we interpret them and synthesize the information to enable the knowledge of past events to be utilized in avoiding future mistakes. It was our desire to present the development of meteorology in the most comprehensive and versatile manner possible.

We hope that this book will captivate your interest and deepen your understanding of the rich history and ongoing advancements in the field of meteorology. Enjoy the journey!

Belgrade, Serbia
Skopje, North Macedonia

Mladjen Ćurić
Vlado Spiridonov

Acknowledgments

Preparing a book like this is a tedious and demanding job. But the authors were not alone on this journey—because many people helped complete one such voluminous work that encompasses this important natural science discipline meteorology. For these reasons, we owe great gratitude to the world scientific community, experts, organizations, and individuals who, with their numerous research studies, published articles, and reports, have given essential momentum and impetus in the preparation of this book. On this occasion, the authors would like to express their great gratitude to a world publishing house such as Springer Nature, and the editorial board in New York, which approved the publication of this contemporary book. The authors express special gratitude to Dr. Aaron Schiller, Assistant Editor of Earth Sciences, Geography and Environment at Springer Nature for his great commitment and contribution to the approval and publication of this manuscript jointly with Henry Rodgers as his collaborator and Nivetha M (Ms.), Project Coordinator. The recommendations, suggestions, and valuable remarks given by the anonymous referees are of the highest appreciation.

Finally, the authors would like to thank the families, for their unreserved support, patience, and understanding throughout the years for this hard and delicate work such as writing a book.

Thank you.

Belgrade, Serbia — Mladjen Ćurić
Skopje, North Macedonia — Vlado Spiridonov

Contents

About the Authors

Mladjen Ćurić A Meteorological Luminary Hailing from the picturesque town of Zabljak, Montenegro, Mladjen Ćurić's meteorological journey has been one of academic excellence and impactful contributions. His quest for knowledge commenced at the hydrometeorological school in Belgrade, where his commitment was rewarded with top honors. Graduating with a degree in meteorology from the University of Belgrade was merely the first milestone in a remarkable trajectory that spanned decades. Continuing his academic odyssey, Ćurić pursued postgraduate studies and successfully defended his Ph.D. thesis at the University of Belgrade. His thirst for comprehensive knowledge prompted him to traverse the globe, engaging in enriching study visits from 1977 to 1980. He graced institutions like Colorado State University, Fort Collins, USA, Imperial College in London, England, Manchester University, and The Meteorological Office in Bracknell, where he imbibed diverse perspectives that further shaped his meteorological insights. From 1972 to 1978, Ćurić's dedication led him to the role of Teaching Assistant at the esteemed Department of Physics and Meteorology, Belgrade University. The culmination of his efforts arrived in 1990 when he was accorded the esteemed title of full Professor at the Institute of Meteorology, Belgrade University, a position he continues to hold with honor.

His leadership capabilities shone brightly during his tenure as the Director of the Institute of Meteorology, University of Belgrade, from 1979 to 1982. Further contributing to academia, he served as Vice Dean in the Faculty of Physics, University of Belgrade, from 1982 to 1984, leaving an indelible imprint on the academic landscape. A global perspective enriched his expertise, as evidenced by his role as a Member of the Executive Committee of the International Commission on Clouds and Precipitation-IAMAP from 1988 to 1996. His commitment to environmental stewardship was evident through his membership in the Executive Committee of the National Association of Environment Protection, which he joined in 1990. His dedication to the academic realm is reflected in his role as Vice Dean at the Faculty of Physics during two pivotal periods: from 1996 to 1998 and later from 2004 to 2007. These administrative responsibilities were seamlessly balanced with his unwavering commitment to his professional and scientific interests. His expertise spans a spectrum of fundamental meteorological subjects, including Atmospheric Dynamics, Cloud Physics, Applied Meteorology, Hydrology, Weather Modification, and Environment Protection.

Mladjen Ćurić's contributions extend beyond the confines of academia. He has authored several influential books and penned or co-authored more than 200 papers. His scholarly endeavors have graced internationally renowned journals such as the *Journal of the Atmospheric Science*, *Quarterly Journal of the Royal Meteorological Society*, *Tellus*, *Journal of Applied Meteorology*, and *Atmosphere-Ocean*, among others. The meteorological community continues to benefit from Mladjen Ćurić's indomitable spirit, profound insights, and enduring dedication. His journey embodies the pursuit of knowledge and the empowerment of future generations in the realm of meteorology.

Vlado Spiridonov Dr. Vlado Spiridonov hails from Skopje and has left an indelible mark in the field of atmospheric science through his illustrious career. His educational journey commenced with primary schooling at "Grigor Prlichev" and high school at "R.J. Korchagin." His thirst for knowledge led him to pursue higher education at the "Institute of Physics," situated within the Faculty of Natural Sciences and Mathematics at "St. Cyril and Methodius University in Skopje" (UKIM). Dr. Spiridonov's academic prowess and dedication to meteorology were evident as he pursued his specialized, master's, and doctoral studies at the esteemed Institute of Meteorology within the Faculty of Physics at the University of Belgrade. His quest for excellence was further recognized in 2004 when he was awarded a prestigious postdoctoral fellowship by the Government of Canada. Even as his international recognition grew, Dr. Spiridonov's commitment to his homeland remained unwavering. In a remarkable confluence of events, he was appointed as the Director of the Hydrometeorological Administration and concurrently as the Permanent Representative of Macedonia to the World Meteorological Organization (WMO) in 2004. Dr. Spiridonov's academic stature continued to ascend. In April 2007, he was elected as the Assistant Professor of Meteorology. His scholarly eminence was acknowledged with successive appointments as Associate Professor of Meteorology in 2012 and 2017, further cementing his reputation as a leading authority in the field. His influence extended beyond the classroom. Dr. Spiridonov's appointment as a representative of the Faculty of Natural Sciences at the National Council for Climate Change in Macedonia reflected his commitment to addressing global environmental challenges. He also became an esteemed member of the Commission on Atmospheric Sciences at the WMO, showcasing his international recognition and influence. The academic world was enriched by his presence as a Visiting Professor of Meteorology at the University of Vienna, Austria, from 2017 to 2019. This tenure not only underscored his expertise but also fostered international academic collaborations. In 2020, Dr. Spiridonov's legacy evolved once again as he was bestowed the honorable title of Visiting Professor at the Faculty of Computer Science and Engineering (FINKI). His meteoric rise culminated in his election as a Full Professor of Meteorology and Atmospheric Physics in 2021, a momentous honor bestowed by the Faculty of Natural Sciences and Mathematics at the Institute of Physics, "Ss. Cyril and Methodius University," Skopje, Republic of North Macedonia. Dr. Spiridonov's impact transcends borders, resonating globally. His dynamic contributions to the advancement of atmospheric science have been exemplified through his participation in international scientific conferences, his organization of global events and symposia, and his pivotal role as a plenary session lecturer and seminar speaker worldwide. His scholarly legacy endures through numerous scientific papers published in esteemed international journals, as well as contemporary meteorological books that continue to enrich the field. This academic journey has been further adorned by awards and recognitions, including the prestigious "Patent of the Year-2000" in Macedonia, a gold medal with distinction at the EUREKA-2000 exhibition in Brussels, and a Genius award from the Hungarian Association of Innovations. His pinnacle achievement, the GRAND PRIX at the Fourth International Exhibition of Inventions in Budapest GENIOUS Prize 2002, attests to his groundbreaking method related to fog scattering. Dr. Vlado Spiridonov's exceptional contributions have left an enduring legacy in atmospheric science, fostering global collaborations, innovation, and a commitment to understanding our past and current weather and climate and preserving our planet's climate system.

Chapter 1
The Earliest Past of the Earth and the Atmosphere

1.1 Introduction

Scientists are continuously refining their understanding of Earth's geological history. According to the latest theory of plate tectonics, our planet's land is viewed as a dynamic and pulsating "living body" that has undergone significant changes throughout the ages. Continents have shifted, oceans have risen and fallen, mountain ranges have formed and eroded away, and the composition of the atmosphere has fluctuated. These changes have led to strong climate variations that have played a crucial role in the development and extinction of various forms of life on Earth.

Over the past few decades, scientists have made significant progress in understanding the underlying causes of geological changes. Specifically, they have identified a layer of semisolid rock called the asthenosphere, which is located between 5 and 70 km beneath the Earth's surface and is responsible for the slow mixing and movement of the mushy rocks in this region. The constant mixing that takes place in the asthenosphere breaks the thin surface layer into rocky parts of different sizes called tectonic plates. Parts of those plates are our continents, whose current forms we are quite familiar with. These huge plates (called lithospheric) float and move on that gently boiling mushy mass. That movement is only a few centimeters per year. With this movement, the plates collide, come together, and separate again. As a result, volcanoes, earthquakes, and the most eruptive (and most disturbing) phenomenon of all appear—the ice age. Strong and sudden changes caused by earthquakes and volcanoes are isolated episodes of limited range. Ice ages lead to global consequences with a strong impact on the entire living world of the planet. In the past, huge ice sheets were formed that covered one-third of the land. When the ice sheets grew and spread toward the Equator, the living world was pushed into the narrow areas. According to the new climatic conditions, the living world moved, adapted, or died out. For the last 3.5 billion years, climate change has caused the complete or partial destruction of the Earth's flora and fauna. About 65 million

M. Ćurić, V. Spiridonov, *History of Meteorology*,
https://doi.org/10.1007/978-3-031-45032-7_1

years ago, the vast land mass of Siberia, North America, and Antarctica moved poleward. Land has a lower heat capacity (can hold less heat) than the ocean. That's why the temperature in those areas dropped. The formed snow and ice cover around the poles reflected solar energy. This caused further cooling. Thus, about 15 million years ago, an ice sheet formed in Antarctica, and 7 million years ago, over a large part of the northern hemisphere, an ice sheet began to grow. This marked the beginning of the Pleistocene ice age about 2 million years ago. The human species began its development about 3–4 million years ago. Since then, the ice sheet has repeatedly descended to the central parts of Europe and North America. Thus, only 18,000 years ago, a significant part of Europe and North America was covered by an ice sheet. The ice age had a strong impact on the living world. Our ancestors successfully adapted to those conditions. He developed both physically and behaviorally. There was an increase in the brain, a bipedal locomotor system developed, as well as hands, which are used to perform more and more complex and delicate actions. It becomes undoubtedly the dominant animal species. In order to survive, ancient people hunted, foraged, and gathered food. The flora and fauna that they used were under the decisive influence of the climate. Human migrations and habitats were also determined by climate. They stayed in one place, until they used up the available resources. Then they would move to another location. They developed a quasi-nomadic way of life. The climate was a crucial factor in the development of early civilization. The generous tropical climate encouraged the development of a more advanced civilization, while the harsh climatic conditions demanded a constant struggle for survival. Seasonal temperature (and generally weather) changes have influenced further social and technological progress. Climate change is influenced by both anthropological and geological factors. Tectonic forces move the continents into different climate zones. They also create mountains around which airflow changes, and soil exposure to solar radiation, all of which produce local climate. Small changes in ocean temperature have an impact on various populations of simple living organisms. They, in turn, change the composition of the atmosphere, which, again, has the effect of changing the climate. Human activity also changes the chemical composition of the atmosphere. All this is reflected in short-term and long-term climate changes. These changes are gradual and are not felt in the weather that we follow from day to day. This part will briefly describe the creation of the Earth, its cooling, and the gradual creation of beneficial conditions that led to the creation of a planet with water and an atmosphere. In those conditions, the living world developed.

1.2 Meteorology of the Universe

Earth was formed from a chaotic, disorganized cloud of dust and gases (Abbe 1910). Like jewels scattered on a black canvas, millions of galaxies shine in the night sky, each one made up of billions of stars. Some are immensely large, others moderate in size. Galaxies can be classified into three basic shapes: ovoid, spherical, and the most visually stunning of them all, spiral. The largest number of galaxies, 80%, is spiral in shape. About 5 billion years ago, on one spiral arm of the Milky Way galaxy, a cloud of cosmic dust floated, created during the explosion of a star. For galactic proportions, this transparent lump rotated slowly, accumulating mass around its center. Over time, this mass increased at the expense of smaller parts from its surroundings, which were attracted by its gravitational force. The central mass grew so large, and with such great density, that its gravity attracted even the most distant particles. The cloud gathered around the central mass. As the cloud gathered, the speed of its rotation increased, similar to a skater who gathers her arms to her body while rotating. As the particles fell on it faster and faster, the central mass flattened its surface. Particles that fell at a lower speed remained rotating around the created nucleus. The application of new material to the growing central mass caused friction, which increased the temperature inside. When the temperature exceeded the critical value, the mass began to glow with a reddish color. Hydrogen atoms that were forcefully ejected into the interior united and created a thermonuclear chain of reactions. The color of the core of this once transparent cloud of dust and gas changed from ruddy, from bright red to bright white. Once, when an explosion occurred, a star, our Sun, was formed, which has been radiating enormous energy into the surrounding space ever since. Several smaller conglomerates that escaped the attractive force remained orbiting the Sun at arbitrary distances from its center. And they continued to collect the remaining material from their vicinity. Thus, lumps with significant mass were created. One of them is Earth.

1.3 The Beginnings of the Formation of Land, Sea, and Air

The strong accumulation of material on Earth was also accompanied by an increase in temperature. But, unlike the Sun, Earth has never accumulated enough mass to start a thermonuclear reaction. However, the increase in temperature contributed to the melting of the inner part. The heavier elements of the molten material, iron and nickel, were attracted by gravity to the center of the planet, while the lighter elements, silicates and carbon, floated toward the surface. Many of them came to the surface in the form of gas. That's how the atmosphere was created. Among other materials, it was composed of water vapor, carbon dioxide, and methane.

Due to the rise of water vapor, a cloud layer up to 20 km thick was formed in the atmosphere. Such a thick cloud layer prevented solar radiation from reaching the Earth's surface. Carbon dioxide and water vapor did not allow the radiated heat from the Earth to go into space. It also rained, but the rainwater would evaporate quickly due to the very warm surface.

Under the pressure of the outer layers and due to the decay of radioactive material, enormous heat was released into the interior of the Earth. Since this heat could not be transferred to space by radiation, the Earth began to glow with a dull red light. Thus, the entire mass of the Earth dramatically melted. The liquid material formed under the influence of gravity into an almost ideal spherical drop composed of liquid rocks that floated above the liquid metal.

The intense release of heat into space, through the proto-atmosphere, started the cooling of the Earth. Thus, from brittle rocks, thinner or thicker islands were formed, which swam in a huge sea of liquid lava. By swimming, some islands merged into larger land masses, thus forming vast continents. During these processes, another atmosphere is formed from water vapor and other gases that left the Earth's interior.

The high temperature of the atmosphere was maintained for a long period of time. A large amount of water vapor and carbon dioxide did not allow the Earth to cool, because they absorbed the heat emitted from the Earth. The surface air temperature could reach between 60 and 90 °C. At higher altitudes, the temperature was lower, which favored condensation. A continuous thick layer of clouds was formed that enveloped the entire Earth. As before, rain fell from the thick cloud layer, which evaporated before reaching the ground due to the high temperature.

Nevertheless, the Earth gradually cooled. Over time, the rains fell on the newly formed Earth's crust. Heavy rains and strong winds associated with thunderstorms crushed and displace soil in higher-altitude areas. Collected water, in the form of rivers, carried minerals taken from the ground to lower areas, thus forming lakes and seas. Hot gases from the interior of the Earth, which were saturated with minerals and nutrients, pushed the underground water through cracks to the bottom of the sea. A rich "soup" of very complex nucleic acids, sugars, phosphates, and proteins was created from nutrients from land and sea, under the influence of heat from the Sun and volcanoes. These nutrients were suitable for the formation of primitive life forms, especially in the shallow waters of lakes and seas. So, after a billion years of "birth pains," the Earth provided itself with the resources needed to create the first forms of life.

1.4 The First Life on Earth

Over a period of a million years, while stormy rains fell, a primordial "soup" of chemicals, minerals, and simple molecules formed in the resulting seas. To this day, it has not been reliably determined how the original life form arose in such a reservoir of ancient elements. There are various theories.

Most researchers agree that our companion, the Moon, formed 5 billion years ago from fragments of a similar composition to Earth. A billion years later, when life is thought to have appeared, the Moon was closer to Earth than it is today. During the Pleistocene epoch, in addition to vast ice sheets covering much of the Earth's surface, the planet rotated faster than it does today. This caused sea waves to last from 2 to 6 hours and tides to cover several kilometers of land. Because of this, the salinity of the sea changed rapidly. It was this change, according to one theory, that caused double strands of DNA-like molecules to join and separate.

At high tide, the salt concentration was low. Double-stranded DNA disintegrates under such conditions, because the charged phosphate compounds on each strand repel each other. But at low tide, the concentration of molecules and salts was very high. This encouraged the development of double strings of molecules, as the high salt level neutralizes the phosphate charge and prevents the strings from separating. Sea currents provided the energy needed to join and separate the polymers.

Many researchers do not believe that DNA was the first molecule to multiply (replicate). There are interpretations that simpler genetic material was created first by crystallization due to the tide. But whatever the first replicating molecules were, they must have existed in a changing environment. By the way, if this theory is correct, life could not have developed on Mars, even if there was water on it. Because, larger than the two moons of Mars, it is so small that the sea moons it causes are only 1% of those caused by our Moon.

For meteorologists, as well as the largest number of researchers who deal with this, the meteorological trigger theory is more reliable for the origin of the living world. According to her, in the gloomy, accumulated clouds of the old atmosphere, constant electrical discharges were carried out. The energy of those discharges in the presence of raindrops generated proteins and simple molecules in the air, which then fell to Earth with rain and accumulated in the oceans. Chemical reactions that took place in the abundance of inorganic substances created biological molecules with more complex nucleic acids, sugars, phosphates, and proteins. They were the "building material" of life. During the abundance of inorganic substances, chemical reactions occurred, producing prebiological molecules composed of complex nucleic acids, sugars, phosphates, and proteins. These molecules served as the fundamental building blocks of life.

Volcanic activity occurred much more frequently 3.5 billion years ago. In the vicinity of the eruption of the volcano, organic matter appeared on the seabed. From them, with the catalytic heat of the volcano, life could be created. Even today, at the bottom of the ocean, on the composition of the tectonic plates, minerals are pumped through cracks. This self-renewing factory of chemicals and nutrients creates an oasis of life with the help of "underground" heat.

Regardless of what served as a catalyst, solar radiation, heat from the depths of the sea, tidal energy, or some other agent, simple single-celled structures—prokaryotic cells—were created from simple organic molecules. The first life forms on Earth originated from there.

1.5 Changing the Composition of the Atmosphere

The original atmosphere was significantly different from today (Brimblecombe 1977; Asimov 1985). It contained a large amount of carbon dioxide and very little free oxygen. This type of composition, no matter how unsuitable it seems today for most of the living world, represented a very hospitable environment for the development of life at that time.

Oxygen is a highly corrosive gas, both in the present day and in the distant past. Oxygen is a highly reactive gas that readily reacts with almost everything it comes into contact with. After its formation, it reacted with minerals in the earth and rocks in the water, leaving little free oxygen and ozone in the atmosphere. The atmosphere at that time did not contain as much oxygen as it does today. This was crucial for the development of the living world, as the embryonic form of life would have been destroyed immediately after creation if the atmosphere had contained as much oxygen as it does now.

The high concentration of carbon dioxide in the atmosphere created a greenhouse effect that led to elevated temperatures in the lower atmosphere and oceans (Allègre 1988). Temperatures were between 60 and 90 °C. That, along with the additional catalyst, did not favor the development of more complex life forms, such as algae and bacteria.

A billion years after the emergence of the first forms of life, the Earth cooled enough, and the uniform thick cloud layer broke up and disappeared, allowing the Sun to shine on the Earth's surface. This created the conditions for the next significant advancement in the development of life on Earth: the emergence of photosynthesis. With the Sun's energy now reaching the surface, early life forms were able to harness this energy and convert it into food, leading to the emergence of more complex and diverse forms of life.

Earlier forms of life, without the presence of sunlight, survived by using nutrients and minerals from water, thus creating sugar, through the process of fermentation. When clouds gave way to sunlight, a type of blue-green bacteria (cyanobacteria) adapted to use sunlight to separate water and carbon dioxide molecules. Thus, high-energy glucose was created, with the release of oxygen in the photosynthetic process.

On Earth, 500 million years later, there was an abundance of cyanobacteria, organisms capable of creating their own food from small molecules, using carbon dioxide from the air and energy from the Sun. Having no natural enemies, the population of cyanobacteria increased without limit. However, the oxygen content increased, so the atmosphere became a poisoned environment for existing life forms, including cyanobacteria.

As the oxygen content increased, the bacteria disappeared or retreated to environments where oxygen did not reach. Thus, cyanobacteria managed to spectacularly change the balance of oxygen and carbon dioxide in the atmosphere but also to destroy their natural habitat. They also changed the conditions of evolution forever.

1.6 New Atmosphere and New Forms of Life

The increased content of oxygen in the air did not favor the survival of the first forms of life (Dickerson 1978). Later forms of life tolerated a more pronounced presence of oxygen, even actively using it in their metabolism. Thus, in a period of a billion years, the interdependence of animals that use oxygen and release carbon dioxide and plants that consume carbon dioxide and release oxygen begins. The ozone layer, formed in the stratosphere, serves as a filter for harmful ultraviolet radiation, making the ground environment favorable for the living world.

By reducing carbon dioxide, the Earth cooled down to a tolerable level. After the next billion years, more perfect life forms developed. Eukaryotic cells appear, based on photosynthesis and an abundance of oxygen, which created previous (prokaryotic) forms. The new cells contained the genetic material that allowed ancestors to pass improved traits on to descendants. More diverse forms of life appear. This makes them more sensitive to changes in the climate, and, in general, the environment.

The most developed organisms require a more stable environment. The country does not provide such stable conditions. Several times in the last 550 million years, catastrophic changes occurred that destroyed almost all eukaryotic life forms, while adaptive bacteria remained almost without consequences.

1.7 The Ice Age and the Development of Primitive Man

1.7.1 Climate Disasters

Of all the disasters that occur in the biosphere (which consists of the lithosphere, the hydrosphere, and the atmosphere), the ice ages leave the greatest devastation (Newson and Richerson, 2021). Volcanic eruptions, earthquakes, and floods are powerful destructive phenomena, but in terms of force and form, the destruction cannot be compared to the ice ages.

The most recent ice age cycle, known as the Pleistocene epoch, began about 2.6 million years ago and ended around 11,700 years ago. During this time, vast ice sheets covered much of the Earth's surface. The continents were not yet arranged in their current positions, and the land masses were gradually moving through the process of plate tectonics. It was formed in the vicinity of the Equator, so it did not experience an ice age. Over time, Pagania was divided into several parts (plates). The largest plates make up today's continents, the appearance of which we are familiar. About 65 million years ago, tectonic plates moved much of the landmass of Antarctica, Siberia, and North America poleward. That part of the land gave off more heat than it received from the Sun. The temperature dropped so much that the accumulated snow during the winter could not melt in the summer. Thus the snow cover increased. Under the weight of the upper layers, the lower layer of snow turned into ice. This begins a long period of cooling, known as the Cenozoic climatic divergence.

The formation of the Antarctic ice cap occurred 15 million years ago, while small glaciers began appearing at higher altitudes in the northern areas of North America approximately 10 million years ago. The growth of the North Pole ice cap began around 7 million years ago, and by approximately 1.5 million years ago, it had expanded to cover Greenland. Huge amounts of accumulated snow and ice reflected the heat of the Sun, so that the Earth was rapidly sinking into the abyss of Pleistocene glaciation.

Since the beginning of the Pleistocene ice age, about 2 million years ago, the Earth has been affected by four extreme periods of glaciation that have left their own traces, which have been well studied by science. It has been established that in Europe, the edge of the glacier has descended further south than the London-Moscow line. Smaller glaciers formed in the Pyrenees and the Alps. From there they spread further north, leaving between the ice only narrow areas of tundra and stunted conifer forests. Similarly, in North America, glaciers descended as far as the Cincinnati-Philadelphia line.

In each period of the onset of glaciers, the existing living world both on land and in the sea retreated, and in the period of melting returned and filled, the areas left by the ice.

1.7.2 Climatological Yeast of Life

People today live almost everywhere on Earth. They even occasionally know that they reside in the vastness of space or in the depths of the ocean. But the predecessor of today's man lived only in Africa. Precisely the climatological factor was the leaven to that life.

The African continent and Europe/Asia began to separate approximately 25 million years ago. At that time, the jungle's lush vegetation extended well beyond the Congo Basin, covering large areas to the north and east. When the African plate assumed its current position, most of the northern continent was under the influence of subtropical high pressure, dominated by the Azores' anticyclone. The warm downward movements of the air did not allow the formation of clouds, except for shallower and lower clouds.

Areas with dry downdrafts were without precipitation. This is how the largest Saharan-Arabian desert complex was formed. This vast open space separated the green forest jungle of equatorial Africa from the Mediterranean Sea and the fertile Iraqi-Iranian plains. Even the Canary and Azores Islands had an arid (dry) climate, with the exception of mountainous areas, where there was precipitation caused by orography.

The Pleistocene climate was cooler and drier. In equatorial Africa, the cradle of our predecessor, large areas of lush vegetation began to wither, to dry up, due to lack of rain. Rivers and lakes dried up, and forests were gradually replaced by grass and bushes. Deserts from the north and south surrounded the area. Plants and animals were forced to adapt due to a lack of water and food.

During this period, a human-like being, in search of food, climbed the mountains of Ethiopia. The forerunner of *Homo sapiens* successfully and adaptively coped with chaotic changes. Pushed out of the forest area into the grassy plateaus of the savannah, it develops both physically and "mentally." Inevitably, there is an increase in brain mass and more and more purposeful and meaningful functions of the hands. After a million years of meta-morphosis, the first "real" man appears and develops: *Homo habilis*.

Homo habilis uses stones as simple tools, makes shelters when hunting, and searches for and collects food in East Africa. It existed for about half a million years when it "gave way" to a much more developed species, *Homo erectus*.

Homo erectus is an extremely successful predecessor of man. It survived for a million years, spreading throughout Africa, including Asia Minor, China, Indonesia, and Europe.

1.7.3 Climatic Migration of Early Humans from Africa

Anthropologists are puzzled and still searching for an answer to the question of why the first man emigrated from Africa. It seems that they are satisfied with the statement that he did it, because he wanted to and could. Wanted—to improve living conditions; could—because climatic factors enabled him to migrate, without whose suitability, despite his desire, migration would not have been possible.

For hunting, searching, and gathering food, a larger territory was needed. Population growth, rivalry, and the struggle for new living space forced nomadic groups to migrate north and south. Some groups of *Homo erectus* lived on the edge of existence along the southern border of the Sahara. They could not migrate, because they knew that even if they brought some food and water, it would not be enough for them to survive on the long desert road.

For thousands of years, erectus has followed how the greening of the desert moves to the north in the summer, due to the summer monsoon rains, and retreats to the south, in the fall, when the rains stop. And they moved like that. Thus, *Homo erectus* was probably the world's first observer and weather forecaster. They carefully followed the signs that would announce when and in which direction they should move.

About a million years ago, the Earth was gradually engulfed by the first of the four Pleistocene ice ages, called the Ginz. As part of these dramatic climate changes, the climate of North Africa changed. Monsoon rains started earlier, lasted longer, and covered a wider area toward the north.

The desert was slowly receding before the coming curtain of beneficial rain. The desert areas, previously covered only in sand and stone, now grow grass and bushes. Oases and dry riverbeds filled up, and Lake Chad reached the size of today's Caspian Sea. The animals rushed to the area rich in herbs, and the people followed them.

Wandering northward, our predecessor, the erectus, encounters the changing of the seasons for the first time. Both summer and winter were colder. Thunderstorms often came from the west. It was impractical and even impossible for them to move to the tropics during the winter season. This forced Erectus to find fire, use clothing, and make shelters. Thus, he became independent of seasonal weather changes.

Trying to figure out the character of the weather to come, our predecessor probably started looking at the sky, trying to figure out the character of the weather to come. Over time, they recognized which clouds do not bring bad weather, and which foreshadow a terrible storm, so they should return home or find a new shelter and light a fire there. Individuals in the tribe who observed in more detail than others came forward. They had the task of observing the weather and issuing warnings of potential danger. They accumulated knowledge and ideas about the weather.

There is no doubt that these, the first forecasters of the weather, suffered from the same problems as modern forecasters. One can imagine what happened when the members of the tribe were warned that a storm was coming and that they should stop hunting, leave some rich bush with lots of fruits, and return home in the middle of the day. Probably even then, there were jokes about the incompetence of people who deal with keeping weather. The jokes could have been something like this: "When the forecaster says it will rain, we will go hunting and gather food, and when he says it won't rain, we will stay in the cave by the fire, prepare clothes, make tools, and draw pictures on the cave walls."

One day a million or so years ago, among the coniferous trees, a small group of people saw an enormous amount of water. They discovered the Mediterranean Sea. A favorable change in the global climate caused a green blanket to be created over the Sahara desert sands on which the early erectus hunted and gathered food and moved to new worlds.

Between the two ice ages, there was a warm period, similar to the one we live in now. Then the glaciers retreated, restoring desert conditions in North Africa, and closing the climatic door to the Middle East. Migration from and to Central Africa was postponed for some time. Animals and people in the face of the threat of the desert in recovery move to the tropics, or north, to the Mediterranean Sea. People who migrated north spread through habitats along the Mediterranean and crossed into the Middle East, Asia, and Europe. A different environment compared to the tribesmen who remained south of the Sahara conditions the somewhat different development of the northern group of people.

1.7.4 Weather and Life in the Ice Age

The third and last species of the human race, *Homo sapiens*, developed in Africa about 100,000 years ago. Then the Earth was engulfed by the coldest part of the Pleistocene, which is called the Wyrm. About 60,000 years ago, in the area of today's Sahara, a completely modern man lived. He found a technical device, called Alteria (first found in El Alter, eastern Algeria).

During the Wyrm ice age, the temperature dropped about 10°, and the ice cap reached enormous proportions. During the strongest phase, a continuous ice cap descended under the Great Lakes in North America, in Europe south of England and Denmark, and icebergs filled the North Atlantic, between Newfoundland and England. In the Southern Hemisphere, the Antarctic ice cap has expanded by over 150 km. Over time, the Antarctic ice merged with the Southern Andes ice in Patagonia.

With intense cooling in the eastern part of North America and western Europe, deep valleys of cold air were established, while a weak ridge of pressure was established over the relatively warm North Atlantic. In America, there were constant intrusions of Arctic air, all the way to the Gulf of Mexico. The mixing of that, cold and dry, with moist and warm air, created explosive processes. A series of storms formed in the Gulf of Mexico. They moved eastward and quickly invaded Spain and North Africa.

During the ice age, especially in the winter, strong westerly winds blew much closer to the Equator than today. Because of this, the Atlantic anticyclone weakened, moving a little further south. As a result, the downward movement over North Africa also weakened, so the storm clouds were very deep and brought a large amount of precipitation to the interior of West Africa. With each passing of such clouds, rivers, and lakes would be filled with water. Vegetation developed deeper in the mainland of northern Mauritania, Mali, and western Algeria.

In Europe, the situation was somewhat different: the Alps, the Pyrenees, and the central mountain massif in France represented an obstacle to the penetration of cold continental air toward the Mediterranean. However, then, as well as today, that prodor was carried out through the Rhône and Carcassonne valleys. Thus, intensive development of cyclones was carried out in that area of the Mediterranean, which during the ice age was more pronounced than now. A series of cold fronts moved from there toward North Africa, bringing rain deep into the deserts of today's Algeria, Libya, and Egypt.

In the Southern Hemisphere, strong westerly winds blew across the Andes, establishing a high-altitude valley east of Rio de Janeiro. At the same time, a ridge of pressure was established over the South Atlantic. The ridge of the subtropical high pressure was weaker, so it was moved somewhat north and east. This caused a stronger flow of moist tropical air across the Equator to North Africa. Weakened downdrafts in the subtropical area of North Africa allowed the monsoons in that area to bring rain to the area of present-day Chad, Niger, Mali, and Mauritania.

Increased rainfall has boosted vegetation in North Africa. At the same time, lower temperatures contributed to a reduction in evaporation from the soil. The soil has accumulated moisture. That accumulated moisture served as an additional source of enrichment for storms coming from the west.

During the Wyrm ice age, cooler and warmer periods alternated. Recent analyses of the ice layers in Greenland have shown that there have been six to seven such shifts in the last 100,000 years. Paleoclimatologists claim that there were similar oscillations in temperature during other ice ages.

Some connections between rainfall in North Africa and the onset of ice ages are speculative in nature. However, archaeologists have discovered that *Homo sapiens* inhabited a large area of the Sahara for a long period of time.

Although it seems paradoxical, the destructive Pleistocene ice age created a motivating force that so markedly improved life on our planet.

1.8 Climate and the Development of Civilization

The predecessor of sapiens was a passive component of nature as long as he lived in suitable areas of Africa, the Middle East, Europe, and Asia. He adapted to the given environmental conditions as much as he could. He hunted, gathered wild food, and acclimated to the climate as best he could.

With the continuous use of hands over a period longer than a million years, "technological" progress was made very slowly. Until the appearance of *Homo sapiens*, a gifted and dynamic man, with his innovative brain, nothing significant changed in the making of tools, shelter, or hunting techniques. Even the Neanderthal used only ordinary tools.

Archaeologists have established that modern man appeared around 120,000 years ago when he began to use more modern technology. Fifty thousand years ago, during the Wyrm ice age, it began to move from North Africa to the Middle East and other areas to the north and east. In the period between 40 and 35,000 years ago, *Homo sapiens* moved rapidly across Europe and Asia. About 25,000 years ago, he changed all previous forms of living.

In the interglacial period, Emien, 130 to 80,000 years ago, the climate was similar to today. As the glaciers receded, life returned to the Eurasian steppes. Taking advantage of the absence of snow and ice, the Neanderthal moved in all directions during that 50,000-year-long period. With the reappearance of the glacier, about 30,000 years ago, it disappeared from Europe. Only one group of Neanderthals remained in the Middle East and lived together with *Homo sapiens* for a while.

Sapiens first inhabited Europe approximately 35,000 years ago. He survived the long cold period of the Wyrm. With his satisfactory social skills, he used natural resources. Using more modern techniques, in the wide-open steppes and tundra of Europe, he hunted from herds of bison, deer, horses, and mammoths. About 26,000 years ago, *Homo sapiens* established more or less permanent communities. They made weapons from bones and left food for the winter.

The conditions of their life changed dramatically when the ice melted, about 18,000 years ago. Water from melted glaciers flooded the open pastures. Some of those Neolithic inhabitants of Europe moved northward, following reindeer herds. Some stayed and changed their diet. They used fish, poultry, and wild herbaceous foods.

When the climate became milder, people grouped themselves into larger communities. They introduced a hierarchy in decision-making, established trade routes along the rivers, cultivated the land, and hunted and fished.

After the end of the Wyrm ice age, about 12,000 years ago, there were already new people who were ready to face the new challenges of the warm Holocene. They replaced the precarious life of hunters with safer farming.

At first glance, it would seem that the area of Asia Minor, the Near, and Middle East is not suitable for farming. The whole area is under the influence of dry continental air flowing from Eurasia. A slight anticyclonic lowering of the air suppresses the development of clouds, and there is no precipitation. Thus, desert and semidesert are maintained on the coasts of today's Syria, Lebanon, Israel, and Jordan. That area is called the Levant.

The Levant region occasionally experiences cyclones from the Mediterranean during the winter season. They are followed by light rains, especially in higher mountain areas. Practically, from April to September, or mid-October, there is no rain. Thus, without water, there is no vegetation, except on the narrow Mediterranean coast.

The Levant region had vastly different conditions approximately 15,000 years ago compared to the present day. There was a persistent area of high pressure over the southern parts of the Eurasian Ice Sheet. This forced the cyclones from the Mediterranean to move much further south than today. Even in the summer, cyclones formed in the eastern Mediterranean, moving across Asia Minor. It was colder and rainier in that area then. Rains also fell in summer, and in winter, there were plenty of them. The ground is green, with an abundance of plants and animals. At that time, a group of hunters living in the Levant, the Natufians, established a high level of organization. They used different methods of using natural resources. They were engaged in farming as well as animal domestication. But at the end of the Wyrm ice age, the thaw reclaimed the deserts.

Squeezed between the mountains, from the north, and the desert, from the south and east, people were forced to engage in production, not food gathering. A suitable location, fertile soil, and warm climate made it easier for them to switch to an agrarian way of life. From there all the way to the Nile Delta and Persia, grains were created from wild grass, and tame sheep, goats, and cattle were created from wild animals.

The mild climate of the eastern Mediterranean enabled the rapid development of plants that could grow even in winter, thanks to the cool and moist air from the Mediterranean. The plants would mature even before the start of the warm wind from the east and Arabia.

In time, the wiser tribes saw the advantage of staying closer to such fertile areas. Thus, around 12,000 years ago, the first permanent settlements were born with stone houses and paved areas for storing and preparing food. The mild Holocene climate was the leaven for the development of civilization. Early civilization did not arise in the higher regions, where a seminomadic way of life was represented. Man, under the pressure of a dry climate, formed cities and developed the first civilization along the rivers. The rivers provided the necessary amounts of water and food. The surrounding soil was not drained but was constantly fertilized by river silt. The four main Eurasian civilizations developed in semiarid areas, on the alluvium of the Nile, Tigris, and Euphrates, and Indus and Van Hoa rivers.

The Sumerians were pioneers in the development of cities, but their rival, Egypt, was also making progress. They invented many new tools and techniques, such as metalworking, the wheel, ships, and, most importantly, a form of writing.

References

Abbe, C. 1910. *The mechanics of the earth's atmosphere*. Washington: Smithsonian Institution.
Allègre, C. 1988. *The behavior of the earth – Continental and seafloor mobility*. Harvard University Press.
Asimov, I. 1985. *How did we find out about the atmosphere*. Walker & Co.
Brimblecombe, P. 1977. Earliest atmospheric profile. *New Scientist* 76: 364–365.
Dickerson, R.E. 1978. *Chemical evolution and the origin of life*. Scientific American.

Chapter 2
Brief General Historical Overview

2.1 Climate as a Development Factor

In accordance with the chosen concept of first exposing the most important factors that enabled the development of science, climate (without any professional bias) surely deserves first place (Lee 1952; Charles 1959; Allman and Wagner 1992; Asimov 1982; Van Doren 1991). Because wherever and whenever the development of the living world began, it was determined decisively by climatic factors. The development and improvement of man and his skills are significantly directed by the hand of the climate.

Despite the difficulties, scientific methods broke through and shed light on the deep darkness of the past (Dampier 1935). Climatic movements of long past times have been reconstructed. According to astronomical theories, primarily by scientist Milutin Milanković, through the analysis of fossil remains and other geological methods, it was established that ice ages alternated on Earth. Milanković showed them on his insolation lines (Fig. 2.1).

A more detailed process of reconstructing the average global temperature on Earth, from the Ice Age to today and the forecasted process for the next 25,000 years is shown in Fig. 2.2. According to the astronomical theory of ice ages, a cooling should occur in the future that would lead to an ice age of 23,000 years. Then global temperatures should be about 6 °C lower than today. This, of course, did not take into account the effects of warming during the interglacials, which is called super interglaciation and is conditioned by the increased concentration of gases from the "greenhouse" group. If this additional warming were to occur, average global temperatures would be several degrees higher than in the past million years. In these conditions, the cooling tendency that would lead to a new ice age would be delayed by about 2000 years, until the aforementioned warming effects are exhausted (Dawson 1994). The loss of this effect will occur due to the depletion of fossil fuel reserves (and thus the accompanying production of carbon dioxide). In addition, the

M. Ćurić, V. Spiridonov, *History of Meteorology*,
https://doi.org/10.1007/978-3-031-45032-7_2

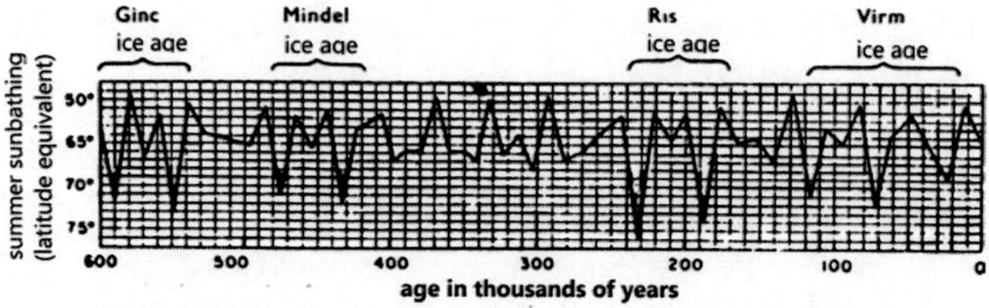

Fig. 2.1 Milanković's insolation curve for the 65th degree of north latitude in the past 600,000 years. The minima on the curve represent the four European ice ages

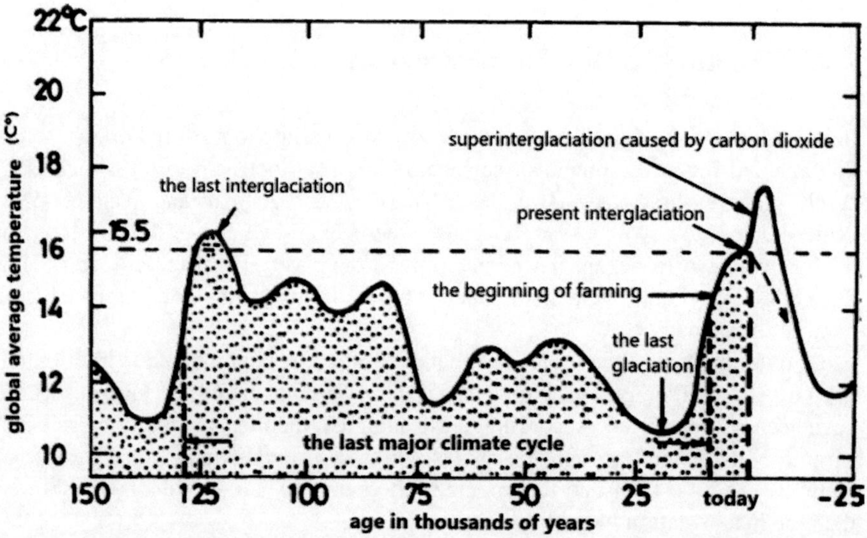

Fig. 2.2 Temperature flow between the last two ice ages and up to today, as well as the forecast for the next 25,000 years

impact of carbon dioxide will last for another thousand years after the use of fossil fuels stops, because that time is necessary for the atmosphere to get rid of the excess of that heated blanket of the Earth.

However, since we are interested in climate conditions in the distant past, we should mention the main characteristics of the Ice Age. 20,000 years ago, when the Ice Age began, the Earth was largely surrounded by thick layers of ice. That ice spread from north to south, burying forests, fields, and mountains. The areas threatened by the slow descent of the glacier will bear the consequences of this conquest for a long time. Temperatures were getting lower, and the surface of the earth in many parts of the world was sinking under thick layers of ice. On the other hand, to create these thick ice deposits, a large amount of water was extracted from the ocean. The sea level dropped by more than 100 m. Large areas of bordering continental rims became dry land. Many parts of the continents were joined by the retreat of water. Then there was a sea connection between Asia and North America.

In Europe, the ice came from Scandinavia and Scotland. It covered most of England, Denmark, and large areas of Russia, Poland, and Germany. A smaller ice cap centered on the Alps buried all of Switzerland and the border areas of France, Italy, Austria, and Germany. Ice also formed in the mountainous parts of the Apennines and Dinarides. The situation was similar in North America. In the Southern Hemisphere, small ice sheets formed over parts of Australia, New Zealand, and South America.

In the Northern Hemisphere, the southern border of the thick ice sheet was bordered by treeless tundra. There, during the short and fresh summers, low hardy plants grew on the swamp land. Reindeer herds and herds of mammoths, which migrated to the south in winter in search of new pastures, grazed them. Stone Age hunters, following mammoths and reindeer across the tundra, could see the southern edge of the ice sheet, and feel the frost and the northern winds that penetrated their reindeer skin clothing.

The ice sheet began to retreat about 14,000 years ago. That withdrawal, up to the present borders, lasted about 7000 years. All that remains of the ice sheet in the Northern Hemisphere is the Greenland ice sheet with a volume of 2,600,000 km^3 of ice (over an area of about 1,736,000 km^2, with an average ice thickness of 1500 m) and a few ice caps in the Canadian Arctic. The largest stock of ice is in Antarctica, and it is about seven times larger than that of Greenland. If all the ice from Greenland melted, the sea level would rise by 6.4 m. Precise measurements carried out in a 5-year period, from 1994 to 1999, showed that the volume of ice in Greenland decreased by 51 km^3 due to melting. The result was obtained by measuring the change in the height of the ice sheet using a laser mounted on the aircraft. The plane flew over the ice sheet on the same route during the winter of 1993/1994 and 1998/1999 years.

The ice sheets are still shrinking. Antarctica is now an icy desert, with an average ice thickness of about 2200 m, with rare snowfall, of less than 60 mm per year (equivalent to 60 liters of water per square meter). Despite its current state, fossil remains show that Antarctica was a tropical region millions of years ago.

The glaciers left behind a very changed landscape. There, in Iowa, where today's farmers harvest corn and in Dakota harvest wheat, 1.5 km high glaciers once rolled across the land, and in the area of today's European forests, valleys without any trees once stretched endlessly.

The traces of the glacier remained incised in the form of deep furrows in the surface rocks, by crushing and smoothing the surface below. This material was transported to the ice fronts, where it was deposited. Evidence left by the ice sheets has often been misinterpreted by modern scientists, especially geologists. Because when the glaciers receded, people's memory of them began to fade. The legacy of talking about them, if it lasted at all, was imprecise, so the world of Stone Age hunters was quickly forgotten. Glacial deposits had to be interpreted. And there were diverse views in the minds of modern scientists. Some strongly supported the thesis that the glacial remains came from the great flood described in the Bible. According to that version, "the Lord God decided to punish the wicked human race." He saved only the righteous Job. He opened the floodgates of heaven, and there was a flood. According to this interpretation, the waters of the flood transported gigantic boulders of stone for hundreds of kilometers.

However, the ice age theory trumped the flood theory. In this sense, the climate change that occurred in the past 10,000 years, the period of the most interesting development of civilization, should be described in more detail. Temperature conditions in the last 10,000 years are shown in Fig. 2.3. This graph shows the course of global temperature arrived at based on geological facts contained in glaciers and fossil plants. About 7000 years ago, at the time of the so-called postglacial climatic optimum, temperatures were higher by about 2 °C, and precipitation was more abundant than today.

It has been difficult to determine how much of a 2 °C drop in mean global temperature has an impact on people's culture and, in general, behavior. The answer to that question is confirmed by the fact that the cooling process is most often accompanied by a decrease in precipitation. This affects the type of agricultural production and thus the distribution of human settlements. So, for example, it was determined that the area of North Africa, which is dry and barren today, had enough rainfall at the time of the climatic optimum and that there were flower fields there, which enabled the development of great civilizations.

On that temperature curve, there is a particularly interesting period in the last 1000 years. That is why in Fig. 2.4 the climate trend in that period is shown. It can be seen that the global cooling from the climatic optimum does not proceed evenly, but there are oscillations of a much shorter period. Of all such oscillations, the most

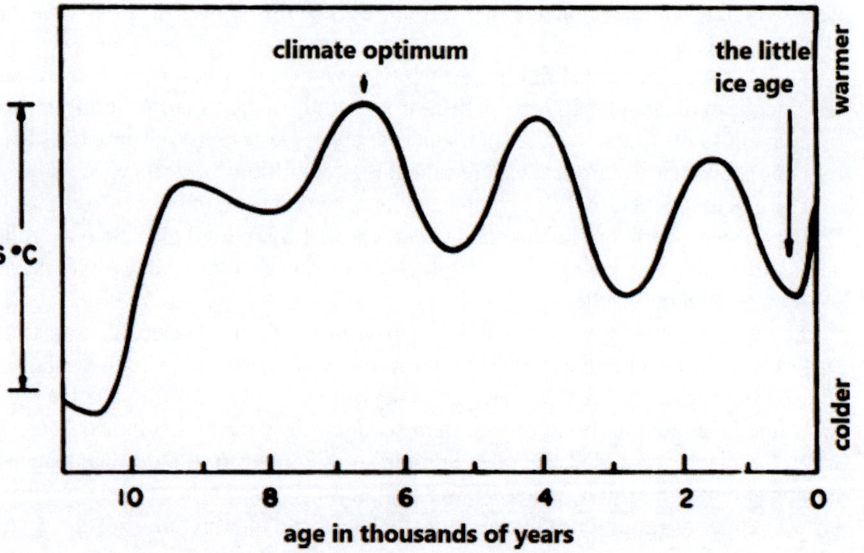

Fig. 2.3 A graph of the average global temperature over the past 10,000 years

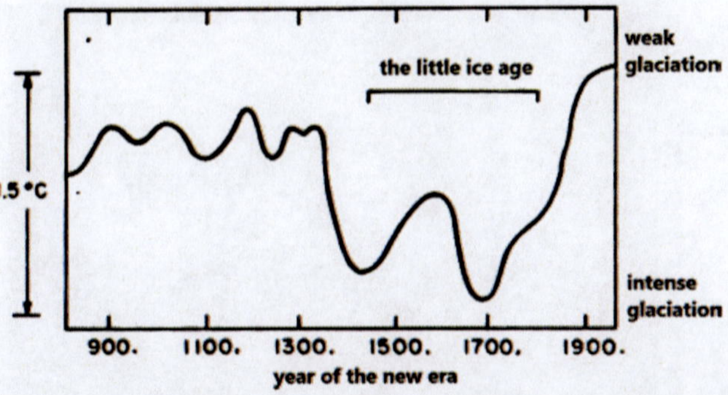

Fig. 2.4 Graph of mean global temperature over the last 1000 years

famous is the so-called little ice age. It appeared sometime around 1350–1850 AD. At that time, global temperatures were about 1 °C lower than today. The images of the time from that era are more faithful and are contained in many written documents. Among other things, there are many records from that period in monasteries. They, along with other works, were diligently collected by Ljubomir Stojanović in books published between 1902 and 1926. These data were further systematized by Prof. Pavle Vujević in his work on the occasion of the International Geophysical Year. The descriptions are, however, necessarily imprecise from a meteorological, quantitative point of view. They are of this type (Vujević 1931):

— In 1597:… it was cold and the earth weakened…

(Written in the Church of the Archangel St. Michael on the Tara River).

— 1605: At that time, there was great mourning in the country and many regions were left desolate: Segechig, Bačka, and Mashkondija were completely devoid of people; because a father sold his child for a piece of bread.

(Note in the record of the Orthodox monastery Krušedol, Irig district).

— 1621: The ice was so thick on Popovo Polje that people walked on dry land from Veličan to the house of Vukojević, on January 30 (February 9, according to the new calendar)

(Note from the gospel of the Orthodox monastery in Popovo Polje).

— 1624: The Kosatica river overflowed in the Mileševi monastery. The flash flood was so wild that it had not been like that since the beginning of the world, it had 4 cells, the abbot's cell, and a guest room. That came on Thursday, June 3 (June 13, according to the new calendar).

(Note in the record of the Imperial Library in Saint Petersburg).

— 1624: Dobrilovina Monastery on the Tara River. At that time, scarcity was great in all countries and life was expensive, amen.

(Note in the record in the National Library in Belgrade. Dobrilovina Monastery or St. George is located 16 km northwest of Mojkovac).

In this brief overview of climate characteristics in the past, it is also necessary to refer to the period of the closest past—the one that contains the most reliable

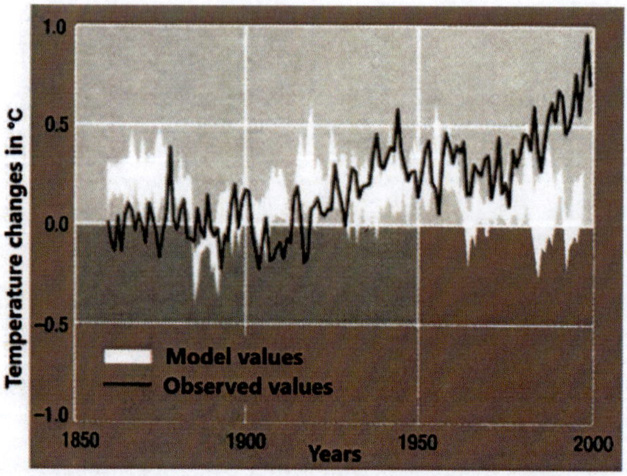

Fig. 2.5 Graph of changes in average measured temperatures in the northern hemisphere, in the past 150 years

evidence of weather and climate (Frakes 1979). It contains instrumental records and quantitative measurements of many meteorological parameters. From measurements at a large number of points around the world, the average air temperature over the last 100 years has been calculated. That course of temperature in the northern hemisphere is shown in Fig. 2.5.

The graph shows that the temperature increased until 1940. Its increase of about 0.4 °C is logical because that period continues to the period when the Little Ice Age ended. From 1940 to 1980, the temperature dropped by about 0.3 °C. From 1980 to today, measurements show the temperature has continued to rise. In that period, the warmest years are represented in many places, since meteorological measurements have been made. Many forecasts say that this trend of rising temperatures should continue in the first half of the twenty-first century.

2.2 General Conditions in Early Antiquity

2.2.1 Paleolithic Age

The period of early antiquity lasts from the earliest times to the middle of the sixth century BC. In that early period, the man was deprived of the simple art of stone carving for a long time. After he mastered it, the development of civilization was much faster. In the later Paleolithic age (the age of rough stone processing, which lasted until about 8000 BC), man also engaged in art. His successors in the Neolithic age (the age in which stones were replaced by metal tools) engaged in agriculture, founded villages, and, soon after that, cities and states. The letter, which appeared then, allowed future generations to learn about the life of those ancient civilizations (Van Doren 1991).

The first "historians" collected memories of family and local traditions. Descendants could also add historical literature to that inexhaustible source, active to this day. In a true historian, there must be a critical spirit that makes him doubt oral traditions, which compels him to seek original records, such as those on clay tablets, stone, bone, papyrus, or parchment. Such records often had the value of an official document. In this knowledge, archeology contributes a lot. Because there were civilizations without any written evidence. The interest is to reconstruct every epoch from which there are any traces of excavation.

Humanity is usually considered to begin when people started using tools. In this sense, it is known that 1,800,000 years ago, there were living beings in East Africa who consciously broke acorns to make a blade. It is not known, however, whether it was a Zinjanthropus, whose skull volume was no more than 530 cm^3 (according to 1100–1800 cm^3, which is the volume of the skull of today's man), or whether it was *Homo habilis* (700 cm^3), which is something closer to the present man. 500,000 years ago Pithecus-anthropic (monkey-man, 1000 cm^3) lived and used fire and stone tools to cut wood. This species was not content to live only in the tropics but also in colder, northern regions.

Neanderthal man (between 100,000 and 35,000) represents a new intellectual advance. The volume of his skull reaches as much as 1450 cm^3. Some representatives of this species buried their dead. But many experts believe that humanity only begins with modern man, Homo sapiens. Its traces have been reliably found since 35,000 years ago. This man perfected the processing of stone, discovered the importance of bone, and made very fine tools and hunting devices. With such a perfected tool, it is easier to hunt game, and tribes are formed. The first art is born—paintings in caves and statues made of clay and stone. In the Magdalenian epoch (about 15,000–8000 years ago), which represents the peak of the Paleolithic era, real cultural progress was achieved for the first time, in the southeast of Europe.

2.2.2 *Neolithic (New Stone) Age*

This age belongs more to history than to prehistory. It is a time that is characterized by the adoption of a new way of life by people due to climate change. As a result of the melting, which occurred in the last interglacial period (beginning in Europe, in the ninth millennium), the Magdalenian culture, based on reindeer hunting in the tundras of southeastern Europe, ended. With the disappearance of meadows, herds of large herbivores disappeared. The hunters had to disperse and hunt small game with bows in smaller groups. They are chasing reindeer that have escaped to today's Arctic steppes (to the north of the Arctic Circle). In the south of the Sahara and in northern China, people from tropical areas moved into savannahs full of wild game, forced by the appearance of large forests. When games became scarce, hunters resorted more and more to picking fruits. They begin to sow the grains and raise the captured young animals. Animal husbandry and land cultivation are slowly suppressing hunting, fishing, and fruit gathering.

Practicing agriculture leads to an increase in the population because they now eat better and live in permanent residences. In the first villages, with a few dozen inhabitants each, there is a division of labor. With the invention of tools for deep plowing, yields are increased by irrigation or watering. Thus, a person becomes attached to one place for a longer time and begins to build a residence from hard material. Later, due to the dry climate, animal husbandry becomes more profitable than agriculture. In the first millennium BC, nomadism appears and horses are used for riding and pulling.

The most significant invention is the use of metal. Man begins to forge pieces of natural, pure metal to make jewelry from gold and silver and tools from copper. Through the work of people settled in one place, wealth accumulates, which attracts smaller groups of people who are still wandering. It conditions conflicts and wars.

After the extinction of the Magdalene culture, cave painting still remains popular, because it is also accepted by nomadic shepherds. Small statues and decorative tomb vases are made. Therefore, religion appears. Agricultural life is accompanied by a religion based on the worship of natural forces, especially those on which the farmer's well-being depends.

The letter was invented by priests in order to have a better insight into the tax paid to the temple by farmers and artisans. That's when the first scribes appeared who had a monopoly on administration and science. The letter is first composed of symbolic drawings (pictograms) which soon turn into ideograms (pictograms to which supplementary, abstract meanings are added by the association of thoughts). An important stage in the use of letters was crossed when the scribes of certain nations (around 3000 BC) came up with the idea of using the ideogram as a phonetic sign, i.e., to represent the voice (usually the voice of the first syllable of the name of the drawn figure). With that procedure, for example, the word kuna would be written like this: Four paws (ideogram of a quadruped) would be drawn, and a drawing of a house would be added, as a phonetic sign for ku, and a drawing of a bracelet—a phonetic sign for na. In the second millennium BC, a syllabic alphabet is created (one character represents one syllable) and then a phonetic alphabet (each character, or letter, represents one sound).

From the third millennium BC, the texts of the most advanced civilizations provide us with numerous data that, unfortunately, cannot be used due to nonunique chronological dating. Namely, in early antiquity, the past year was first named after some significant event. Years were counted by the accession of a new ruler to the throne. The chronology of very old civilizations, which did not know about writing, is done by applying the theory of radioactivity. Thus, more reliable records are obtained than for the historical era. For the periods of the deep past, the protactinium-thorium and argon-potassium method is used, and for the end of prehistory, the carbon 14 (C14) method is used. This radioactive element is found in all living things in a constant ratio with ordinary carbon. After their death, carbon 14 disintegrates, and its ratio constantly decreases, so it is possible to use a Geiger counter to determine the time of death of the person, animal, or plant from which the remains originate.

2.2.3 *Stone and Copper Age Civilization*

This era lasts from the ninth to the third millennium BC. Back then, people lived in permanent settlements. Cultural differences were created between individual areas, among other things, depending on climatic conditions. Areas with a more favorable climate and better soil enabled faster development of technology and social organization. In the third millennium BC, three such civilizations developed.

Copper tools and farming appeared first in the valleys of the Middle East (from Greece to Iran). There was great mineral wealth and a favorable distribution of rainfall throughout the year. Later, man cleared and cleared the great wetlands of Mesopotamia, Egypt, and the Indus Valley (Wainwright 1938). They become the three focal points of civilization. Their influence is strongly spreading to the surrounding countries, Syria, Iran, Africa, Anatolia, and Greece, to areas whose relief is fragmented or threatened by drought in order to create a state and culture in them. At that time, technical devices were transferred from the coast of the Eastern Mediterranean to Europe and the Eurasian steppes. The Yellow River countries, Mexico and Peru, however, are creating their own agriculture.

Heavy rains, which in the area of Mesopotamia coincided with the exit from the last ice age, made the land around the Tigris and Euphrates inaccessible. Residents of the surrounding hilly areas (Kurdistan, Zagros) are engaged in agriculture, ceramics, and making copper objects. The climate, which gradually becomes drier, allows the valleys to be inhabited. This is where the differences between individual areas arise. The hills in the northwest (future Assyria) are quite cold and wet (there is now 300–600 mm of rain per year). In the southeast (the lands of Akkad and Sumer), due to the hot sun and little precipitation (today there is about 100 mm of precipitation), the land must be irrigated, which is made possible by the proximity of two rivers. The rest of Mesopotamia is gradually drying up.

The first significant progress in development occurs from 3600 to 3000 BC, with the Uruk civilization. A large architecture appears in the lowlands. Pottery winches, roller seals, and letters are found. This progress is attributed to the invasion of the Sumerians, who appear to be the creators of the Mesopotamian script. The inhabitants live in cities above whose modest houses rise tall temples. Obviously, those Sumerian temples reminded them of the mountains of their homeland, where heaven and earth meet and where the gods descend. Uruk was the metropolis of the lowlands for a while.

The invention of the letter was encouraged especially by the importance of calculation. The first evidence of this is provided by the lists of foodstuffs that the managers of Uruk received and distributed. Scribes who use over 2000 characters (ideograms) cut with reeds write on fresh clay, which is then baked or dried in the sun to keep it stable. Roller seals are used, with which every respectable citizen could be immortalized. Stamps are introduced for economic reasons. There is an imprint of a small scene on them. The impression of the seal was placed on the clay stopper of the liquid vessel to guarantee its inviolability. Later, contract drafters printed it on contracts.

Afterward, the painters used a script with characters far from the primitive pictorial script. It is called cuneiform (made of wedges).

Iran, whose mountain ranges maintained high humidity for a long time, was the cradle of agriculture. The fertile valley of Suzia has seen progress. The great Elamite city of Susa also developed, thanks to trade and irrigation, and represents a link between the urban Sumerian civilization and the mountain kingdom of Elam. The proto-Elamite script was used there (around 4000 BC). It is similar to Sumerian and goes beyond pictorial writing. Around 2250 BC, it was replaced by the cuneiform script.

Separated by high mountains and a still barren ocean, India lagged behind the Middle East in development. The Indus Valley Civilization arose around 2400 BC. People cultivate barley, wheat, rice, and cotton. They also use bronze. Two large cities are developed with a checkerboard plan and a system of sewage and drainage. The script used to be written on stone and copper plates is used. To this day, that letter has not been understood. The wealth of the great valley led to the creation of the original civilization.

In the part of Africa where excavations were carried out (in the north and east), there was the same civilization until the end of the Paleolithic. There are no deserts on that continent then because of the abundant rainfall. The terrain, which is not crossed by mountains, means that everywhere there are the same tools and the same pictures on the walls where they lived. People engaged in stone polishing, ceramics, animal husbandry, and agriculture. Droughts in the Sahara became more frequent after 3000 BC. This forced its farmers to move south and north. During this period, the fertility of the silt around the Nile allowed the development of Egypt (Craig 1909).

In the fourth millennium BC, a general civilization is born, thanks to the established river navigation between villages. Navigation was done using boats made of papyrus. A copper tool appears. Artisans produce beautiful vases from hard stone. Small kingdoms are created. The beginning of the new Egyptian progress was conditioned by the arrival of the Asians, who came from Mesopotamia, around 3300 BC, brought new art forms. Then the Egyptians found hieroglyphs. It is a letter that had the appearance of an exact drawing, with the help of which the inscriptions were carved into the stone.

We learn about the civilization of those eras from the excavations of temples and tombs. Between 3100 and 2700 BC, the famous intellectual and technical values that pharaonic Egypt used throughout its history were formed. Fairly accurate mathematical formulas (it was found that $\pi = 3.1605$) are created for the needs of cadastre and architecture. Priests observe the sky to know which direction to turn the monuments and to determine lucky days and hours. Thanks to them, two calendars are used. One is lunar, for determining holidays, and the other is official, solar, with 365 days a year. For administrative purposes, the cursive hieratic script was introduced, as well as the material on which to write, papyrus. The religious zeal of the Egyptians can explain the enormous dimensions of the pyramids of Cheops and Chefren (built in 2550 BC). Their heights are 145 and 143 m.

From the eighth to the third millennium BC, the rapid development of civilization also took place in Anatolia, Cyprus, and the Aegean islands (Bowra 1963). Cyprus, the island of copper, was inhabited by Ana-dolci and Syrians, who were attracted by its land and forests. Around 5700 BC, villages developed, but technical progress was very slow. Copper ore on the island was found only in the middle of the third millennium. Crete, unique in the Aegean for its forests and fertile land, attracts Easterners. Only in the old Minoan era (named after Minos, the legendary Cretan king) from 2600 to 2200 BC, the original civilization was created. Large families who were engaged in maritime trade at that time raised tholos and made elegant objects (baked ceramics, vases made of mottled stone, and seals made of ivory).

On the Hellenic peninsula, from Macedonia to the Peloponnese, in the seventh millennium BC, villages appear almost everywhere, whose builders use first glued wattle, and then batting and bricks. Cultural development was particularly pronounced from 2500 to 2000 BC. The use of copper is spreading; cities are being built.

Small groups of peasants from Macedonia or Anatolia reached today's Hungarian border. There they find a fertile black forest that was created by the decomposition of steppe grasses. In the north of the Balkan Peninsula, they created the Starčevo civilization (from 5400 to 4200 BC) as well as the first Danube civilization (from which the famous decorative motif with meanders and spirals originates). Their influence is felt even in the Ukrainian steppes and Italy.

2.2.4 Bronze Age

It is the era between 2300 and 1200 BC. This time is characterized by various migrations. The old centers of civilization were permanently affected by it. Mesopotamia becomes easy prey for invaders. Egypt, being naturally better protected, repels the Asiatic barbarians and regains its traditional prosperity. The Aegean world, dominated by the Cretans and then the Mycenaeans, represents the link between the eastern kings and warlike European tribes. These tribes, as victims of various climatic adversities, and under the pressure of steppe inhabitants, end their migrations by conquering the Middle East. In the Far East, only the Chinese civilization is advancing rapidly.

Migrations are well known only in the east, as written monuments inform us. Elsewhere they are nameless barbarians of whom only traces of ash remain, ending their cultural activities.

In the cities of Mesopotamia from the twentieth to the seventeenth century BC, dynasties appear, mostly originating from illiterate Semites. The Sumerian race, which had already begun to disappear, overwhelmed by the constant influx of Semites, finally merged with them. Nevertheless, in that age, science is progressing. The Akkadian tablets show that the Semites, as disciples of the Sumerians, had more of a scientific spirit than is usually attributed to the scholars of the ancient East. Scribes of the seventeenth century BC use Euclid's postulate and the Pythagorean theorem and are able to calculate the sides of a rectangle if they know its area and diagonal. The progress of the Mesopotamians in geometry and astronomy is explained by the discovery of the positional value of numbers, which the Sumerians came up with.

The work of Hammurabi of Babylon (1723–1680 BC) is particularly well known from that period. Leaving the Mesopotamian Javanese to be exhausted by mutual wars, Hammurabi, after several decisive battles, imposed his rule on all the cities of the great plain. As a good administrator, he succeeds in merging the cities of Sumer and Akkad into a unified state, Babylonia. Since then, its capital, the Babylonian language (as a local form of the Akkadian language), and its gods have had no competition. Of all the works of the great king, the most memorable is his Code, which is actually a collection of judgments. The famous text, written in clear language, goes beyond the Sumerian laws. The king can no longer be satisfied with demanding a fine from the guilty party. He applies terrible punishment on an eye-for-an-eye, tooth-for-a-tooth system.

In Anatolia, in the second half of the second millennium BC, there were palaces notable for having hot-air heating devices as well as wall decorations. Bronze objects, beautiful ceramics, and, especially significantly, thousands of pieces of the famous Cappadocia tiles were found. The writers of these texts are Mesopotamian merchants from the nineteenth and eighteenth centuries BC who imported cloth and tin from Mesopotamia and sent silver, copper, and lead to their country from Anatolia.

Around 2200 BC, Cretan civilization experienced a sudden rise. On the rich fields in the middle of the island, cities develop, bronze metallurgy spreads, and Cretan merchants sell seals, jewelry, and ceramics. Around 2000, palaces appeared in Knossos (ancient Minoan culture), which were built in a very significant way, with sculptural columns whose base is wider than the top. There were huge jugs in the shops and clay tablets dried in the sun in the archives. Inhabited places in Crete around 1700 were destroyed (probably by an earthquake), but the palaces were rebuilt very quickly.

2.2.5 Early Iron Age

This period lasts from the twelfth to the sixth century BC. Iron, a new metal, is spreading due to a terrible invasion of people from the sea. Egypt managed to stop them but never recovered from it. From the tenth to the eighth century BC, the Israelites, who created the moral law, and the Phoenicians, who spread the alphabet of their country, come first. Then comes the age of warrior empires (from the eighth to the fourth century BC): Assyrian, Babylonian, and Persian. The Hellenic peninsula is slowly creating its own Greek culture that surpasses the Eastern models. The Phoenicians and Greeks inhabited the western shores of the Mediterranean Sea and had a favorable influence on the progress of Europe. Thus, the Italian peninsula is gaining importance. The Jews, at the time of their migration, were called "the people of the Bible." Later, when their state existed (eleventh to sixth century BC), they were called Israelites after the patriarch Jacob Israel (who fights with God). Finally, when Judea (the land of the tribe of Judah) became the center, they are called Judeans. They did not leave many archaeological traces. They took their place in history according to the holy book, the Bible (from the Greek word Byblos, which means book). The volume of words is called only the Old Testament. This work shows the mentality and culture of the Middle East. It expresses a religious thought that is significantly different from other religions of that time. Yahweh, the god of the Jews, has no divine consort. Man cannot make him do something with magic. The happiness he promised his people would be eternal.

The value of the Bible as a document depends on the circumstances under which it was written. The book is a reconstruction of the past with a specific goal. The Old Testament was compiled slowly, from the tenth to the second century BC. Its writers used a wide variety of documents, oral (family traditions, epic poems, laws, court decisions) and written (chronicles, official acts, ritual books, works of ancient writers). Having previously cleaned these sources, the Bible writers combined them into one work, with the intention of giving the Israeli people a theological, moral, political, and social basis. The Bible is precious, because it reveals to us the material and spiritual life of the Jewish people. It mixes literary genres (taken from Mesopotamian myth), epics, and history. The book presents historical events only in an instructive form.

The first book of the Bible is Genesis. It begins with stories (the creation of the world, sin, the first people, the flood) that form the basis of Israelite theology. It then talks about Abraham, a shepherd leader from Ur, who will settle in Palestine with his tribe. Then the Bible tells about the life of Isaac and Jacob, Abraham's son and grandson. Joseph, one of Jacob's sons, becomes Pharaoh's top chieftain and brings his father's tribe to Egypt.

According to Exodus, the second book of the Bible, the Jews stayed in Egypt for 400 years. At the end of that age, Pharaoh begins to persecute all foreigners. One of them, Moses, retreats to the desert, where he experiences a revelation. In the "unburnt Blackberry," the exile recognizes the god of his father and asks for his name. He receives the answer "I am" (Yahweh). Moses then receives the duty to convey God's command to Pharaoh to let the Jews out of Egypt. Because Pharaoh refused, ten punishments came upon the Nile Valley. However, this forced him to relent, and the Jews left Egypt. When Pharaoh changed his mind, he began to pursue them all the way to the Red Sea. The sea parted and the Jews crossed it on foot, then the waves of the sea came together and covered the Egyptian army. After their liberation (at the end of the eighth century BC), they spend 40 years in the desert that separates Palestine and Egypt. The events of that period are written in the book of Exodus. Moses is portrayed as an outstanding personality, a leader of the people, and a deep thinker.

2.2.6 The World in the Age of Greek Civilization

From the middle of the sixth to the end of the third century BC, the civilized world stretches from the Pillars of Hercules to China. Relations between the countries it occupies are loose. The Persians in the continental areas manage to form a relatively unified empire. They are in contact with the Greek world, which has been centered around the Mediterranean Sea for centuries. In the east, the Greeks reached as far as India. In the west, Greek, Phoenician, and Etruscan coastal cities battled with Celtic tribes.

At the time when Greece was in conflict with the Persians, Greek culture reached its peak. The first philosophical systems appear. Gradually separating from religion, their creators turn to matter and explore the basic elements that are hidden behind the emergent, apparent. Thus, they significantly contribute to progress in the field of knowledge. Such is Thales of Miletus and his students Anaximander and Anaximenes. Heraclitus builds his own teaching on standards and the existence of the world, while Pythagoras of Samos found a school in Croton. The two great centers of culture are Ionia and Athens. Temples are built in stone, based on two styles: Doric style, very geometrically shaped, and Ionic, less strict and more elegant.

But Greece is also torn apart by the internecine war. Athens, in addition to wars, is affected by an epidemic of plague, which affects thousands of people, including Pericles (429 BC).

At the same time, Confucius' philosophical system is currently in China. It emphasizes two terms: fairness (ji) and altruism (yen), which mean respect for oneself and others—purely social virtues. Confucius' ideal is a virtuous man. Like Socrates in Greece, Confucius teaches man to know himself, so that he can perfect himself. As Greek thought, Confucian wisdom and humanism made Chinese culture play the same role in Asia that Hellenism played in the Mediterranean and Western world.

2.2.7 *The World in the Age of Roman Civilization*

This period includes the end of the third century BC until the end of the fourth century AD. It begins with a great victory over Carthage, in 201 BC. Since then, Rome occupies the Mediterranean world and at the same time succumbs to the spiritual supremacy of Hellenism. This is how the Roman civilization was born, which radiated from the Pillars of Hercules to the borders of Mesopotamia. In the Far East, Chinese civilization, thanks to the powerful Han dynasty, has the same reputation and spreads under similar conditions.

This period is interesting in that it saw the establishment of two large communities, each achieving full unity, both politically, socially, and economically. Two phases are observed. The first is the fulfillment of efforts for unification, conquest, and organization, both in the Far East and in the West. In the second phase, the people of that time could believe that they had achieved the unity of the world and that they were waiting for the golden period. It is the age of Roman peace or Roman peace, that is, of Chinese peace or Chinese peace. It remained in people's memory as extremely happy because of the long five-century Roman peace because of the general stability, power, and cultural importance of both of these empires.

In Rome, literature and art developed, thanks to patrons, among whom the most famous was Maecenas (around 69–8 BC). Rhetoric, which reaches its peak with Pliny the Younger (first century AD), is also cherished, and philosophy, from Seneca to Marcus Aurelius, has several famous names.

References

Allman, W.F., and B. Wagner. 1992. *Climate and the rise of man*. U.S. News and World Report.
Aristotle, trans. H.D.P. Lee. 1952. *Meteorologica*. Cambridge, MA: Harvard University Press.
Asimov, I. 1982. *Biographical encyclopedia of science and technology*. Doubleday.
Bowra, C.M. 1963. *The Greek experience*. Mentor Books.
Charles, S. 1959. *A short history of scientific ideas to 1900*. Oxford: The Clarendon Press.
Craig, J.I. 1909. Meteorological conditions controlling the Nile flood. *The Quarterly Journal of the Royal Meteorological Society* 35: 141–143.
Dampier, W. 1935. *A history of science*. New York: The Macmillan Company.
Dawson, A.G. 1994. *Ice age earth – Late quaternary geology and climate*. Routledge.
Frakes, L.A. 1979. *Climates throughout geological time*. Elsevier Scientific Publishing Co.
Van Doren, C. 1991. *A history of knowledge*. Balentine Books.
Vujevic, P. 1931. *Old Serbian records. International Commision for the study of climate*. Belgrade.
Wainwright, G.A. 1938. *The sky religion in Egypt*. Cambridge: The University Press.

Chapter 3
Early Development of Meteorology

3.1 The Beginnings of Meteorology

3.1.1 Introduction

As with other sciences, it is not possible to find the exact beginning of meteorology. In this case, it must be taken into account when talking about meteorology as a science and when about meteorology as a field of knowledge (Abbe 1906, 1907; Hellmann 1908; Clarke 1910, 1927; Heninger 1960; Frisinger 1977). Meteorology as a scientific discipline is younger than meteorology as a field of knowledge, the beginnings of which go back to the distant past of human civilization (Crew and DeSalvio 1939; Rackham 1937).

Hunters and farmers of the early period were strongly influenced by the weather. This forced them to observe the phenomena of the weather and to find signs that would tell them what awaits them in terms of weather in the future. Those collected signs of weather were passed down from generation to generation, taking the form of short sayings, making remembering them easier.

The first great ancient civilizations developed around the largest rivers of Africa and Asia: the Nile in Africa, the Tigris and the Euphrates in Western Asia, the Indus and the Ganges in Central Asia, and the Huangpu and the Yangtze in Eastern Asia.

Even before 3500 BC, the ancient Egyptians had religious practices related to the sky, in the form of rituals for invoking rain (or artificial stimulation of rain, as we would say in our modern language). All ancient religions believed that atmospheric processes were under the control of the gods.

The Babylonian civilization developed around the Tigris and Euphrates Rivers 3000–300 BC. In the absence of suitable plants, such as those from which the Egyptians made writing papers, the Babylonians used clay tablets. Various interesting records were written on such plates. The plates would then be baked, so that the record would remain permanent, etc. (Fig. 3.1). Thousands of such plates were used

M. Ćurić, V. Spiridonov, *History of Meteorology*,
https://doi.org/10.1007/978-3-031-45032-7_3

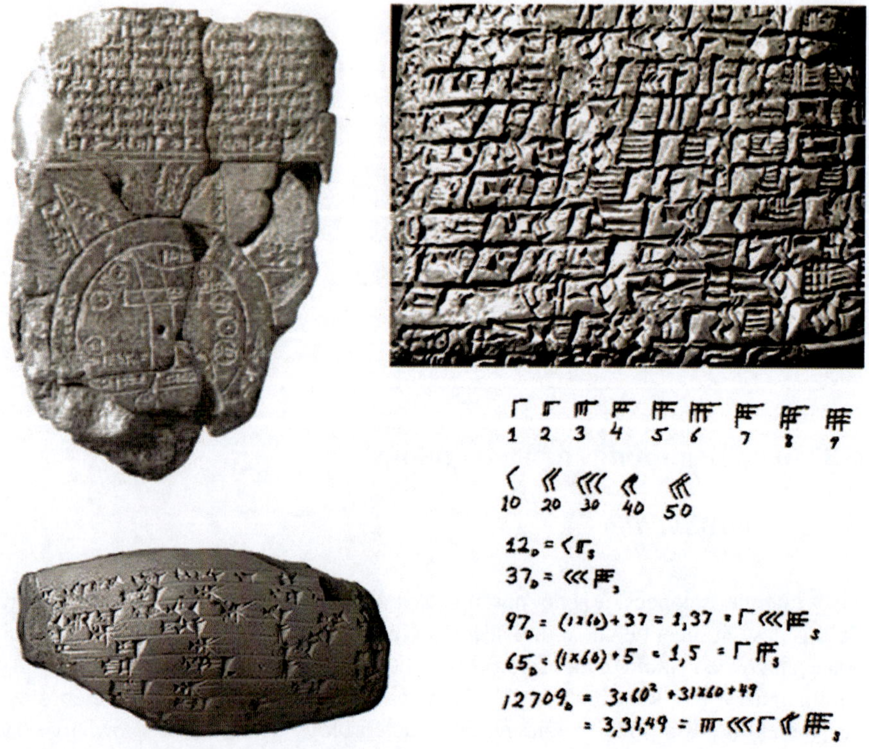

Fig. 3.1 Inscriptions on Babylonian baked tablets and the numerical system

to record mathematical, astronomical, commercial, and other works. From those plates, it was possible to read that meteorology was an important part of the culture.

In order to connect atmospheric phenomena with the movement of celestial bodies, the Babylonians introduced astrometeorology. She dealt with weather forecasting based on certain signs of the heavenly bodies. Predictions are of this type: "When a dark halo is seen around the moon, that month will be rainy or clouds will gather," or "When dark clouds rise in the heavens, the wind will blow." They also used clouds, winds, and thunder as omens of good and bad events. Thus we find: "When it thunders on the days of the waning moon, the harvest will be good and the market will be stable."

The Babylonians used a wind rose in eight directions. For that, they used four basic directions and sides of the world: south, north, east, and west (sutu, iltanu, sadu, amur). Combining these words with "u" made a name with four intermediate directions. Thus, they called the southeast "sutu u sadu," and the northwest "iltana u amura." Until recently, this method of marking intermediate wind directions was thought to be based on four basic more recent dates.

3.1.2 *Some of the Earliest Records of the Weather*

Even in the earliest period of life, man sought to distinguish between natural phenomena that bring discomfort and those that do not (Dufour 1943). Poor shelters did not provide good enough protection from bad weather, and their crops were often destroyed by drought or heavy rainfall. The monks of primitive religions told them to believe in the gods of those elements against which man cannot fight. Thus, the first gods of all nations were the gods of the Sun and Moon, thunder, lightning, wind, and sea. So has the god Osiris in Egypt, the Scythian sun god Otosirus, Poseidon in Greece, Indra the Thunderer in India, and Vulcan in ancient Rome. A little later, in the Taoist religion, an entire divine ministry was formed with a clearly arranged hierarchical order, arranged according to the intensity of meteorological processes. This divine administration includes the God of thunder and lightning; the wind counts; the rain lord and their apprentice Yun Tun, as well as the cloud boy, whose job it is to keep the floating tank (the cloud) constantly full of water. The ancient Slavs revered the god Perun, who was the god of the Sun and lightning. He was also called "Dah-God."

You can learn about the weather in those old eras from poems and various philosophical works. From the depth of the weather details shown, it can be concluded that the writers were dedicated weather observers. Here are some examples from various countries and cultures.

In the Odyssey, book five, Homer describes the winds that blow, and how north and west replace south and east. In the Iliad, book 24, the beauty of a rainbow is described, the lower part of which sinks into the sea.

In the work "Tao Te Ching" (circa sixth century BC), which was previously attributed to the Chinese philosopher Lao Tzu, it is written: "A strong wind cannot blow all morning and a heavy rain cannot last all day."

In the Indian epic poem "Mahabharata," the summer monsoon's arrival in India is vividly described in vivid colors. Thus, in one part, it is written: "And then (the god of thunderclouds and lightning)… so worshiped by Kadru, covered the entire firmament with massive clouds… And those clouds, illuminated by lightning, constantly roared at each other in the heavens, boiling an abundance of water… And as a result of the countless waves caused by the heavy downpour, the strong roar of the clouds, the lightning, the fierce wind, and the general boiling, the sky seemed like a rapturous dance. The sky becomes completely covered with clouds and the sun's rays and the moon disappear as a result of that upheaval… And water covers the earth all over."

In the same poem, a dust storm in India is described: "Ga-ruda (the legendary king of birds) … gathered his wings and ascended to the sky." He immediately fell down because of Nishada… "And bowed down overcome by Nishada, who raised a huge dust-rail and scattered it across the heavens."

We find similar descriptions in the Koran. Thus, in Surah 30, it is written: "Allah alone sends the winds so that they carry clouds that spread across the sky just as He wills."

It can be seen that the idea that God rules over all nature and people is represented everywhere. It distanced people from finding the true laws of the functioning of nature. It was not easy to fight against that. Thus Pythagoras (born 570 BCE) tried to limit divine power by asserting, "God always works in such a way as to obey the rules of geometry."

In meteorology, the first rules by which nature behaves were those by which the weather changes during the year. Old Slavic legends speak of a constant struggle between Good and Evil, winter and summer, and light and darkness. These opposites are personified through the gods. "Belobog" represents Good, and "Cernobog" represents Evil. The same motifs are often found in the traditions of other nations. So, Hesides (eighth century BC) writes in the work "Works and Days" that the Greek peasant was dependent on the movement of the Sun and other heavenly bodies all his life.

In the more recent past than the one in which the abovementioned works were created, meteorology as a factor of survival entered people's lives in a more "serious" way (Webb 1963; Gershenson and Greenberg 1964). Thus, in the age of Meton (around 433 BC), calendars with the characteristics of the weather in that year were placed on the squares of Greek cities. Those calendars contained, among other things, the first annual weather forecasts. They were called "Parapegma" (*παραπηγμα*), which has its root in the word hang. Some of these parapegmas have been preserved, thanks to the great astronomer Ptolemy (born around 90 AD). Most of the information on those calendars is related to wind, precipitation, cold weather, and some phenological phases. Thus, for example, an Alexandrian parapegm mentions very often the occurrence of the south wind (*νοτοζ*) or the west wind (*ζεφμροζ*), which is not in accordance with what is now known about the prevailing wind (northerly winds now prevail in Alexandria). And then, as now, storms occurred in Alexandria, mainly in the winter period. It is recorded that rain appeared about 30 times a year, and thunderclouds appeared in all months (which, again, does not agree with what is happening today, because today Alexandria has dry summers). It is also recorded that fog appeared quite often in the summer, which is not the case today. Obviously, either the weather has changed significantly since then or the parapegmas cannot be considered as complete climatological representations in the sense of comprehensiveness as they are today.

In Chinese classical literature, there are data on phenological phases, which can be used in the interpretation of the weather of that era. For example, in the book *The Book of Users*, the author Li Qi devotes an entire chapter to the agronomic calendar, which includes data up to the third century BC. In the book of Chu Kung, which was written at the very end of the pre-Christian era, it is stated that then the peach blossomed in Shanghai on March 5th, and now it blooms around March 25th. It also says that the swallows were returning to Ningpo on March 21st (now they return around mid-March), while they migrated on September 21st, and now they are rarely seen in September (they moved away earlier). All these various data show that it was a warmer period then than it is now. In Chinese chronicles, there are many records of frost, snow cover, floods, and droughts. Droughts were particularly common in the fourth, sixth, and seventh centuries AD. The date of the last snowfall (averaged over a 10-year period) was April 1 during the Sung Dynasty (1131–1260), which is about 16 days later than in the decade from 1905 to 1914.

The first attempts at weather forecasting based on regional characteristics were made a long time ago. In the book *Shih Ching* Book of Signs, which refers to the Chu period (1122–247 BC), it is written: "If a rainbow is seen in the west at sunrise, this means that rain will immediately begin to fall." We find very similar signs of the weather in Greece. So Theophrastus of Eros (ca. 372–287 BC) writes: "…We have described the signs for wind, rain, storms and clear weather so that they can be understood. Some of these signs I observed myself, and some I took from other reliable sources." According to him, a reliable sign that it will rain is a golden-ruddy color on the clouds before sunrise. A dark red sky at sunset or a wisp of mist in the mountains are also signs of rain. Many of Theophrastus' predictions are related to observing the behavior of birds and animals. Theophrastus in his work "Signs of Weather" listed a large number of prognostic rules. Thus, he lists 18 signs that predict rain, 45 signs for wind, 50 for storms, and 7 signs for forecasting the weather for a period of about a year.

In India, a country where there are regular seasonal changes, anomalies in these changes have been monitored, with the aim of forecasting the weather well. It is not known exactly when the first attempt was made to predict whether the summer monsoon would be good or bad (i.e., whether there would be plenty or famine).

In more recent times, the Indian astronomer Varaha-Mihira (fifth century) in the 21st chapter of his book *Brihat Samita*—The Great Compendium—systematically presented various signs that could be used to predict the arrival and abundance of monsoon rains well in advance. He grouped these signs according to the Hindu lunar months. According to him, the signs of good rain are in October and November (their months do not correspond to ours); there is a red sky in the morning and in the evening, halo, clouds, and the absence of severe cold; in November and December, red sky morning and evening, clouds and halo, and only a little snow; in December and January, strong wind, the very cold, gloomy appearance of the Sun and Moon and thick clouds at sunrise and sunset; in January and February, strong, dry gale-force winds, thick, flat-based clouds, a torn halo, and a copper-colored Sun; in February and March, clouds accompanied with wind and snow; and in March and April lightning, thunder, wind, and rain.

According to this, if the signs are as described, then the following number of days with rain will occur (by rainy months): May 8, June 6, July 16, August 24, September 20, and October 3. There is no data to verify these rules, even though they were set so long ago. There are only some fragmentary representations. Thus, the Indian meteorologist Sen found that during the strong monsoon in 1917, the number of rainy days was significantly less than those from the "calendar" signs, and there were 5, 6, 12, 15, 13, and 5 days in the corresponding months with rain.

Ancient Jews also paid great attention to meteorological phenomena. This is described in the Old Testament when the Jews faced a dry climate when they left Egypt from slavery. Rain was also important for them in Asia Minor. And no matter how often rain is mentioned in the Old Testament, there is nowhere about how they imagined the nature of rain. There is a hint in one place in the book of the prophet

Amos (around the eighth century BC) that he understood the water vapor cycle. But the Bible is a grandiose human document and not a textbook of cloud physics. That's why you shouldn't expect answers of that kind.

A lot of information about weather and climate can be found in the book *History of Jermaine* by Moshe from Korin (fifth century). Mention is made of humid air and frequent fogs in Azharia, snowfall, strong winds, and snowstorms in the higher regions of Armenia. At the end of the book, when enumerating the reasons why the country's power is declining, he cites the inhospitable climate "... Winds, bringing dry summer (warm, dry wind) and diseases; clouds throwing out lightning, hail, rain, merciless fall at the wrong time of snow storm, frost...".

3.1.3 Learning About the Weather of the Ancient Greeks

Ancient science achieved its greatest development, systematicity, and clarity in Greece, especially in Athens (Cohen and Drabkin 1948). At the beginning of the sixth century BC, Greek colonies were scattered all along the Mediterranean and the Black Sea. In that period, through their colonies, the Greeks were familiar with the culture of the entire Western world. They were able to take a lot from their predecessors, the Egyptians and the Phoenicians, and to improve the earlier relatively fragmented knowledge into a science of the modern type.

The Greeks paid great attention to facts or data that had been gathered earlier. They had a talent for penetrating the essence and finding the simplest and most significant explanations. They had the ability to think abstractly. For the ancient Greeks, the natural sciences were closely related to the philosophy of life. How much importance they attached to meteorology can be best served by the record of Aristophanes:

Two Greeks, Strepsiades and Aminias, are talking. Strepsiades said: "Tell me, does Zeus constantly bring new water to the earth with rain, or does the Sun raise the old water from the earth and use it again." Aminias replies: "Neither do I know, nor am I interested." Then Strepsiades says: "And how do you expect to become rich if you don't know anything about meteorology?"

Meteorological observations of the oldest ancient people, and their successors, the Greeks, led to thinking about the laws of nature, which were later "seized" by physicists as physical laws. Indeed, they thought about ordinary concepts: cold, warm, light, and dark, and their alternation and interdependence. It imposed the first concepts of physical events. Thus, meteorology was initially the driver of the scientific approach. Over the centuries, physics has taken over that driving force in many things.

It can be said, therefore, that the Greeks were the first to start regular meteorological observation of the weather and that they established some theoretical rules. Thales of Miletus was one of the first important Greeks in this business as well.

In the earlier history of meteorology, thunder and lightning attracted the most attention from weather phenomena. Thus, two followers of Thales, Anaximander (ca. 611–547 BC) and Anaximenes (ca. 585–528 BC), had similar theories about the origin of lightning and thunder. They claimed that thunderstorms are caused by the collision of air and clouds, which as a result of that rush of fighting produces thunder. It also ignites the lightning fire. This theory implies that there is a substance of fire in the atmosphere. This belief prevailed in meteorological theory even for the next 2000 years.

Thales of Miletus (620–540 BC)

He was born in Miletus, one of the Ionian Islands near the coast of Asia Minor. He was one of the seven sages of that age. He is a philosopher, mathematician, astronomer, and also meteorologist. He was interested in meteorological phenomena. Thales tried to connect the weather phenomenon with the movement of the heavenly bodies. In particular, he observed a group of stars called the Hyades. He connected the arrival of rain with the position of this constellation in relation to the Sun. Thales believed that water is the basic element from which everything is composed. Water is the basis of everything, and by moving it makes a circle, coming from the sky to the Earth, moving through all the living world, and returning back to the sky. He knew unequivocally that clouds were composed of water, but there is no evidence that Thales understood the process of condensation and cloud formation.

He is the first person who knew how to apply meteorological knowledge in practice. Through careful observation, he noticed the weather trends, which enabled him to predict an abundant harvest of olives one summer. Fully convinced of this, he decided to buy all the olive presses he could get his hands

(continued)

(continued)

on. The forecast really came true, and with his presses, he was able to press a rich harvest. That's how he got rich. He could devote the rest of his life to science. And he explained various things. Once, while traveling in Egypt, he became familiar with the problem of the Nile flooding every year. Afterward, he gave an explanation of this phenomenon. According to him, it is not the increase in the amount of water in the Nile that is responsible for the flooding, but the northerly wind, the Ethesia, which at that time blows in the basin area, preventing the waters of the Nile from flowing into the sea. This problem occupied the attention of scientists for more than 300 years after Thales.

Anaximander was a fellow citizen and friend of Thales. His book *Peri Physios*—On Nature—was published near the end of his life. Only a small part of it has survived to our day, but a beautiful account of it was given by Theophrastus. He had strong mental abilities that he used to observe the phenomena of weather. He was the first to define wind as the "flow of air." As you know, this is also today's scientific definition of wind. But, surprisingly, this was not generally accepted by later natural scientists.

Anaximenes, the third Miletian, accepted Thales' theory of the "basic" elements but thought that air, not water, was the basic element, because his observations showed the need for air to sustain life on Earth. In his theory, it is claimed that the air contains something significant that he called "pneuma"—breath. He believed that it sustains all life in the universe, just as air sustains human life. Despite the fact that air is the basis of the substance, he believed that through condensation (coagulating) and rarefaction, various forms of matter pass into one another. Thus, when water is heated, the air is created, and when it is heated, fire is created, which is nothing but heated air. He also believed that the earth was formed by the coagulation of water. Traces of this understanding of the four basic elements—earth, water, air, and fire—have taken such a prominent place in science.

He claimed that the air we breathe out through a wide open mouth is warm and when we let it out through a small opening on the lips, it is cold. That's an entirely accurate observation. However, his interpretation of this is wrong. He believed that when the air is compressed, it cools (thinking that the air is compressed in the mouth when the lips are almost closed), and when it thins, it heats up (corresponding to the open mouth). Therefore, he did not notice that the air expands only after leaving the mouth. This expansion is more intense when the mouth is slightly open than when it is completely open, and hence faster cooling in the first case.

The last of the group of the celebrated Ionian school was Anaxagoras (about 499–427 BC). He exhibited his works in Athens, from where he was eventually banished to his native Ionia for his materialistic views and for not agreeing with the divine character of heavenly events.

Among other interests, he studied meteorology. His sound scientific approach was also visible in meteorology. His approach was that one should first carefully observe nature and then carry out experiments to test some hypothesis when the observations are not enough. It is essentially the same approach to science as it is today. The first meteorological phenomenon he considered was the appearance of hail in the summer months. This confused all the natural scientists who said that water cannot freeze in warm summer weather.

He concluded from his observations that the air temperature decreases with increasing altitude and that the cloud contains moisture. From there, he drew the conclusion that even in summer the water in the clouds can freeze at high altitudes. The question arose as to what forces the cloud to rise to the height where icing occurs. Anaxagoras found the answer easily, since he knew that heat causes air to rise, creating convective currents in the atmosphere. Therefore, the heat of a summer day drives the moist cloud to such heights that the moisture freezes and returns to the ground in the form of hail.

The decrease in air temperature with increasing altitude Anaxagoras explains that with increasing altitude, the intensity of reflected sunlight from the ground decreases. This thermal effect, according to him, occurs only up to a certain height in the upper space. Above that point, the atmosphere has a different composition. It consists of a substance it calls "ether." Because of this, in the upper atmosphere, the temperature would start to rise, becoming so hot that it starts to burn. But this happens above such high altitudes that it is not significant for meteorology. According to this, we see that Anaxagoras described the exact change of temperature with height for completely wrong reasons. It is interesting to note that this exact change of temperature with height was not accepted in scientific circles until the nineteenth century.

Anaxagoras introduced "ether" into the upper atmosphere to explain the cause of thunder and lightning. As Aristotle later wrote, Anaxagoras believed that there is also fire in the clouds, which is a part of the ether from the upper universe that has descended into the lower atmosphere. The lightning was due to the flash of this fire through the clouds, and the thunder was a screeching noise created by extinguishing the fire with water from the clouds. Anaxagoras did not answer why the warm substance "ether" descended into the atmosphere instead of rising, as he pointed out in the theory of the origin of the summer city.

Thales' and Anaxagoras' theory of the basic element (water or air) encouraged other "scientific rivals" to think in the same sense. Thus, Empedocles from Argentum (around 492–430 BC) introduced the four basic elements with the Pythagoreans—air, water, earth, and fire, with four properties—hot, cold, wet, and dry. This understanding was represented in meteorology for 2000 years. Since water extinguishes fire, Empedocles concluded that these are two opposite elements.

Empedocles' interest in meteorological phenomena included an explanation of the origin of thunder and lightning. His theory is essentially the same as that of Anaxagoras, except that he claims that the fire in the cloud is actually a trapped ray of the sun. He seems to have been the first to suggest that lightning originates in a cloud.

Applying his concept of the universe of four elements, Empedocles tried to explain the opposite climatic characteristics of summer and winter. The basic elements of fire and water are constantly against each other in the atmosphere. When it's warm, dry fire reigns, and summer begins. When it is wet, cold water prevails, and winter occurs. Since these two basic elements change in a random way, he did not explain why these two seasons alternate regularly.

The next, which in the fifth century BC dealt with the phenomena of weather, was Democritus (about 460–370 BC). He is known as a geometer and proponent of atomic theory. He was a great traveler. He wrote: "I wandered over the greater part of the Earth more than any man of my age, inquiring about distant things. I observed many climates and countries and listened to many learned men…". This was probably during his time in Egypt when the problem of the annual flooding of the Nile was considered. Like Thales, he emphasized the importance of the north wind of Etesia. But Democritus' interpretation is more complex and includes the consideration of a larger number of atmospheric processes. He takes into account the melting of snow in the northern part of the world, around the summer solstice, and the water runoff from that. Then the formation of vapor clouds, and their movement toward the south, toward Egypt, pushed by Etesia. It produced violent storms that filled the lakes and the Nile.

It is important to note here that Democritus, like Anaximander, viewed wind as a current of air. More importantly, he introduced the concept of storm movement, which was not discussed until the eighteenth century. It was considered that the storm does not move from one place to another.

Democritus applied his atomic theory to the definition of wind, that is, he gave the cause of air movement. He claims that wind occurs when there are many particles (atoms) in a small space, while on the other hand, there is a calm state in the atmosphere where there are few particles in a large space.

Thunder and lightning also occupied the attention of Democritus. Based on the atomic theory, he explains thunder and lightning as a unique mixture of particles that produce strong movements from or within the cloud. It is concluded that both of these phenomena occur simultaneously, but we feel them separately, because sight is faster than hearing. This important statement about the simultaneous occurrence of lightning and thunder was not mentioned by the followers for the next 2000 years.

Certainly, Hippocrates of Kos (about 460–375 BC) cannot be left out when mentioning learned Greeks who focused their thoughts on meteorology. He is considered the "father of medicine." He sees the understanding of nature as one of the important parts of his medical doctrine. He believed that one cannot be a good doctor without knowing meteorology. In his treatise "On Airs, Waters and Places," Hippocrates analyzes the different climatic parts and their effect on the health of the inhabitants, as well as the diseases that prevail in locations exposed to specific winds. It is obvious from the title of the discussion that he considered that there are several types of air, with special characteristics, related to that place. It can also be seen that he did not try to look for the causes of those local meteorological phenomena. He only observed the influence of weather on people's health, which is still poorly known even today.

The last known Greek before Aristotle's period who studied meteorology was Plato's student Eudokia (around 408–355 BC). He was believed to be the author of the work "Ceimonos Prognostica"—Forecasting of Bad Weather. He worked on problems related to the periodicity of meteorological phenomena. He noticed this periodicity not only in the wind but also in other features of bad weather. According to his conclusion, the periodicity was four years, and that period always begins in leap years, at the rising of the star Sirius.

In this earlier phase of the development of meteorology, most naturalists paid considerable attention to meteorology. However, that study was based on observation and mental reasoning, devoid of quantitative indicators, because meteorological instruments were mostly found in the seventeenth century. Whether a theory was accepted or not depended on speculation not on actual established facts. Nevertheless, all these theories form the basis of Aristotle's meteorology, which was the undisputed authority on meteorological theory for the next 2000 years (Alhazen 1572).

3.1.4 Aristotle's Meteorology

Meteorology included a significant part of the total knowledge about nature. At the beginning of the first book, Aristotle says: "We will consider the contents which the earlier writers called meteorology." This shows that this science got its name much before him. In fact, even Homer (around the eighth century BC) used this name, but by it, he meant everything that is above the ground. Aristotle divided that space into two parts: the first, which extends from the Earth to the Moon, and everything that happens there belongs to meteorology, the second, above the path of the Moon, and that is described by astronomy. This determined the content of Aristotle's Meteorology.

Meteorology is written in four books. The first book describes phenomena that occur, according to Aristotle's division, in the upper atmosphere (comets, shooting stars, etc.) but also hydrometeors, wind, rivers, and springs. According to Aristotle, the upper atmosphere is dry and warm, while the lower is moist. In the second book, we talk again about winds, earthquakes, lightning, and thunder. The third book describes storms, eddies, and light phenomena in the atmosphere. The fourth book covers mainly the chemistry of the atmosphere, i.e., the "Theory of the Four Elements."

The content of Meteorology shows that the Greeks in Aristotle's time knew most of the most important meteorological phenomena (Freeman 1953; Heath, editor. Archimedes 2009). They were such careful observers that they were not even unfamiliar with the northern or polar light. He knew that the city appeared more often in the spring than in the summer and much more often in the fall than in the winter. He knew that in Arabia and Ethiopia, it rains in summer and not in winter, as in Greece, and that lightning precedes thunder. He noticed that the colors of the rainbow are

Aristotle (384–322 BC)

He was born in Stagira, a Greek colony located a few kilometers from the current Hilandar monastery complex on Mount Athos. At the age of 17, he came to Athens as Plato's student. When Plato died, in 347 BC, he settled in Lebos. His reputation as a scientist was very great, and he became the teacher of the young prince Alexander of Macedonia. When he returned to Athens, he was the most famous teacher, writer, and philosopher of that time. He wrote the first book on atmospheric phenomena, titled: "Meteorologica"—Meteorology.

always very similar and that in the outer, weaker rainbow, the sequence of colors is opposite to that of the primary, inner rainbow, and that the dew occurs in light winds.

Aristotle wrote Meteorology guided by two basic principles. The first is that the universe is considered to be spherical in shape. He accepted Eudoxus' system of the motion of the stars and planets in concentric spheres. Earth is at the center of all spheres. Aristotle divided the universe into two regions: the region of the stars, which is outside the region of the Moon's path, and the region of the Earth, or the sublunar sphere (Fig. 3.2).

His second principle is based on Empedocles' theory of the four elements, with the fact that Aristotle distributes them in the area of the Earth as shown in Fig. 3.2. The spheres of earth, water, air, and wind are not regular and solid, as shown in Fig. 3.2; they already permeate each other in some places. Thus, dry land rises above the water, and fire often burns on the ground. That permeation is constantly

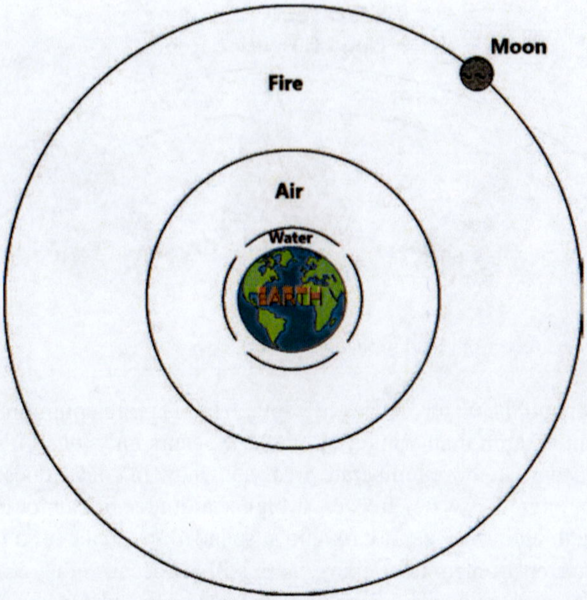

Fig. 3.2 Aristotle's concept of the universe of the sublunar sphere

changing in form. When the heat from the Sun reaches the Earth's surface, it mixes with cold and moist water, creating a new substance that is similar to air but warm and moist. By mixing that heat with the cold and dry earth, another substance similar to fire, which is hot and dry, is created. Thus, the Sun creates two types of vapors. One is mostly moist and in the form of warm vapor. Phenomena such as clouds, rain, etc., are created from it. The other vapor is hot and dry, and it represents the ingredient for wind, thunder, etc.

Aristotle's atmosphere contains two movers, air and fire. Within the sphere of air, he makes a further division. Thus, according to his theory, clouds are not formed either above the tops of mountains or near the ground itself. Because there is fire above the tops of the mountains, and next to the ground itself, the reflected heat from the ground prevents the formation of clouds. The area of cloud formation is shown in Fig. 3.3.

In Aristotle's Meteorology, there are many facts collected from predecessors, natural philosophers, historians, poets, and many significant experiences. Many of the prognostic rules of the weather were taken from the Egyptians and Babylonians, especially those about wind classification. The content of Meteorology was everything that was known about meteorological phenomena up to that time. At the same time, it should be emphasized that Aristotle himself developed a large number of meteorological theories. Here it will be stated how he explained the processes related to hail formation.

"When observing the processes that lead to the formation of hail, one must take into account facts that are easily explained but also those that seem inexplicable."

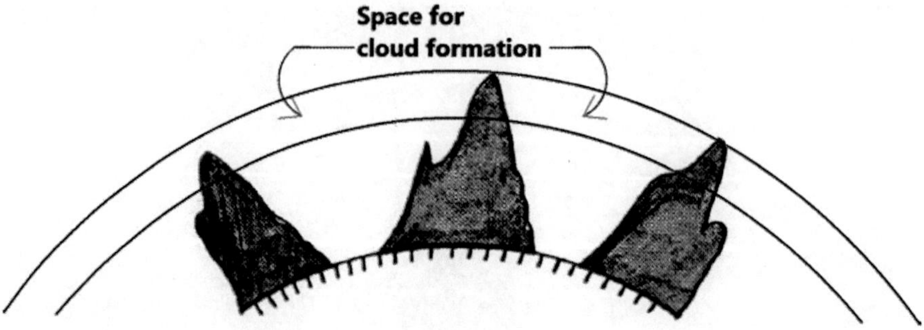

Fig. 3.3 Aristotle's theory of cloud formation

The hail is ice, and the water freezes in winter. Hail is more common in spring than summer and in autumn than winter, in winter it occurs only when it is not cold. In general, hail occurs in more temperate areas and snow in colder ones.

It is also strange that water freezes at higher altitudes, it cannot be frozen until water is formed, and water cannot remain suspended in the air for a long time. But we note that water droplets float above us in still air because they are small, much like tiny particles of earth or gold can float on water. Thus, the water will float in the air until a large number of tiny droplets combine to form a large falling drop. This cannot happen in the case of hail, because frozen drops cannot coalesce like when they are liquid. It is clear that the suspended drops must be of the required size, because their freezing could not create such a large hail.

Some think that the cause of its formation is as follows: When the cloud is forced to rise in several areas, the temperature is lower up there, because the reflected solar rays from the ground do not reach a great height, and the water freezes. That's why hailstones occur more often in summer and in warmer areas because the heat forces the cloud to rise further from the ground. But, in places of high altitude, hail rarely occurs, and according to their theory, this would not be like this. We know that snow falls in places with high altitudes.

Forms are often seen descending close to the ground with a great noise, causing fear, from what is seen and heard, that this is a sign of some greater disaster. But sometimes, even when such forms appear without noise, hail falls in large quantities, with large grains of irregular shape. The reason is that the grains formed close to the ground and do not fall from a great height. Smooth grains fall from a great height and are polished as they fall.

We now know that there is an interaction between hot and cold (which is why deeper in the earth is cold in warm weather and warm in frosty weather). Such reactions must also exist at altitude. Thus, in the warmer season, the cold is concentrated within the warm environment. This often results in the rapid formation of cloud water. This is why we have larger drops on warm days than in winter and heavier precipitation. Precipitation is stronger when it is more intense, and it is more intense when there is rapid condensation… Hail is less frequent in summer than in spring and autumn, because the air is drier in summer and more humid in spring, while it becomes more humid in autumn. Because of this, hail sometimes occurs in late summer.

If the water is preheated, it will contribute to its faster freezing and faster cooling. Because it is known that many people when they want to cool water quickly, first keep it in the sun. It is similar to the inhabitants of Pontus when they encamp on the ice to fish. They catch fish through holes they make in the ice. To get their sticks well and quickly fixed in the ice next to those holes, they pour hot water on the ice around the stick. Then the water around the stick quickly freezes. That's how they fix sticks in the ice. For the same reason, water condenses faster in the air in warmer areas and warmer seasons.

For this reason, in Arabia and Ethiopia, it rains in summer and not in winter, and it often rains several times a day. "Clouds cool faster due to the reaction with the large heat from the land."

From this part of Aristotle's text, which is taken from the twelfth chapter of the first book, it can be concluded what Aristotle's discussions are like. He first introduces the theory, showing what others think about it, and then refutes it.

Anaxagoras and others before Aristotle used the inductive method in explaining weather phenomena. Their theory was based on observations. Aristotle uses a deductive approach. Instead of using data to develop a theory, he first develops a theory and then uses the data to support his own position.

Aristotle rejected the opinion of Anaximander and others about wind as flowing air. His theory starts from the fact that the Sun creates two vapors from the Earth. According to him, the wind is a warm and dry vapor. He explains the origin of the wind analogously to the origin of the river. The water of the river gradually collects in the higher mountain areas and descends to the lower areas. The wind is also created due to the gradual accumulation of dry and warm vapor from the Earth. Aristotle explains that the wind flows horizontally and the vapor rises vertically, "because the whole air that surrounds the Earth follows the movement of the heavens."

Aristotle believes that there are two types of wind: north and south. Northerly winds originate in cold regions, in the region below the constellation of the Big Dipper. It is the upper limit in the north where people live and where it is cold. Those from the south come, not from the South Pole, but from the tropics, where the southern border is where people live, because beyond that it is too warm for life. Since they come from warm areas, southerly winds are hot.

In Aristotle's time, the Greeks had poorly developed symbols for determining the sides of the world. That is why Aristotle determined the direction of the wind according to astronomical directions, such as the position of the sunrise at the equinox, the winter sunset, the position of the sun at noon, etc. He divided the entire area of 360 degrees into 12 equal areas (Fig. 3.4). This division is reminiscent of the Babylonian one.

What is characteristic of that era is that it was considered that the direction from which the wind blows automatically determines other characteristics of the weather. So, certain winds are always accompanied by the same type of weather. Regarding this, Aristotle writes: "Apaktias, Trakias, and Argestes (which are roughly equivalent to the winds that blow from the corners of the world: north, north-northwest,

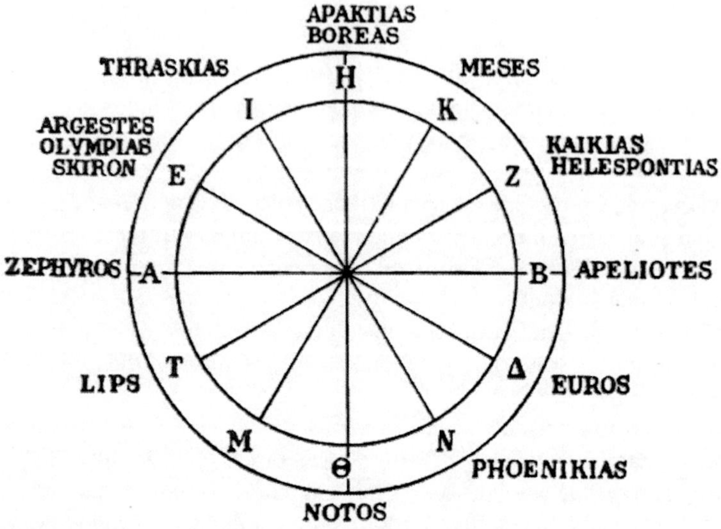

Fig. 3.4 Greek wind rose

and west-north-west) disperse thick clouds, bring clear weather, or in every in case the clouds are not too compact. The effects of these winds are different if they are cold and not so strong. Then they cause vapor condensation instead of cloud dispersion. Argestes and Euros (east-southeast wind) are dry winds when they start to blow and become wet towards the end. The moon (north-northeast wind) and especially Apaktias bring snow because they are very cold. Apaktias also brings the city, as do Thrakias and Argestas. Notos (south wind), Zephyros (west wind), and Euros are hot. Kaikas (east-northeast wind) covers the sky with thick clouds, while the clouds brought by Lipa (west-southwest wind) are not so thick…"

Aristotle tries to explain these properties of winds. The explanations are as follows: "… There are more winds coming from the northern countries than those from the south. Far more rain and snow are brought by those from the south, because they lie below the Sun and are located in its path."

It should be noted here that these statements of Aristotle are in contrast with the author of the "Song of Songs" from the Bible, who claims that: "in Palestine, the north wind brings bad weather. This discrepancy can be easily explained. In an arid (dry) area, to which Palestine belongs, the penetration of a warm front may be without accompanying clouds, while the penetration of a cold front is accompanied by clouds, rain, and thunder. Along the northern shores of the Mediterranean, on the other hand, tropical air that brings moisture from the south is sometimes accompanied by snowfall in winter and heavy spring rains".

This strong feeling that the wind "rules" the entire type of weather is also expressed artistically in a building called the "Tower of Winds." The tower was built in Athens by Andronikos Kirestes in the second century BC (Fig. 3.5).

Fig. 3.5 The Tower of the Winds in Athens with an isolated detail

The tower is in the shape of an octagonal prism. At the top of each page are mythical figures that characterize a certain type of weather. At the top of the tower is a mounted metal pointer that shows where the wind is blowing, a weather vane. The weather vane is a bronze statue of the god Triton turning toward the wind, so that the staff he holds in his hand indicates the direction of the wind. Thunder and lightning were the subjects of speculation by all thinkers, including Aristotle.

He argued that dry vapor in the atmosphere that is trapped in a cloud is forced out when the cloud condenses and when the cloud collides with another cloud. Then there is thunder. Because the cloud is not uniform, different sounds occur. The ejected "wind" burns in the form of a "fine and mild fire, which we call lightning." Contrary to most of his predecessors, Aristotle wrongly claims that lightning appears after thunder. Aristotle also misinterpreted the strong winds in hurricanes, with a combination of dry and moist vapors (Book III, Chap. 1).

Aristotle's treatises show how great mistakes were made in reasoning due to the lack of experimental methods. Meteorology was written by a philosopher of nature, not a scientist of natural science, meteorology. Regardless of the misconceptions expressed in it, Meteorology is of great historical value. It is the earliest attempt to present knowledge about meteorology in such a comprehensive way.

3.1.5 Theophrastian Signs of the Weather

Before moving on to describing the next period, let's focus a little more on an important Greek scientist who outlived Aristoteles by about 40 years, and in terms of his interest in weather forecasting, he surpasses all his predecessors.

We mentioned earlier Theophrastus (373–287 BC) and his "Signs of the Weather." His signs of the weather are not the result of summing up sayings and experiences from the past. He makes forecasts based on some general principles, which are based on the following:

The first rule is that the forecast period, e.g., years, divide in half. The year is divided in half by the principle of the rising of the Pleiades. The first part of the year is from its sunset to its sunrise. That period is further divided into two parts, the solstice and equinox points. According to this, whatever conditions prevail in the atmosphere when the Pleiades sets, such conditions will prevail until the winter solstice. If some changes happen, they will happen after the solstice. If the changes do not occur at the solstice, the weather will be the same until the vernal equinox. The same principle applies to the second part of the year, from the vernal equinox to the rising of the Pleiades. So, what the meteorological conditions are like on the day of the vernal equinox, they will be like that until the summer solstice, and from then until the setting of the Pleiades.

Similarly, each month is divided starting from the full and new moon. See it on the fourth and eighth day after the full moon. The rule applies to the annual forecast. As the weather is when the moon is full, it will remain so until the fourth day. That's when changes usually happen. If there is no change, then the same time will be until the eighth day.

The day is divided according to the same principle. The turning points are sunrise, midday between sunrise and noon, midday, mid-afternoon, and sunset. The night period is similarly divided. According to the weather in these moments, it is forecast whether it will be good weather or stormy. If there are changes in the weather type, then they occur at one of the specified breakpoints.

Applying this principle, and the principle of the general balance of weather in the course of the year, Theophrastus makes the following general forecast: "If much of the rain falls in the winter, the spring will usually be dry. If the winter is dry, the spring will be cold."

The behavior of animals and birds has served as an indicator of the future for centuries. Even today, it is a widespread means of forecasting among the people. Some of such rules are: "If a cow licks its front hoof (hoof) there will be a storm or rain." Or, "If a dog rolls on the ground there will be a strong storm." "Early mating of sheep in autumn means that winter will start early."

Observation of the state of the sky and the Moon is used by Theophrastus for forecasting. Thus, if many shooting stars are seen in the sky, there will be rain or wind: "If the Moon has the color of fire, that month will be cool with a light wind"; "If the moon is dark it will be wet weather."

Theophrastus also tried to explain many meteorological phenomena, but he always referred to Aristotle's views (Hort 1948). However, there was a big difference between them. Aristotle was an extensive theoretician, while Theophrastus, with short treatises, was a great practitioner. Thanks to him, the oldest signs of the weather have been preserved.

About 10 years after the death of Theophrastus, Eratosthenes was born (about 274–194 BC). He, among other contributions, finally solved the cause of the flooding of the Nile. He said that one should go to the area where the Nile rises. It was found that the rains in that area are the cause of the increase of water in the Nile bed.

3.2 The Dark Ages

3.2.1 Introduction

The period after Aristotle was significantly marked by his student Alexander the Great. He discovered new areas through the military conquests he undertook and faced a different climate. His generals described for the first time the monsoons they faced in India. Science moved from Greece to Alexandria. After that came the Roman era. The Romans did not care much for science. It was significant for them only if it was practically applicable.

One of the important naturalists of that time was Poseidon (135–50 BC). He also dealt with meteorological speculations that were close to Aristotle's theory. And he claimed that thunder is actually the burning of dry vapor that is trapped in a cloud, while most ancient thinkers claimed that clouds occur up to a height of about 150 km. Poseidon said that wind and clouds occur up to a height of about 10 km. Above that, the air is clean, fluid, and completely transparent. Drugi značajni naučnik iz toga perioda je.

Claudius Ptolemy (about 85–165 AD). He was also involved in weather forecasting. In the section "Tetrabiblos," he lists several astrological weather forecasts.

"One must monitor the state of the moon three days before and after the new moon, the full moon, and its quarter. When it is thin, clear and there is nothing around it, it means that the weather will be nice. If it is thin and red, and the entire disk is covered by some dark disturbance, there will be a wind that will be directed in the direction of those disturbances. When the moon is dark, or pale, and full, there will be storms with rain."

We should also mention Hira from Alexandria, who around 100 BC studied the elasticity of air. He also invented small air pumps. He observed the thermal expansion of air and water vapor.

3.2.2 From Seneca and Pliny to Descartes

The Roman period was significantly marked by two names, Seneca (third year BC–65 AD) and Pliny the Elder (23–79 AD). They were not scientists but engaged in copying and commenting on the theories of Greek scientists. Thanks to them, many of these theories have survived.

Toward the end of his life (63–54), Seneca (see Clarke 1910; Gorcoran 1971) wrote the work *Quaestiones Naturales*—questions about nature. He mainly deals with astronomy and meteorology. In the discussion, he combines the findings of Roman scientists with Greek, Babylonian, and Egyptian. It includes all meteorological phenomena, from wind, thunder, and lightning. He always cites earlier theories about this and presents his conclusion, which is a sort of compromise between earlier views. Thus, the wind is considered to be not only a flow of air (according to Anaximenes) but also evaporation from the earth (according to Aristotle).

Seneca was a careful observer of the weather, which shows that he possessed a certain degree of scientific spirit and imagination. He was first and foremost a moralist and secondly a scientist. He believed in fate, and since the atmosphere is between the earth and the heavens, atmospheric phenomena, such as thunder, are tied to fate. Here we are faced with facts that place us, today's people, among those who distinguish the ancient Greek thinkers so much in relation to Seneca and his contemporaries.

Pliny's most important work is the *History of Nature*, composed of some 2000 works, of which 146 were by Roman and 326 by Greek authors (most of these works are now lost). The second book, or chapter, deals with meteorology. Pliny states that more than 20 Greek scientists dealt with meteorology from the earliest times. They published meteorological observations of various meteorological phenomena. He discussed various theories of earlier authors but did not make any personal contribution.

Pliny discusses the wind in particular detail, as he states, "having in mind hundreds of sailors – navigators, who depend on it." His views are interesting: "The first spring opens the sea to sailors: the Favonius (west wind) softens the severe winter six days before the end of February…" Pliny claims that the Favonius blows on February 22 and from February 28 to March 8, Solana (east wind, Fig. 3.6) blows on May 10, when the Pleiades rises, Aquilos (northeast wind) blows from July 10, etc. It is obvious that Pliny when discussing the wind uses Aristotle's descriptions. But, nevertheless, he points out quite clearly that these properties of winds depend on latitude. Thus he states: "They have two winds which by nature change when they enter another country." In Africa, Auster (south wind) brings warm weather and "Aquilo brings clouds," (which is the opposite of what happens in Italy). And Pliny with his *History of Nature* saved earlier works on weather from oblivion.

With the fall of the Roman Empire, another period begins without any significant development in meteorology. But, nevertheless, in the period between 400 and 1100 AD, the interest in meteorology did not die completely. In some places, the spirit of Greek and Roman learning broke through alone. It mostly happened inside

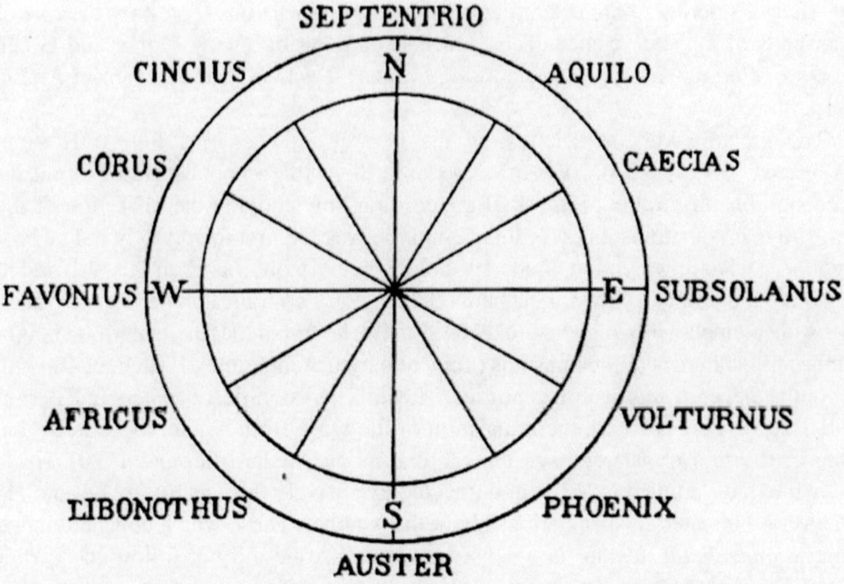

Fig. 3.6 Roman wind rose

the church. One of the greatest medieval church scholars was Brad (ca. 673–735). He was the first Englishman to write about the weather. He is called the founder of meteorology in England.

In his De Natura Rerum, written around 703, Brad devoted a chapter to the atmosphere, wind, thunder, lightning, clouds, and snow. In this chapter on meteorology, he summarized the knowledge available to him mainly from classical sources. It talks about already known theories about wind, as well as about wind as a disturbed movement. It also describes thunderstorms created by the collision of clouds driven by the wind. Brad does not cite Aristotle's theory, since it was not yet known in Western Europe until the twelfth century. Brad's discussions of meteorology were not without their superstitions. Thus, he claims that a thunderstorm with a westerly wind brings "a very great pestilence – a plague." Nevertheless, looking at the whole, he tried to present meteorology less in a philosophical and more in a scientific sense. There are some signs of progress.

Brad was not the only medieval scientist interested in meteorology. At the beginning of the seventh century, Isidore (ca. 570–636), a bishop from Seville, appeared. In his work *Etymologiae, De Ordine Creaturarum*, he devotes a considerable part to meteorological questions. Similar to Breda, Isidore also disputed the prevailing theological view of science at the time. However, he expressed some strong thoughts when discussing weather phenomena such as frost, rain, hail, and snow. His theory on this was reasonable. He thereby gives the possibility that common sense can enable the development of meteorology.

After this period, there is a big upheaval. Arabs—Muslims—pre-take the role of guardians of ancient science. They translate into Arabic many Indian and Greek works, including Aristotle's meteorology. Later, it was all translated from Arabic to Latin.

The greatest Muslim scientist in the natural sciences was Ibn Al Hai-tam (Alhazen) (c. 956–1039). He is best known for his work on optics. He also contributed significantly to the progress of meteorology by studying the refraction of the sun's rays in the atmosphere. In this regard, he was the first to correctly define twilight, as a state in which the Sun is 19° below the horizon. Based on this, he tried to calculate the height of the atmosphere. Using a very complicated geometric procedure, the elements of which he took from Euclid, he found that the maximum height of the atmosphere is 52,000 pasums (units of length at that time). It is about 400 km.

In the eleventh and twelfth centuries, Muslims had significant centers in Palermo and Toledo. This is where the translation of the work from Arabic to Latin began. One of the first translators was the English monk Adelard (around 1120). He is known to have studied in Spain and traveled extensively throughout the Levant. He was very interested in speculations about the weather. Those which appeared in his *Quaestiones Naturales* are in a sense new, because he did not follow the earlier Greek theories completely. For example, when discussing wind, he says, "Wind is air in a state of motion, and dense enough to have a motive power; That's why I think that wind is a kind of air." It also explains why the wind travels around the Earth, and how it achieves such tremendous force.

And his theory about thunder and lightning is original. He believes that thunder is created by the breaking of ice during a collision in the cloud. In summer, it is conditioned by the melting of the ice that collides in the cloud. As for lightning, Adelard believed that with every strong collision of bodies, the lightest parts are separated from them first. Ether in the form of fire is the lightest substance in the air, and it is released and expelled from the clouds when the ice collides. That's how lightning is created.

In the thirteenth century, thanks to translations, Aristotle was again at the forefront of meteorology. His systematically developed theory was superior to others. For the next 400 years, his theory enjoyed almost undivided support. In that period, the work on meteorological problems was reduced practically only to commenting on his work. It is estimated that by 1650, 156 commentaries on Aristotle's meteorology appeared. In the thirteenth century, the most famous "commentators" were the Dominican theologian Thomas Aquinas (1225–1274), de Beauvais (about 1190–1264) from France, and Albert Magnus (1206–1280) from Germany.

Research in the Middle Ages mostly took place in libraries. This was especially pronounced in meteorology. When any problem had to be discussed, between the choice of going out into nature and observing phenomena, and going to the library, scientists would choose to go among the books. Thus, their own experimental record was suppressed by the written positions of authority in the books. Before meteorology could take a new path of development, this condescension toward the authorities had to be broken. It was a slow process, the originator of which was the most famous English scholar of the thirteenth century, Roger Becken (1214–1294).

Becken is a great Franciscan scientist, who is often called the father of modern science in Europe. He vigorously advocated the experimental and mathematical approach in all scientific works, including meteorology. In his work *Opus Majus*, Becken follows Aristotle's theory of the composition of the atmosphere of water, air, and fire, arranged concentrically around the earth. He also analyzed Ptolemy's views on climate. He completed his definition of cold being found in the north by saying that cold can also be found at higher altitudes in the mountains. He thus observes the influence of orography on the climate. In the section "In Meteora," Becken presents a large number of comments on Aristotle's Meteorology. He points out that experiments and observations are more important in science than believing in the views of authorities. It was undoubtedly the first significant step that freed meteorology from the shell of Aristotle's theory. That shell was finally broken only in the seventeenth century.

3.2.3 Forecasters: Prophets of Weather

After Becken, there was a complete standstill in the development of meteorology for 400 years. In the middle of the sixteenth century, meteorology began to develop in two opposite directions. Some were dealing with Aristotle's theory, and others were developing pseudoscience. They prognosticated or, better said, they predicted the weather. Especially popular was the forecast based on the "signs" of the heavenly bodies. It was astrometeorology. Thus, a strong connection between astrology and meteorology emerged. Astrometeorology had strong support from the church and the government.

Among the astrologers who made weather forecasts were prominent cultural and scientific workers such as John Miller (1436–1476), Leonard Digges (around 1550), and Johann Kepler (1571–1630). They gave legitimacy to that pseudoscience.

Those prophets of weather from the Middle Ages can be divided into four groups. They were the first to deal with forecasting based on celestial phenomena, including the moon. Others made predictions based on the weather prevailing on certain significant dates. A third predicted based on the behavior of animals, and a fourth predicted based on the current state of the weather.

The methods of forecasting in the Middle Ages of those of the first group are similar to the methods developed in antiquity. This was discussed in Sect. 3.1.3. It will be added here that a lot of attention was paid to the forecast of late spring frosts. It was thought to be related to the moon. Because, during clear spring nights, the moon can be seen, and the temperature drops due to the radiation of ground heat. One of the widely held notions was that the weather of the entire month would be as it was on the third, fourth, fifth, or sixth day of that month. That's the rule: "Quintus, Sextus qualis tota luna tails."

Another way of forecasting is related to the weather on certain dates. The dates are usually some church holidays. Those rules depended on the region in which they were created. Large geographical differences also contributed to different meteorological conditions on a given date. Thus, in Russia, it is said: "Pokrov (October 14-bar) heats our house without wood"; "Alexei (March 30) brings streams from the mountain." December 7 is called "moist Catherine"; etc. In France, for example, says: "On St. Blaise's day (February 24th) winter often loses its teeth," or "If it snows on St. Ambrose (April 4th) there will be eighteen more days of cold weather after that."

Forecasters—prophets made forecasts for a longer period based on the weather and also from a longer period. Thus, in the Arkhangelsk region, it is said: "If there is less snow in the winter, there will be little rain in the summer." It is also said that "in the seventh year, winter turns into summer, and in the seventh year, summer turns into winter." This would mean that every seventh year is extreme, i.e., summers are extremely cold, and winters are extremely warm. Many people today and in our country consider the repetition of this saying to be top knowledge in meteorology.

Of the forecasts of this type, the most important rule is the weather forecast for the whole year. It outlived all the rules of forecasting, because it was easy to follow and covered such a long period. That rule was based on the type of weather during 12 consecutive days counting from Christmas. Thus, Leopold of Austria (thirteenth century) stated this rule in his "Derivations." It states "that the old people claimed that the weather on Christmas Day and during the next eleven days will be the same for twelve months of the year." The fact that different calendars were in use (so Christmas falls on different dates) was not taken into account.

Predictions of the third type were based on the behavior of animals, birds, and other living beings. They are very old and were collected by Theophrastus. Such forecasts were mentioned in 1340 by William Merle in his book *De prognosticateonis aeris*—about weather forecasting. Also, later (in 1554) A. Mizo lists 46 indicators of bad weather, 42 of which refer to the behavior of animals, birds, and insects. Even today, we have such prophets who claim that before the rain, swallows fly low above the ground. It is difficult to confirm and deny this rule without systematically conducting research.

Forecasts of the fourth type, based on the weather characteristics themselves, are more significant than others. Those rules are centuries old and still interesting. May they still fit well into the scheme of short-term local weather forecasts.

Konard of Megenburg published a book in 1340: *Buch der Natur*—The Book of Nature. In it, he states that a rainbow is a sign of rain that will soon begin if the bow of the rainbow expands and contracts, "because the vapor becomes denser and a cloud is formed." This would correspond to the weather following the arrival of a warm front.

In ancient sayings, we have seen the wind was often taken as a sign of the weather. Thus, in some areas, it was claimed that the wind from the direction of the summer sunrise, east-northeast, brings frost in winter and good weather in summer.

The daily changing of the wind (local circulation during the day and night) was taken by many peoples as a sign of good weather. Today it is understandable. Because when local air circulation occurs, it means that there are no "penetrations" that would disturb the stable situation in which local winds occur. In those areas, it is said that "the wind that follows the movement of the Sun brings good weather." The prophets also took the clouds as signs of the weather. Thus it was said: "Striped clouds (obviously meaning cirrus clouds) bring rain and warm weather in winter." Or, "if the clouds in the form of flocks of sheep, or larger masses, gather together, there will be rain." "Thunder in the morning heralds windy weather, rain at noon, and a big storm in the evening."

The audibility of the ringing of church bells was also used in prophesying the weather. It was said: "If the sound of the bell is clearer, it will be clear and cold in winter, and if the sound is muffled, it will be rain or snow." The first part of the saying is true, since good audibility is related to the temperature inversion at the ground, which occurs on cold winter days. The second part of the statement is more difficult to explain, although it is known that there is attenuation of sound through the medium with precipitation.

A large number of sayings were collected in the English book *The Shepherd of Bambury's Rules*, which was printed in 1740. It has precise rules that forecast various detailed types of fog (rules 9 and 10), different types of rain (rule 14), and lots of facts describing the penetration of cold air accompanied by thunderclouds (rule 6). All such detailed records of the nature of the process have a scientific value.

3.2.4 Some Weather Records

The most valuable and interesting materials from the Middle Ages have reached our days in the form of various records—chronicles. These records were made by individuals or official institutions. These records differ from others in the length and precision of the description. With various historical events, there are descriptions of storms, floods, snowstorms, etc. These events are recorded year after year. Although these records are not actual "weather diaries," they are clear and sometimes artistic descriptions of weather phenomena. A good example of these records is the *Saxon Chronicle* and *Holinsheed's Chronicle*.

Certainly, not all chronicles can be shown. Some will be selected that show the weather in the best way. The Moscow Chronicles contain the following record from 1092: "This year there were many signs in the heavens; a large circle appeared in the middle of the sky. It was very clear that year, so the land was dried up and a lot of forest was burnt." In the same chronicle for the year 1164, it is written: "… this year there was a great flood in Galicia, and by God's will the rain suddenly intensified both day and night, so that great water poured down the Dniester… and more than 300 people drowned."

In the Armenian Chronicle for the year 1275, it is written: "18. In April of this year, a sign appeared on the Sun: colored arcs were seen around it, dark blue, green, yellow, bright red and purple, and in its center a cross."

It was recorded in 1230 "that a severe frost destroyed all the winter crops, so that after that a great famine prevailed in all of Russia, except for Kyiv." And for the year 1371, he writes: "This year, a thick fog descended that lasted for about two months, and a man could not see two sagens (an old measure of length, 1 sagen = 2,134 m) in front of him." Even birds could not fly but landed on the ground and walked on it. "In the Moscow Chronicle for the year 1406, it is written:" "On St. Peter's Day (July 12), Nizhny Novgorod was hit by a storm and a whirlwind so that a man with a horse and cart was carried away and disappeared." "Only the next day they found the car on the other side of the great Volga, hanging from a big tree… but the man was never found."

In the Moscow Chronicle, there are a large number of records about various weather phenomena. Since 1740, there are 350 records. Storms, electrical discharges, early snows, early and late frosts, halo, and drought are mostly described. There are descriptions related to the aurora borealis. Unusual events are, e.g., the snow that fell on April 26, 1498. It is recorded that "the snow fell to the middle of the knees and remained on the ground for seven days." In order to understand the significance of such a record, one should know the typical events for that phenomenon in that area. There are mentions of great droughts that occurred in 1024, 1060, 1092, 1124, 1161, 1193–1194, 1223, 1298, 1325, and 1365, which was the driest and when a large part of Moscow burned down in flames.

In addition to these chronicles, those from China are also interesting, which are detailed, and cover the period of 1500 years. Here, a summary of the number of droughts and floods will be displayed. That presentation by centuries is given in Table 3.1.

The table shows that from the fourth to the seventh century, there was a relatively dry period. It is known from historical sources that many people emigrated from China between the seventh and twelfth centuries.

The Chi Cheng Encyclopedia of China, they have three chapters dealing with severe frosts and periods of extreme cold. The number of those harsh winters in China and Europe is shown in Table 3.2.

Obviously, there is a similar trend in these secular temperature changes between the two separate parts of the world.

In the following example, some records of Jules Jasmin from Liège will be cited. He cites an exceptionally warm spring and summer in 1473, when roses bloomed in

Table 3.1 Number of floods and droughts in China in 15 centuries

Century	II	III	IV	V	VI	VII	VIII	IX	X	XI	XII	XIII	XIV	XV	XVI	XVII
Floods	18	15	5	18	10	13	31	24	36	41	56	43	57	24	43	67
Droughts	35	24	41	37	41	43	41	43	64	69	58	77	60	54	84	82

Table 3.2 A number of cold winters in China and Europe

Century	VI	VII	VIII	IX	X	XI	XII	XIII	XIV	XV	XVI
China	19	11	9	19	11	16	94	25	35	10	14
Europe	–	–	–	11	11	16	25	26	24	20	24

the gardens on Easter (April 18) and the harvest ended on St. John's Day (June 24). It is recorded that in 1480, the frost started on December 26 and lasted continuously until February 6 of the following year. The frost was accompanied by heavy snowfall. That year, horses and carts moved on ice, wine froze in cellars, and water in wells. Until then, there was no saying that would tell about such an extremely cold winter.

In England, William Merle kept notes on the weather from 1337 to 1343. He published these data in one of his papers. It is interesting that he also observed the wind. He arranged the data in the form of a table, on the basis of which the fraction of winds from the eight directions for London could be drawn. He found that the resulting wind was from a direction of 250°. Similar winds were blowing as in the period 1901–1930 (the resulting direction is 239°, southwest).

There were other individuals in Europe who recorded the characteristics of the weather. Thus, Kilian Leib from Bavaria kept a diary about the weather from 1513 to 1531.

From all these activities in the Middle Ages, it can be seen that there was almost no progress in meteorology, although certain phenomena attracted attention. This interest in weather was with individuals who made individual notes, or systematic records over shorter periods of time. There were also attempts to give weather forecasts based on some rules.

3.2.5 The First Hints of the Progress of Meteorology in the Middle Ages

All scientists between the thirteenth and seventeenth centuries supported astrological weather forecasts. One who doubted the validity of that work was the famous mathematician and commentator on Aristotle's Meteorology, Nikola Orezm (1323–1382). He is one of the first to notice the problems related to weather forecasting and that astrologers cannot solve them. He believed that weather forecasting was possible but based on some rules that were not yet known. Despite conflicting opinions similar to those of Oresmus, the influence of astrology lasted until the beginning of the eighteenth century, when it lost its status as a scientific activity.

As we have seen, Aristotle's Meteorology, maintained from ancient days to the seventeenth century, slowed down the progress of science. The theoretical speculations of Robert Ricardo (1510–1608) extended their life. One of those who tried to break with such meteorology was the famous algebra scientist Djirolamo Cardano (1501–1576, Fig. 3.7). In his treatise *De Subtilitate* (1550), Cardano devotes a significant part to meteorology. He discussed wind, clouds, rain, and lightning. And in that work, you can see the strong influence of Aristotle's teachings. But when Cardano speaks of air, he divides it into two parts: "free air," which destroys inanimate things and sustains living things, and "unclosed" air, which has opposite properties. Cardano omitted fire as the fourth element.

Fig. 3.7 Djirolamo Kardano

In the series of those scientists who tried to break the ruling point of view of classical authorities, the first place certainly belongs to René Descartes (1596–1650). It can be said that with him the era of speculation ends and modern meteorology is born (Descartes 1668).

In his book, printed in 1637, *Discours de la Methode* (Fig. 3.8), Descartes presents four methods that should be used in research.

1. Never accept anything as true until it becomes absolutely clear that it is so.
2. Divide each difficult problem into smaller parts, and solve them part by part.
3. Always solve first simple and then more complex problems, looking for a connection between them.
4. It is necessary to be as broad and deep as possible in scientific research, not allowing prejudices to rule.

In the appendix of this book was the section "Les Meteores." There, Descartes applied these four principles to meteorology. Descartes, like Aristotle, used the deductive method. He tried to explain the nature and causes of all meteorological phenomena through the basic laws of nature, which were not yet fully known. Descartes first analyzes the nature of earthly bodies and then the vapor that rises from the earth. After that, he explains the formation of clouds and wind and the way in which rain, hail, and snow are produced. Thunder and lightning are described in his order. He finished the description with an explanation of the influence of light on the formation of rainbows and other light phenomena in the sky.

DISCOVRS
DE LA METHODE
POVR BIEN CONDVIRE SA RAISON,
& chercher la verité dans les Sciences.
PLVS
LA DIOPTRIQVE, LES METEORES;
LA MECHANIQVE,
ET LA MVSIQVE, Qui sont des essais de cette METHODE,
PAR RENE' DESCARTES.
Auec des Remarques & des éclaircissemens necessaires.

A PARIS,
Chez CHARLES ANGOT, ruë saint Iacques,
au Lion d'Or.
M. DC. LXVIII.
AVEC PRIVILEGE DV ROY.

Fig. 3.8 Title page of Descartes' book *Discours de la Méthode*

René Descartes (1596–1650)

He was born on March 31, 1596, in La Haie, France. His Latinized name is Renatus Cartesius. After studying at the University of Poitiers in 1618, at the beginning of the Thirty Years' War, he served in the army under Prince Maurice. Later, he was reassigned to the Duke of Bavaria's army on the Danube. He did not take part in the war skirmishes but worked on his Cartesian geometry. On November 10, 1619, allegedly after some mystical revelation, he came to a radical position in research. He concluded that if he wants to discover something, he must carry the whole program inside himself, doubting everything except his existence of himself (his famous motto "Cogito ergo sum"—I think, therefore I exist). In 1649, Descartes traveled to Spokosoli, where he was offered the position of court philosopher to King Christian of Sweden. He died in Stockholm the first winter after his arrival, on February 11, 1650, from pneumonia. His headless body was returned to France. The skull remained in Sweden until 1809.

In his discussion, the hypothesis was first put forward that air, water, and other bodies on Earth are composed of tiny particles, between which there is very "fine matter." He hypothesized that the particles of water were long, smooth, and slippery, "like little eels that twist around each other and are so attached that they cannot easily be separated." Particles of solid substances are attached to each other, so they have an irregular shape. If the particles are smaller, then there is less space between them for the fine matter to move, and then air or oil is formed. All further Descartes explanations of meteorology are based on these hypotheses.

Although he rejected Aristotle's teachings, Descartes could not avoid some of his views (Hellmann 1897). Thus, according to Descartes, the wind is conditioned by several things. First, vapors are created by the Earth's particles under the influence of the sunrise. Then, the falling clouds displace the air that was below them. If the cloud suddenly descends on the cloud below it, thunder occurs. Lightning occurs due to the presence of flammable vapor between two clouds. Descartes' explanations of rain are quite modern. He argued that clouds are composed of water droplets or small ice particles. Droplets are created by the collision and pouring of small vapor particles. They are round, unless the shape changes under the influence of the wind. "When they become large enough that the air can no longer contain them, they fall as rain, as snow if the air is not warm enough to melt them, or as hail if the melted ones meet a colder wind that freezes them."

Descartes demonstrated the superiority of his method in *Les Meteorites* applied to meteorology, compared to the previous ones, primarily Aristotle's method. As with his predecessors, Descartes' views are flawed. They are primarily conditioned by the lack of any other means of knowledge except visual observation. Given the deductive method of explaining the phenomenon of weather, an error in one step is reflected in others.

It can be confidently argued that Descartes' enthusiasm for meteorology served as a catalyst that restored meteorology to the ranks of true sciences in the seventeenth century.

References

Abbe, C. 1906. The present condition in our schools and colleges of the study of climatology as a branch of the geography of meteorology as a branch of physics. *Bulletin of the American Geographical Society* 38: 121–123.

———. 1907. *The progress of science as illustrated by the development of meteorology*, 299. Smithsonian Institution Annual Report.

Alhazen. 1572. *Opticae thesaurus*, ed. Federico Risnero, 286–287. Basileae: Per Episcopios.

Cohen, M.R.I., and I.E. Drabkin. 1948. *A sourcebook in Greek science*. New York: McGraw-Hill Co. Inc.

Descartes, R. 1668. *Les meteores, discours de la methode…*, 227–230. Paris: Charles Angot.

Dufour, L. 1943. Les grandes époques de l'histoire de la météorologia. *Ciei et Terre* 59: 357.

Freeman, K. 1953. *The pre-Socratic philosophers*. 3rd ed. Oxford: Basel Blackwell.

Frisinger, H.H. 1977. *The history of meteorology to 1800*. New York: Science History Publications.

Galileo, G., trans. H. Crew, and A. DeSalvio. 1939. *Dialogues concerning two new sciences*, 64–68. Chicago: Northwestern University Press.

Gershenson, D.E., and D.A. Greenberg. 1964. *Anaxagoras and the birth of scientific method*. New York: Blaisdell Publishing Co.

Heath, T.L. 2009. *The works of Archimedes*. New York: Dover Publications, Inc.

Hellmann, G. 1897. Die altesten quecksilber thermometer. *Meteorologische Zeitschrift* 14: 31–32.

———. 1908. The dawn of meteorology. *Quarterly Journal of the Royal Meteorological Society* 34: 221–232.

———. 1927. *Die entwicklung der meteorologischen beobachtungen bis zum ende des XVIII*. Berlin: Jahrhunderts.
Heninger, S.K., Jr. 1960. *A handbook of renaissance meteorology*. Durham: Duke University Press.
Pliny, trans. H. Rackham. 1937. *Natural history*. Vol. 1, 247 str. London: Heinemann.
Seneca, trans. John Clarke. 1910. *Quaestiones naturales*. London: Macmillan and Co., Ltd.
Seneca, trans. T.H. Corcoran. 1971. *Natural questions*. Vol. 2, 273. London: Heinemann.
Theophrastus, trans. Sir A. Hort. 1948. *Enquiry into plants and minor works on odors and weather signs*. Vol. 2, 395. London: William Heinemann Ltd.
Webb, W.L. 1963. Missile range meteorology. *Weatherwise* 16: 100–107.

Chapter 4
Beginnings of Quantitative Meteorology

4.1 Introduction

Until the end of the sixteenth century, meteorology was based on the thoughts (speculations) of naturalists such as Aristotle. Attitudes about meteorological phenomena were the result of thinking and visual observations. Then it became quite clear that some thoughts were wrong, due to the lack of sufficient data about the atmosphere (De Luc 1784; Hooke 1665). Descartes himself saw the growing need for knowledge about the atmosphere. He is a leader in the effort to gather quality meteorological knowledge. In order to do that, it was necessary to have meteorological instruments. Fortunately, this shortcoming was quickly compensated.

In the seventeenth century, three basic instruments in meteorology were found: a thermometer, barometer, and hygrometer. For the invention of the first two instruments, the famous English meteorologist, Sir Naper Shaw (1854–1945), said: "The invention of the barometer and the thermometer marked the birth of a serious study of the physics of the atmosphere, quantitative studies without which it was impossible to form an accurate conception of their structure".

Data on air temperature were immediately interpreted in the right way, while data on air pressure were not. For a long time, people thought and made mistakes about what barometric data actually represents, why it changes with time in one place, with altitude, etc. The development of basic meteorological instruments will be discussed in this part.

M. Ćurić, V. Spiridonov, *History of Meteorology*,
https://doi.org/10.1007/978-3-031-45032-7_4

4.2 Thermometer

4.2.1 The First Thermometers

Although the thermometer was invented at the end of the sixteenth century, the basic principle on which its operation is based was known much earlier (Réaumur 1731; Crombie 1953; Middleton 1966). The Greek Philo (third century BC) wrote the work *De Ingeniis Spiritualibius* (On Printing Machines). In it, he described an experiment proving the expansion and contraction of air.

"Let us take a lead ball of moderate size which is empty and closed. It should not be too thin, so that it can be broken easily, nor too thick. For the experiment to succeed, it must be completely dry inside. Through the opening, a bent tube enters the ball almost to the bottom. The other end of the pipe enters the container with water, again close to the bottom (Fig. 4.1). I have found that when the sphere is left in the sun and heated, the air enclosed in the tube escapes. This can be seen as air from the tube flows through the water, moving it, and producing air bubbles, one after the other. If the ball is placed in the shade, or in any other place where the sun does not penetrate, the water in the tube will rise and flow into the ball. If the ball is returned to the sun again, the water will return to the ball… The same effect is obtained if the ball is heated with fire or hot water is poured on it…".

Later, Hiro of Alexandria built a similar device, a kind of steam engine that works on the principle of jet propulsion. Heron called such a machine "aeolipile," from the Greek word "Aeolus" (god of the wind) and the word "pila" (ball). This machine consists of a hollow ball that can rotate around the horizontal axis (Fig. 4.2). It is connected through that hollow shaft to the water tank, which is located under

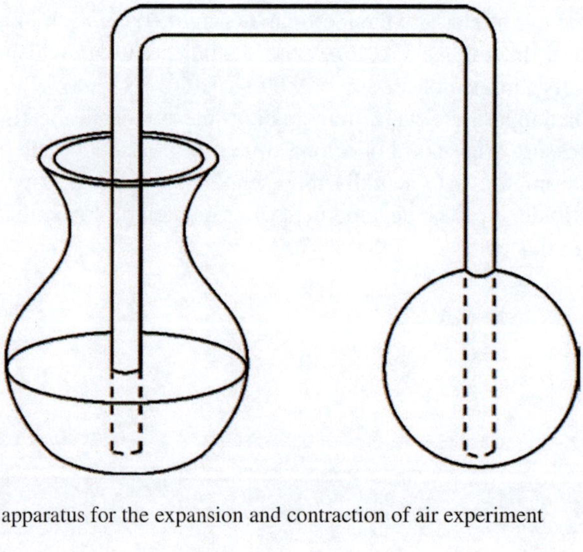

Fig. 4.1 Phil's apparatus for the expansion and contraction of air experiment

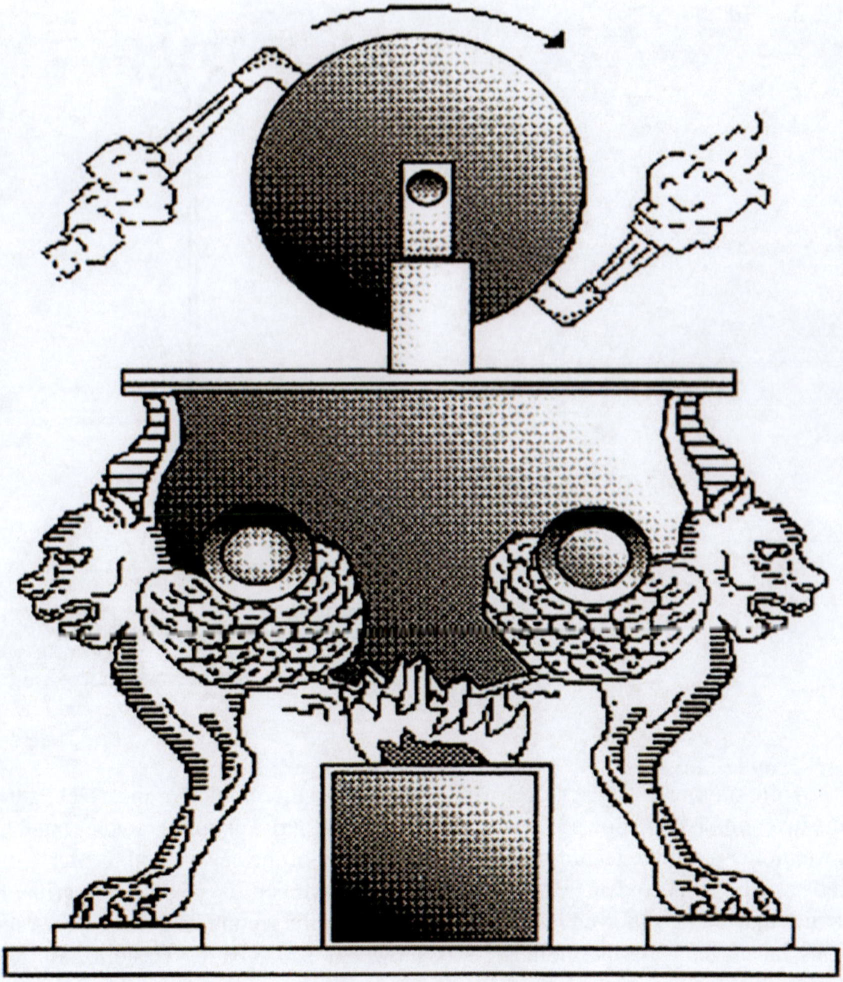

Fig. 4.2 Heron's "aeolipile" machine

the ball. When the water is heated, steam escapes through two bent tubes placed on opposite sides of the sphere. The escape of steam pushed the ball to rotate. It is the forerunner of the steam engine.

Both Philo and Hiro experimentally determined the property of air to expand when heated and contract when cooled, but they did not see that this could be used to measure heat (only Joseph Black (1728–1799) distinguished between heat and temperature). But that possibility did not escape the famous Italian scientist Galileo Galilei.

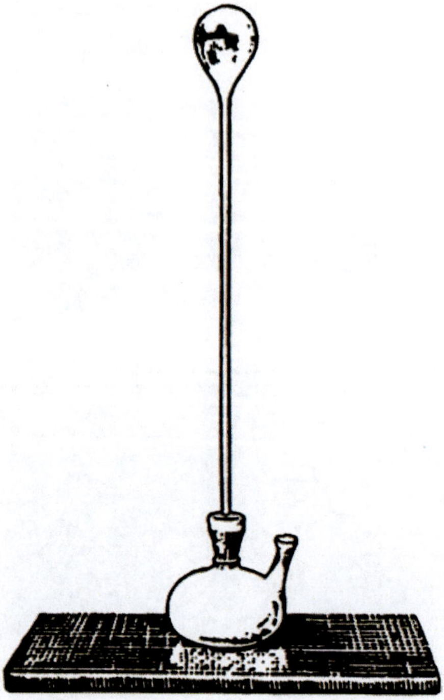

Fig. 4.3 Galileo's thermometer

While staying in Padua (1592–1610), he found a thermometer. The exact date of the invention is not known, because Galileo did not write about it. He presented his invention at a public lecture right at the beginning of the seventeenth century. In a letter sent by his student, Father Castelli, on September 20, 1638, to Cesarina, he writes that Galileo showed his thermometer at a public lecture he gave 35 years ago. This means that the thermometer was shown in 1603. Father Castelli described Galileo's thermometer as follows: "Galileo took a container the size of a hen's egg (Fig. 4.3). At the end of the straw-wide and two-hand long tube, he slipped a glass ball. He warmed the ball with his hands and turned the tube downwards, plunging it into the water. As soon as the ball began to cool, the water rose up the tube to a height of one hand from the water level. He used this instrument to measure the degree of heat and cold[1]".

[1] Where we are in terms of science at the end of the sixteenth century, when Galileo invented the thermometer, let the fact that at that time, in 1597, the first Serbian primer was printed in Venice. The author was the icon of Sava, a native of Paštrović, a monk of the Dečani monastery. It is written in the Serbian-Slovenian language. Only in 1827, Vuk Karadzic published *The First Serbian Primer*, written in the vernacular in Vienna.

Galileo Galilei (1564–1642)

He was born on February 15, 1564, in Padua. He studied and later taught at the University of Pisa. He was an energetic man with reddish hair. His sayings were powerful, and in this way, he differed from many of his university colleagues in Pisa. He moved to Padova in 1592, where he spent 18 years. There he married the Venetian Maria Gambole, with whom he had two daughters and a son. He moved to Florence in 1610, to the court of the Grand Duke of Tuscany, Ferdinand II de' Medici. There, in 1632, he published the work *A Dialogue Discussing the Two Principal World Systems Proposed by Ptolemy and Copernicus*. It was written in the form of a dialogue between two men, one of whom supports Ptolemy and the other Copernicus. He wrote that "dialogue" with the approval of the newly elected Pope Urban VIII, whom Galileo knew personally. Despite this, the infamous Roman Inquisition condemned Galileo for supporting Copernican ideas. As an influential man of science, he was allowed to "repent." Before the greatest theological authorities of the Catholic Church, he had to kneel down and declare that his teachings were false and contrary to the Holy Bible. After that, standing up, he muttered "Epur si muove" (Yet he turns) and lived under house arrest. He was blind for the last 4 years of his life. Grand Duke Ferdinand II wanted to erect a large monument to him, but he was warned not to do anything that could provoke a reaction from the "Holy See."

More than a century later, the Vatican admitted the mistake. Galileo's "Dialogues" were removed from the list of forbidden works, which the church declared in 1757. Galileo was eventually exonerated by Pope John Paul II in 1992, 359 years after the accusation.

Galileo used a thermometer to determine the relative difference in temperature of different places and in the same place at different moments in time. His thermometer did not have a scale, which is why it is more correctly called a thermoscope. Another student of Galileo, Sagredo, made measurements in 1613 with a similar thermometer. It is believed that his thermometer had some kind of graduated scale.

Porta, a well-known scientist interested in unusual experiments, wrote a book in 1606 entitled *Pneumatico-rum libri tres* (*Three Books on Pneumatics*). In this book, there is a description of the thermometer, but it does not say who invented it. Because of this, some have mistakenly claimed that he invented the thermometer.

There is, however, evidence to show that, independently of Galileo, Dane Cornelius Drabl (1572–1634) invented the thermometer. He was a professional inventor (he built the first submarine). It is not known exactly the year, nor the circumstances in which it happened. He mentions a thermometer in his book *Elements of Nature* in 1619. The Italian Gaspar Enns wrote about his thermometer in 1936.

This new instrument did not immediately receive the name "thermometer". Somewhere it was called a "glass calendar." The word "thermometer" was first used by the Jesuit priest Leremont in 1624 in the book *Recreation Matematic*. Until 1665, Boyle called his thermoscope "weather glass" (Boyle 1969). Boyle's thermometer (thermoscope, or closed glass for weather) is shown in Fig. 4.4.

The championship in the discovery of the thermometer was also given to the famous Venetian doctor, professor of Medicine Santorini (1561–1636). He described his instrument in the work *Commentaria in artem medicinalem Galeni* in Venice in 1612. Santorini was the first to use a thermometer to measure the rise in body temperature when a person has a fever. Galileo's new instrument for measuring temperature quickly spread throughout Europe. After Italy, the thermometer was transferred to Poland in 1657, to France in 1658, and to England in 1661.

There were many who tried to improve it. One of them is the French theologian Father Marin Mersen (1588–1648) Fig. 4.5. At that time, communication between European scientists was done mainly through personal visits (there were no scientific journals). Probably the most active in this sense was Father Mersen. He was in constant contact with such famous people as Galileo, Descartes, Huygens, and Roberval.

Fig. 4.4 Boyle's thermometer

Fig. 4.5 Marin Mersen (1588–1648)

Mersenne played a major role in the dissemination of Galileo's thermometer. By the way, he is known for being the creator of the original method of stimulating science by asking reward questions. He often repeated experiments that others had set up. Thus, experimenting with the thermoscope, he tried to improve its sensitivity.

In a paper published in Paris in 1644, Mersenne described an instrument with a narrow tube and a large ball at one end and a small one at the other end. The small ball had a narrow hole (opening). He gently heated the large ball, while the small one was submerged in water. After the heating of the big ball stopped, the water began to float along the pipe in the form of a short column. The pipe was graduated (provided with a scale), so that it served as an indicator of temperature change, Fig. 4.6.

Ferdinand II, Grand Duke of Tuscany, patron of scientists, stood out by improving the characteristics of the thermometer. Around 1641, he devoted his attention to improving the thermometer. His thermometer is similar to Morse's, Fig. 4.7. He poured some water into it and sealed it hermetically by melting glass tubes. The closed instrument was supplied with a scale in the form of short spikes. It was the first thermometer that showed the temperature independently of the air pressure.

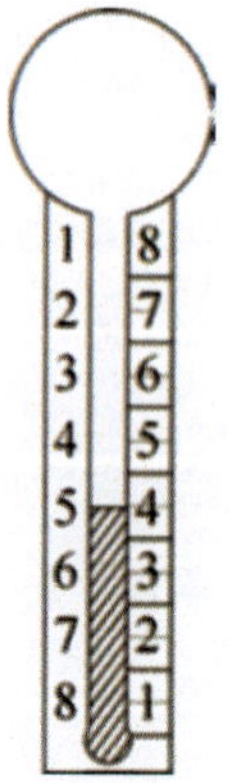

Fig. 4.6 Mersenne air thermometer with scale

Fig. 4.7 Ferdinand closed thermometer

Ferdinand II di Medici (1620–1670)

Ferdinand II is the son of Grand Duke Cosimo II, ruler of Tuscany. His family, the Medici, ruled Florence continuously since 1420. Ferdinand II was not only a great sponsor of science but also a hobbyist scientist. He supported the establishment of the "Accademia del Cimento" in 1657 (founded by his brother, Prince Leopold) and dissolved it in 1667, upon the intervention of the church. The society measured temperature, pressure, humidity, wind direction, and the state of the sky, and recorded it in special forms, which were collected from various places in Florence. In Fig. 4.8. Ferdinand II is shown as a participant in an experiment in 1657 that was supposed to show whether cold can be reflected using a mirror. The ice in the bowl was the source of the cold. Of course, the cold could not be reflected.

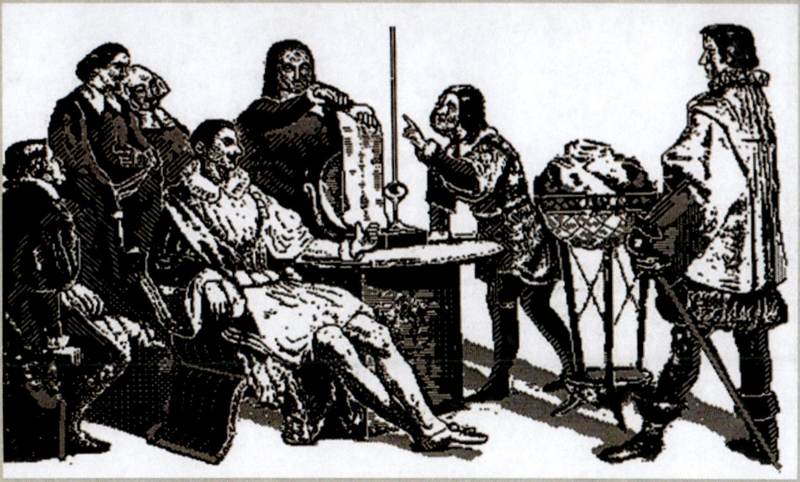

Fig. 4.8 An attempt to reflect the cold through a mirror

Ferdinand's thermometer was used to measure the air temperature within the famous "Accademia del Cimento,", whose establishment he supported in 1657. Instead of water, they also used alcohol, and they also mentioned that mercury should be used. There is no doubt that the most interesting in the development of thermometers is the one constructed between 1660 and 1662 by the German Otto von Gerik (1602–1686). This particular instrument was about 6.5 m tall. It consisted of a large copper ball, painted blue with gold stars (Fig. 4.9). The ball is connected to a long copper tube, 2.5 cm in diameter, in the shape of the Latin letter "U." A certain amount of alcohol was poured into the tube. The shorter leg of the pipe was open. A float was placed in that part, which floated on top of the alcohol. The float was made of a thin brass plug. A rope is tied to the float, which is passed over a reel attached under the ball. A figure of an angel was attached to the other end of the cone. Next to the pipe was a scale used to measure the angel's position. On one side of the large ball was a valve, through which air could be drawn using an air pump. This allowed the alcohol level to be adjusted. This huge thermometer was attached

Fig. 4.9 Gerikov termometar

to the outer wall of the house which was in the shade. The intention was to use it to measure "the hottest and coldest weather throughout the year."[2]

The use of mercury has been contested because it has a low coefficient of mass expansion. This was claimed by the famous scientist, Halley, in 1693. During this period, Newton constructed a thermometer that used linseed oil. Fahrenheit (1686–1736) significantly improved the thermometer technically. He used alcohol. Unlike other thermometers, Fahrenheit ones showed equal values (Fahrenheit 1724). Around 1715, he started using mercury thermometers, whose readings coincided with alcohol thermometers.

4.2.2 Thermometer Scales

The biggest drawback of the thermometers at that time was that the data could not be compared with each other. This problem existed even when thermometers were supplied with their own scale. The scales were different. One of the first who tried to overcome this problem was the English scientist Robert Boyle (1627–1691). He put a lot of effort into constructing a closed thermometer with a standardized scale, which allows the comparison of data among themselves.

Robert Boyle (1627–1691)

(continued)

[2] It should be noted that entomologists (scientists who study insects) believe that crickets can be used to measure temperature. They are cold-blooded creatures, and their activity depends on the air temperature. They can be used as a sign of high temperatures, but what about situations when they are not?

(continued)

He was born on January 25 in Lismar, Ireland. He was the youngest of fourteen children in the family of Richard Boyle, the first Earl of Cork, and a great thinker. He was first interested in chemistry. Nevertheless, he published a book on heat, thermometry, and color. Boyle is one of the founders of the Royal Society. From 1662–1663, he dealt with the questions of whether experiments lead to evidence, i.e. whether real truth is determined by experiments. He claimed that experiment is one of the key factors of truth. He wrote one of the first historical novels in England, *The Martyrdom of Theodora*, which describes the conflict between love and religious obligations. He never married. He died on December 30, 1691, in London.

He thought that the melting point of ice depended on latitude and that therefore onyza seed oil should be used to obtain one fixed point on the scale. An alcohol thermometer tank was placed in that oil, allowing the oil to freeze. He determined the height of the column of alcohol in the tube when the oil began to coagulate. He tried to calculate the absolute spread of alcohol and to divide the scale into ten, or some other number of parts. That didn't turn out well. It was necessary to take two fixed points instead of one.

Robert Hooke (1635–1703)

He was born in Freshwater, England, the son of a minister. He was a sickly child, and his parents did not hope that he would survive. After his father's death, Robert was taken in by Richard Busby, headmaster of Westminster School, to educate him at the age of thirteen. Hooke studied Latin, Greek,

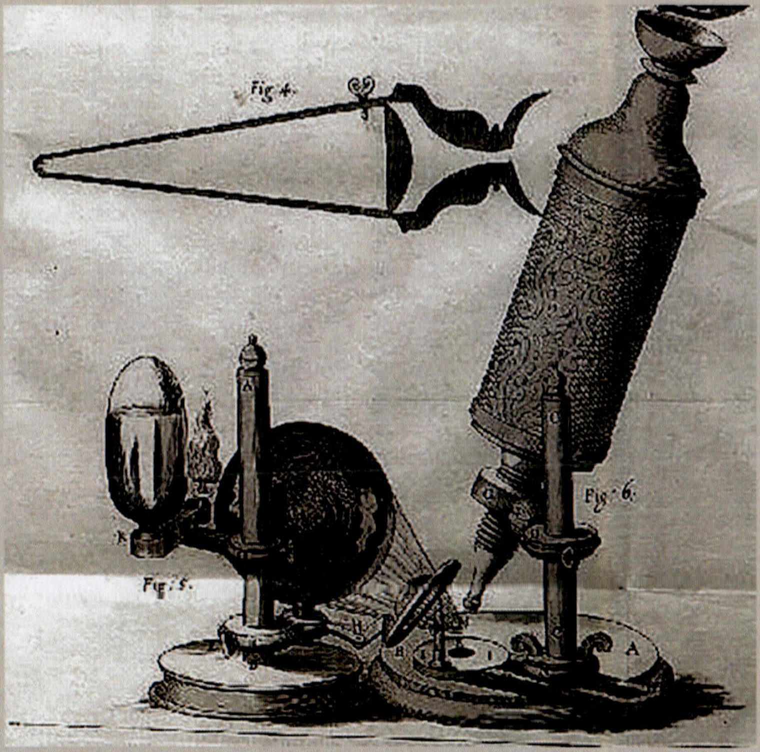

Fig. 4.10 Part of the equipment Hook was experimenting with

Hebrew, and mathematics. He also learned to play the organ. He moved to Oxford in 1653, where he obtained his master's degree. He was a permanently defective, crooked figure. This is probably the reason why his original paintings are rare. He was assigned to Robert Boyle as an assistant. In 1658, Hooke constructed an air pump and began working on a chronometer with a spring (Fig. 4.10). In the same year, he constructed an anemometer. When the Royal Society was founded in 1662, he was appointed as the official experimenter. He had to do two to three experiments for each week-long meeting of the Society.

It contributed to the intellectual rise of the Society. In 1665, Hooke published the book *Micrographia*, dedicated to microscopic measurements. When Christian Huygens constructed a spring clock in 1674, Hooke wept, accusing Henry Oldenburg, the Society's secretary, of betraying the secret. He called him "a trafficker among the intelligentsia." When his niece Gracie, first his guardian and then his longtime mistress, died, he couldn't get over it. He lost interest in science and died on March 3, 1703, in a room at Gresham College in London, where he lived for almost 40 years.

Robert Hooke attempted to standardize the scale. In October 1664, he published *Micrographia*, in which he described the work on closed thermometers. He wrote: "After several attempts I managed to make a pipe over 120 cm high, in which the liquid expands to near the top in summer, and to near the beginning of winter when it is coldest." He filled the thermometer with the purest high-quality wine alcohol. He divided the scale by marking the position of the column of alcohol with zero when the tank is placed in the distilled water that is freezing, and the other divisions above and below that are proportional to the expansion of the liquid.

While Hooke used one fixed point on the scale, Christian Huygens chose another point, the one where the liquid level is when the tank is placed in boiling water.

Carlo Renaldini proposed in 1694 that the length between these two points be divided into 12 parts. Unfortunately, this position did not find significant support. Gabriel Fahrenheit (1686–1736), a German who was born in Gdańsk (a city belonging to Poland until 1793 and again after 1945), spent most of his life in the Netherlands and proposed three fixed points on the scale. That proposal was made even earlier by the Danish astronomer Reomir. One point, marked zero, is determined by the behavior of the liquid when the tank is a mixture of water, ice, and sea salt. The second point on the scale is marked 32 and is defined by the temperature of the mixture of water and ice. The third point is determined by the temperature of the human body and is marked with 96. Until 1717, Fahrenheit used alcohol for the thermometer liquid, and then mercury. Mercury had the advantage that it covered a wider range of temperatures and that it does not evaporate at ordinary temperatures. After Fahrenheit's death, it was customary to take the freezing point, designated 32 °F, and the boiling point of water, designated 212 °F.

A professor at the University of Uppsala, Sweden, Anders Celsius (1701–1744) proposed a different scale. He marked the first fixed point with zero (melting temperature of ice) and the second with 100 (boiling temperature of water). In the original version, these two points are marked opposite to this one (with 100 and zero). This proposal was presented by Celsius to the Swedish Academy of Sciences in 1742 under the title "Considerations about the two fixed degrees on the thermometer." Celsius finally made the transition to the mercury thermometer as the basic instrument (Celsius 1742). He showed the shortcomings of existing scales and calibration methods, especially Reomir's scale. Celsius originally marked the boiling point with zero and the freezing point with 100. He divided that interval into 100 equal parts. In this way, he wanted to eliminate the use of the "minus" sign for winter temperatures. In 1745, at the suggestion of the famous botanist Linnaeus (1707–1778), the Council of the Academy adopted a new thermometer, marking the freezing point with zero and the boiling point with 100. This thermometer was known as the "Swedish thermometer" for a long time.

In Lyon, France, a year after Celsius and independently of him, Christine constructed the "centigrade" thermometer, with zero at the freezing point and 100 at the boiling point.

It can be seen that the development of an accurate thermometer, suitable for scientific measurements, took 100 years, from the time of the invention of the first graduated Florentine thermometer to the Celsius thermometer. Meteorologists of the eighteenth century, and partly of the nineteenth century, worked on comparing the results of measurements with different scales. In 1779, in the work *Pyrometric*, Lambert described 19 different scales.

About 60 years ago, the unit of the Celsius scale was called centigrade (from the Latin word centum, meaning 100, and gradus, meaning degree). The scale, prior to an international agreement, is called Celsius. It is used all over the world, except in the USA, where the Fahrenheit scale is still more commonly used. The connection between them is:

$$t_C = 100/180(t_F - 32) = 5/9(t_F - 32),$$

respectively,

$$t_F = 9/5 t_C + 32.$$

The Englishman William Thomson (Lord Kelvin, 1824–1907) proposed the absolute scale. The temperature in it is marked with the capital letter T and is defined by

$$T = t_C + 273.15.$$

These three scales are shown in Fig. 4.11. So far, more than 60 different scales have been in use. It can be seen that it was not easy to make all the measured temperatures uniform.

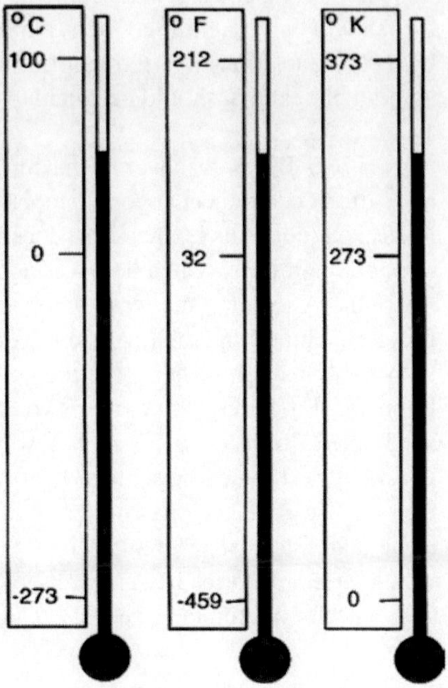

Fig. 4.11 Celsius, Fahrenheit, and Kelvin scale

4.2.3 Temperature and Heat

In ordinary speech, the word "heat" is often replaced by the word "temperature" (it has the same meaning). In reality, in science, they mean completely different things. Heat is the factor that causes something to become hotter or colder. Temperature is a quantity that shows the measure, effect, and degree of that cooling or heating.

For centuries, the heat was a great mystery to natural thinkers. We have shared how this term was explained in ancient times. From the writings of the Roman author Lucretius (95–55 BC), it can be seen that heat was viewed as a type of matter in fluid form. The word "caloric" (from the Latin "calor" meaning "heat") was used to denote this special, unfathomable fluid. Lavoisier, the famous French chemist, even included it in the list of elements.

With this understanding of the caloric, it could not be explained why equal volumes of water and mercury, heated in equal vessels and in the same way, change their temperatures by different amounts. Mercury increases its temperature almost twice as much as water. Similarly, equal masses of iron and water placed in a heated furnace will result in the iron immediately becoming too hot and the water remaining acceptably warm. Nor could it be explained why a mixture of equal masses of water and mercury has a temperature closer to that of water than that of mercury before mixing.

The first scientist who noticed the need to distinguish between the meaning of the words "heat" and "temperature" was Joseph Black (1728–1799). In 1764, he wrote that we use a thermometer to measure the intensity of heating, and that heat is a quantity, i.e., the amount of energy required to heat or cool a substance. Black noticed that the concept of specific heat must be introduced, as the amount of heat that should be brought to a gram of a substance in order to increase its temperature by 1 °C. Based on this, the calorie is defined as a unit of heat. A calorie is the amount of heat that should be added to a gram of water to raise its temperature from 14 to 15 °C.

In 1761, Black discovered that under certain conditions heat transferred to a body did not cause a change in temperature. He observed that when water is cooled to freezing point, its temperature does not decrease despite continued cooling (heat removal). Similarly, when heat was added to water, its temperature rose to the boiling point. After that, the temperature did not change, although heating continued. Black concluded that melting, cooling, evaporation, and condensation are connected with the transition of heat into the so-called latent heat. Latent heat ("latent" is a Latin word meaning "hidden") is released or absorbed during ice-water-water-to-vapor phase transitions, in such a way that the temperature remains unchanged. Black also estimated how much heat would have to be removed from the water to freeze it. He got 82 calories per gram of water. He also found that 82 calories needed to be added to melt a gram of ice. That agrees well with the more recent value, which is 79.9 calories per gram. Black did not determine the heat of vaporization. Only later was it found to be 597.3 calories per gram.

4.2.4 Heat and Energy

Benjamin Thomson (1753–1814)

He was born on March 26 in Woburn, Massachusetts, USA, in the family of a small farmer. Everything he learned was with the help of a local priest. He moved in 1772 to Concord (formerly Rumford), New Hampshire, and married a wealthy widow, 14 years his senior. He had one child with her. After 3 years, he divorced. After the fall of Boston, in the American Revolutionary War, he emigrated to Europe. In England, he managed quickly, so he became under-secretary of state for the Ministry of Colonies and was knighted by King George III. He later moved to Bavaria, where he was appointed Minister of War. In that position, as a military commander, he proved to be a great innovator. For example, he made a study on the insulating properties of fabrics and fur. He showed that a vacuum prevents the passage of heat. He analyzed the nutritional properties of food and wrote studies on recipes for healthy meals. He researched which drinks could replace alcoholic drinks. He strongly advocated drinking coffee instead of alcohol. He constructed many devices for preparing coffee. In 1793, Thomson became Count of Bavaria, under the Bavarian Duke Maximilian. Then he called himself Count Rumford. He was elected an honorary member of the Royal Society in 1799. He moves to Paris and marries the widow of the beheaded Lavoisier. He was not a gallant man. The marriage was not a happy one. All of Paris was talking about their public, violent fights. The marriage was dissolved 2 years after the wedding. Thomson died on August 21, 1814, in Othello, France.

American Benjamin Thomson (Count Rumford) in 1798 noticed that a large amount of heat is released due to the friction that occurs when drilling cannon barrels with a drill. Thomson carried out an experiment in a gun factory in Munich. He was drilling a metal roller, whose temperature at the beginning was 15 °C. The steel drill was 10 cm long and 1.5 cm in diameter. The drill turned 32 times per minute, using the power of two harnessed horses.

After half an hour of work and 960 revolutions, the temperature of the punched roller was 50 °C. Then the weight of the roller, and especially of the taper, was accurately measured. The weight of the roller was also measured before drilling. The ground roller weight was 113.13 pounds (1 pound = 0.454 kg). The shavings were 1/948 part of the mass of the roller. Rumford did not believe that calories from such a small amount of sawdust could produce such a large rise in roller temperature.

In the next experiment, he drilled by placing the roller in two gallons of water (1 American gallon = 3.785 liters), the temperature of which was 15 °C. After an hour of drilling, the water temperature was 75 °C. After 2 h and 20 min, it was 85 °C, and after 2 h and 30 min, it started to boil. He calculated that the heat released by drilling was greater than the heat released by melting 10 candles 3/4 inch in diameter (1 inch = 2.5 cm). That's a lot of heat. Rumford concluded that the caloric cannot be material in nature, and rejected the caloric theory.

To discover the nature of heat, James Prescott Joule (1818–1889), from Manchester, England, did the following experiment: He poured water into a cylindrical metal vessel. He lowered the wings attached to the vertical shaft into the court, Fig. 4.12. On top of the part of the shaft that protruded above the court was a roller on which a rope was wound. The other end of the string is passed over a reel and tied to a weight. When the weight was released, he turned the shovel. Due to the friction of the paddles against the water, the temperature of the water increased. The experiment showed that the mechanical work performed by the weight during the fall was partly converted into heat and partly into the energy of rotating the fluid. With this experiment, Joule showed that heat is not a substance, but a form of energy. His result was in great doubt. The work on heat and energy, which was prepared for publication, no one wanted to print in professional journals. The suspicion came from the fact that he was not a famous name in the world of science, but a brewer's merchant. Nevertheless, he managed to hold public lectures in 1847, which were popular. Among others, 23-year-old William Thomson, later known as Lord Kelvin, listened to him. Thanks to Thomson, Jules' work was recognized after 3 years, i.e. in 1850.

Later, Joule was honored, and the unit of energy was named after him, Joule. Joule's experiment showed that the mechanical equivalent of heat is 1 Joule = 4186 calories.

Heat is shown to be transferred from one place to another in three ways: conduction, convection, and radiation. Conduction is the transfer of heat by air molecules. Convection is the transfer of heat through the flow of a heated substance. Radiation is the transport of energy by electromagnetic radiation. Count Rumford conducted experiments that showed the characteristics of convective currents in liquids. He

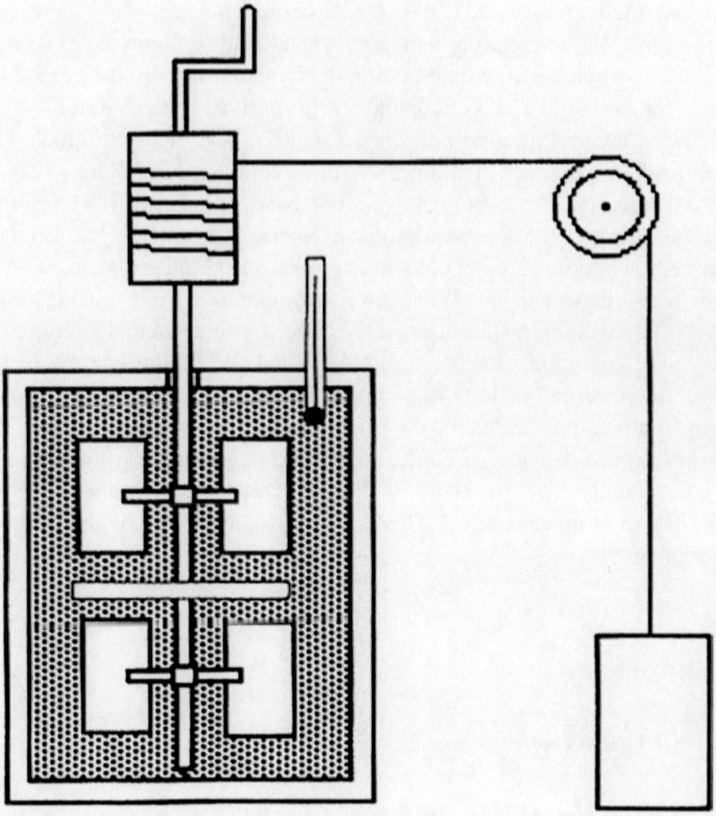

Fig. 4.12 Joule's apparatus shows the connection between mechanical work and heat

took a thin glass tube with a cylindrical opening at the top. On the other side, the pipe was closed. He poured a few teaspoons of yellow amber powder into the vertically placed tube. Then he poured in mixed distilled water and alcohol (so that the powder could float in the fluid—be suspended in it). By heating the pipe from the bottom side, all three types of heat transfer occur. First, a thin layer of solution at the bottom is heated by thermal conduction. Then the absorbed heat is transferred through the solution by convection and is emitted as radiation. Part of the energy is transferred over the top into the air by evaporation, in the form of latent heat.

4.2.5 Adiabatic Temperature Change

With the invention of the thermometer, the temperature could be accurately measured. It was already well known that the temperature of the vazouch increases when heat is added to it and that it decreases when heat is taken away from it. However, John Dalton (1766–1844) noticed that the air temperature changes even when it is isolated (separated) from the environment, i.e., when there is no heat input or output. Air temperature increases when the air is compressed (compressed) and decreases when the air expands (thins). He, therefore, discovered the adiabatic change of temperature. The word "adiabatic" is derived from the Greek word "adiabatos," which means "impossible." occurs without heat input or output. This word was first used by the Scottish engineer William John Rankin (1820–1871).

The adiabatic process is part of a more general law that in 1881 formulated by Henri Louis Shatier (1850–1936), known as Shatier's principle. He formulated it as: "If a system is exposed to any change from the outside, the system will rearrange itself to resist that change." When air is compressed from outside, its volume decreases. The system (air) opposes this change by increasing the temperature. An increase in temperature causes the air to expand. When the volume of the air is increased by external action, it counteracts this by lowering the temperature. Lowering temperature causes the air to recondense. When dry air rises in the atmosphere, it expands, because it comes to a place of lower pressure. In doing so, it cools adiabatically. This lowering of the temperature is constant and amounts to 1 °C per 100 m of upward travel. This is the well-known dry adiabatic vertical temperature gradient.

4.3 Barometer

4.3.1 Introduction

It does not happen often in the history of science that an instrument is made for a certain purpose and later serves other needs. In particular, it does not happen that it completely serves another purpose later. That's exactly what happened with the barometer. It was not found by having the idea of air as a substance that has weight so that weight had to be measured. When the barometer was invented, it was not air at all, but a vacuum. The vacuum occupied the attention of thinkers from ancient times to the seventeenth century.

As we have seen, even Thales, Plato, and Aristotle asked questions about whether a vacuum can exist in nature, and whether it can be created. Aristotle said that it cannot exist, because empty space has no dimensions, neither down, nor up, nor east, west, north, or south. And light cannot pass through a vacuum. Many thought this way in the seventeenth century. Confusion also reigned over whether air has weight and whether it exerts pressure on something exposed to it.

At that time (seventeenth century), a large group of people dealt with this issue. The largest group was in Italy. Some names should be mentioned: Giovanni Balliani, Gasparo Berti, Isaac Beckmann, René Descartes, Athanasius Kircher, Raphael Maggiotti, Emmanuel Mangan, Jan Ray, Vincenzo Viviani, Niccolò Zucci, Michelangelo Rizzi and, of course, Evangelista Torricelli and Galileo Galilei. These names are given to show that the barometer was not found by the work of "one hand."

There is a mystery surrounding it because proving the existence of a vacuum was blasphemy, a heretical work. The Catholic Church, through the famous Roman Inquisition, severely punished heretics. Giordano Bruno was burned, Galileo Galilei barely made it out alive, and he lived under house arrest. In that century and in England, the church deprived people of their lives lightly. It was enough to declare someone a witch, and the sentence was death. More destinations of people were burned. How can science be free and public in such conditions?

4.3.2 Forerunner of the Barometer

In most books, the invention of the barometer is linked to the year 1643. There are those who think that it happened in 1644. But that will be discussed later. One activity of this kind that is generally omitted, and which preceded the invention of the "real" barometer, should be highlighted here.

Fifty years before Torricelli's experiment, there was an instrument called "Dutch time glass," or "weather glass." Namely, there is a document from 1619, in Ghent, Holland, which shows that the Dutch engineer Jijsbrecht Donker found an instrument that "shows the several tempers of the times." This instrument is believed to have been made well before this date.

A contemporary reproduction of Dutch glass from the beginning of the seventeenth century is shown in Fig. 4.13. The colored water in the vessel is raised and lowered in an open pipe, bubbling, depending on the atmospheric pressure. However, the change in temperature also contributes to this movement of the water level, due to the contraction and expansion of the air above the water in the main part of the court. This, of course, could not be a precise instrument, but in the historical sequence of events, it is very important to know. This time glass is now called a "sympiezometer"—a barometer that has air or some other gas above the column of liquid, instead of a vacuum above mercury, as in ordinary barometers. Air or other gas above the liquid compensates for the temperature. This instrument is used for weather forecasting. The principle was not known, due to which the state of the instrument was linked to a state of time. That was left for Torricelli to explain.

It should also be noted that the "glass for storms, or weather" had different shapes and different liquid compositions. A very common type consisted of a glass tube about 25 cm high and 3 cm in diameter, filled with a mixture of water, camphor, alcohol, ammonium chloride, and calcium nitrate. The tube is hermetically sealed. Such a device appeared at the beginning of the eighteenth century. Some say it was produced by alchemists. The appearance of the content in the tube changes over time. Beautiful camphor crystals appear, which change appearance as they fall towards the bottom. Sometimes piles of little stars, or snow crystals, float on top. Sometimes the liquid becomes dark, and sometimes it is completely transparent, with many shades of turbidity. These different liquid appearances were interpreted as harbingers of the type of weather to expect.

Fig. 4.13 Dutch time glass

This device was used by Admiral Fitz Roy when he traveled (about 1830) on Darwin's ship, for research. Daniel de Fau, in 1703, in his book *The Great Storm* gives instructions on how to use "camphor glass." (De Fau is best known for the book *Robinson Crusoe*, published in 1719). This "weather glass" is often placed next to the mercury barometer.

FitzRoy inquired with the scientist Faraday if he could counter-sword the display of this device. Faraday clearly emphasized that the state of that mixture is determined only by temperature, and nothing more. In 1965, the well-known contemporary English scientist Sutton pointed out "that the 'storm glass' is nothing more than an interesting toy." "It is not a barometer, and at best it is some unpredictable qualitative thermometer. It is not a device that has any prognostic value". Despite this attitude, it should be emphasized that the first author of this book, during many visits to the English Meteorological Service, saw that there was still "storm glass" at the entrance to the main building in Bracknell.

4.3.3 Creating a Vacuum

Pressure and vacuum are two basic concepts that are now well-understood by almost everyone. But it wasn't like that before. Even the wise ancient Greeks denied the possibility of a vacuum. Aristotle emphasized that "nature fears a vacuum." And Galileo did not have a clear idea about it at the beginning. To show that there is a vacuum, he constructed a simple device (Fig. 4.14).

He poured water into a glass container with a long neck, but not to the top. He closed it with a wooden stopper and turned it over. He attached a weight to the cork. The plug was gradually pulled out. He noticed that the water was also separating from the bottom of the vessel. An empty space, a vacuum, was created there.

In his work *Discourses*, published in 1638, he tried to explain it. It was also necessary to explain why the Florentine well diggers could not pump out the water with suction pumps above a height of about 10 m. The suction pump (Fig. 4.15) consists of a cylindrical pump body in which there is a movable piston P. On the upper side of the piston, there is a cover O, which opens upwards. The cover S, which is moved

Fig. 4.14 Galileo's vacuum creation experiment

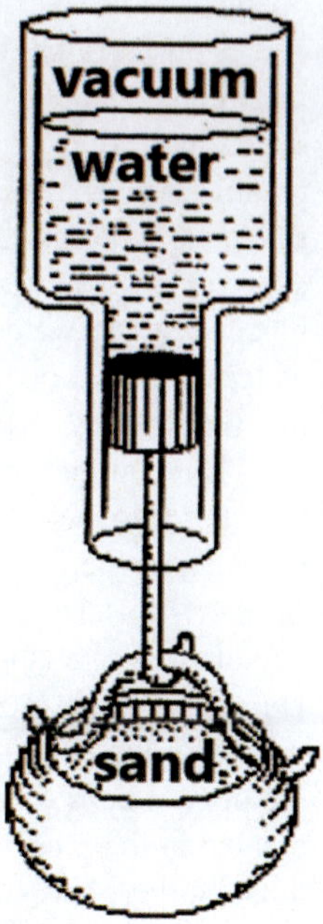

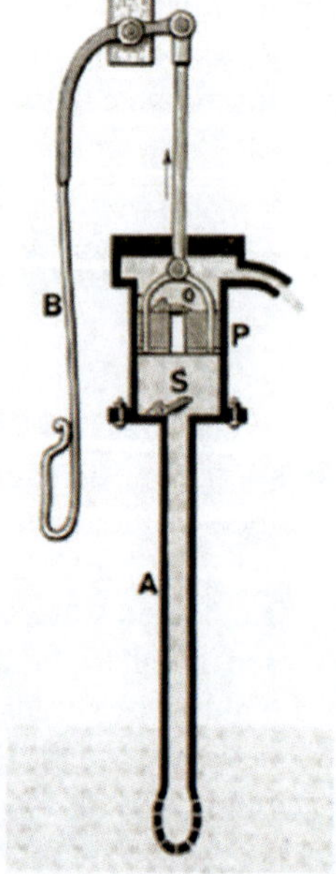

Fig. 4.15 Suction pump

downwards, closes and opens the suction pipe A, which is immersed in the liquid to be raised. When the piston is raised by lever B, O closes and S opens. Atmospheric pressure pushes the liquid into tube A. When the piston descends, S closes, and O opens. Fluid comes above the piston. The next time the piston is lifted, this liquid comes out through the drain. The pump functions well only when it is full of liquid, i.e. when activated. The Florentine well diggers noticed that the water could not be raised more than 10 m. Due to the penetration of air through inevitable cracks, in practice, water can never be raised to a height greater than 7–8 m. This stimulated scientists to think about why there is a barrier of about 10 m. Only Torricelli's experiment gave a complete answer to this question. The air that maintained the balance of the water column weighed so much, pressed, and because of this, the water could not be ejected to a greater height using this method. It was only later that pressure pumps were found, where this limitation does not exist.

When Galileo's *Discourse* came to Rome in December 1638, in the hands of the scientist Raphael Maggiotti, it stimulated him to make an experiment on the vacuum. The experiment was performed by Gasparo Berti (1600–1643) in June 1641 (Fig. 4.16). Bertie used an apparatus consisting of a lead pipe about 12 m long and bent downwards at the top. At both ends, there were valves for closing and opening. During the experiment, the tube was filled with water. Both ends with valves were submerged in containers of water. In the beginning, the lower valve was closed and the upper one was open. Then the water in the pipe went down and a vacuum was created in the pipe. When the upper valve opened, air entered the pipe with a clear noise, filling the space.

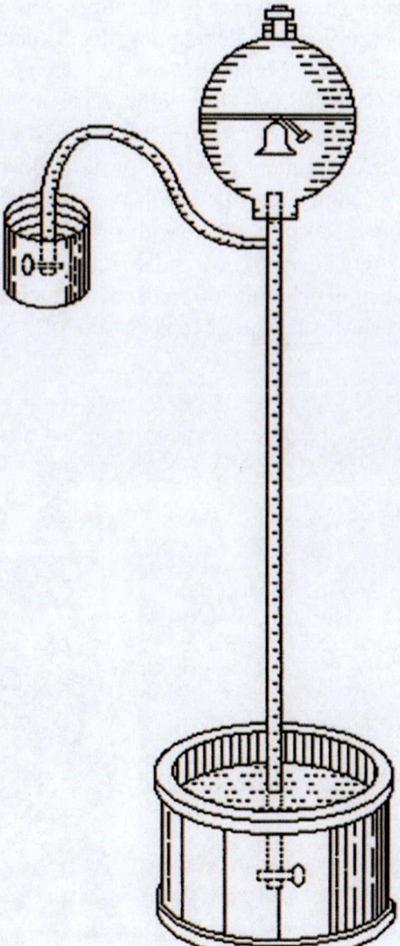

Fig. 4.16 Bertie's experiment

Later, he placed a ball with a bell in it on top of the tube. He repeated the experiment in the same way. When the sphere was emptied, a vacuum remained in it. He moved the bell hammer from the outside with a magnet, but no sound was heard. Sound does not travel through a vacuum.

4.3.4 The Invention of the Barometer

In a letter to Evangelista Torricelli (1608–1647), Raphael Maggiotti described Bertie's experiment. In the letter, he emphasized that when seawater is used instead of ordinary water, the water level in the pipe is lower. This prompted Torricelli to use mercury, which is 13.6 times denser than water, instead of water. At that time, mercury was called "quicksilver" (because of its silver color and fluidity). There were mercury mines in the Tuscany area.

The famous "Torricelli" experiment was performed by his assistant Vincenzo Viviani (1622–1703), in 1643 or 1644 (it is not known for sure when it happened). The experiment consisted of the following: A glass tube, closed on one side, about 1 m high, was filled with mercury. It is closed and turned on the other side and immersed in a vessel with mercury. Then the lower side is opened. The mercury slowly began to descend down the pipe. The top of the mercury column stopped at a height of about 760 cm from the mercury level in the vessel. This found the barometer. All subsequent repairs were of a cosmetic nature.

Evangelista Torricelli (1608–1647)

He was born on October 15 near Faenza, Italy. He was left an orphan as a boy. He was raised by his uncle, the learned monk Jacopo. Torricelli first studied with the Jesuits in Faenza, before moving to Rome in 1627 to study with Benedetto Castelli, Galileo's friend. It was immediately obvious that he was a talented mathematician. He was appointed Galileo's assistant in 1641 when he returned to Florence. Galileo died 3 months after his arrival. Then Grand Duke Ferdinand II appointed Torricelli as Galileo's successor as court philosopher and mathematician in Florence. In 1644 Torricelli published his major work *Opera Geometrica*. He died on October 25 after a short illness, probably from typhoid fever.

Torricelli concluded: "We live submerged at the bottom of an ocean of air" The total weight of the layers of air in the "50-mile-deep atmosphere" (so he used to say) **creates a pressure that keeps the 76 cm tall column of mercury in balance.**

Torricelli never published a paper on vacuum and barometer research. One of the reasons is that he knew what consequences he could face from the church (like his predecessor, Galileo). Another reason may be his preoccupation with research in mathematics. He researched cycloids. Then (in 1644) he published his main work "Opera Geometrica". Nevertheless, he described the experiments in detail in two letters that he sent to Rome in 1644 to his friend Michelangelo Rizzio. According to the enthusiasm for the results of the experiment, one could guess that the experiments were completed in 1644, and not a year earlier, in 1643, as is usually stated in the literature. Rizzi wrote a letter to Mersenne in Paris, in which he informed him in detail about Torricelli's experiment (Middleton 1964). Mersenne, known for his good connections, spread the news of this throughout Europe.

Everyone was delighted with the new instrument. They repeated the experiment in their performance. Already in 1645, Torricelli's experiment was repeated in Rome, with Cardinal Giovanni Carlo (brother of Ferdinand II). Pipes of various shapes and sizes were used and placed at various angles from the vertical position. The vertical height of the top of the mercury was always the same in relation to the level of mercury in the vessel. This means that it depended only on atmospheric pressure.

There is some ambiguity regarding Descartes' contribution to the invention of the barometer. Descartes wrote a letter on June 2, 1631, to his student Renery. In it, he explained atmospheric pressure and vacuum. There, Descartes used wool as an analog of air. In his explanations, he used his famous attitude about vortices. There he reasons correctly that the atmospheric pressure keeps the mercury in the tube in balance, but he does not agree that there is a vacuum in the tube above the mercury, but that this space is filled by ether when the mercury descends.

It can be seen from the letter that Descartes knew the idea of how the barometer works, therefore, much earlier than Torricelli did. From that, one could guess that he also carried out the measurement. After careful reading, however, it can be concluded that there are some omissions in the explanation, which show that he did not perform the experiment. However, Descartes wrote a letter to Mersenne on December 13, 1647. It shows that Descartes made a barometer scale, so he sends a paper copy to Mersenne so they can compare the measured values. If the barometer is considered to be an instrument with a scale, then it can be said that Descartes was the first to make a barometer.

4.3.5 Improving the Barometer

More than 20 years have passed without practically anything being done to improve the barometer. In that period, the barometer that the Academia del Cimento used since 1657 was used. Robert Hooke made an outstanding contribution in 1665, with the invention of the "wheel barometer". It was mainly used for domestic rather than professional needs, so it will be discussed in a separate chapter.

The word barometer was first used by Robert Boyle in 1665 (Day and Ludlam 1972). Until then, they were talking about Torricelli's tube. Namely, in a letter addressed to Oldenberg (Secretary of the Royal Society), Boyle gave a brief description of his balance baroscope and a comparison with the height of mercury in Torricelli's tube. This is where he uses the word "barometer" for the first time.

Boyle was interested in both the thermometer and the barometer. The barometer was used in many investigations of atmospheric phenomena, including the question of the existence of a vacuum. He never expressed a clear position about the vacuum (perhaps due to religious convictions). In 1669, he wrote the paper "Continuation of New Experiments," in which he describes the first completely portable barometer. On June 4, 1668, the Royal Society proposed that such barometers be sent to several parts of the world in order to measure pressure. In particular, they believed that he should be sent to distant English areas (colonies) "to the plantations of Bermuda, Jamaica, Barbados, Virginia, New England, Tangier, St. Helena, the Cape of Good Hope and Cameroon." This grandiose plan was not realized at the time, due to the fear of whether Boyle's portable barometers would "survive" the long journey.

Attempts have been made to increase the sensitivity of the barometer. The first such attempt comes from Descartes. He constructed a barometer with two liquids, mercury below and water above (Fig. 4.17a).

The idea is to increase the sensitivity with the upper, thinner liquid (water, pure alcohol). With the mercury below, the total length of the barometer tube is reduced, and with the upper, lighter one, that length is increased, thus increasing the sensitivity. Water as the upper liquid is not good because water vapor in the upper part of the tube makes the reading very temperature dependent.

Fig. 4.17 Barometers: (**a**) Cartesian, with two liquids; (**b**) Hooke's with two liquids; (**c**) Hooke's, with three liquids

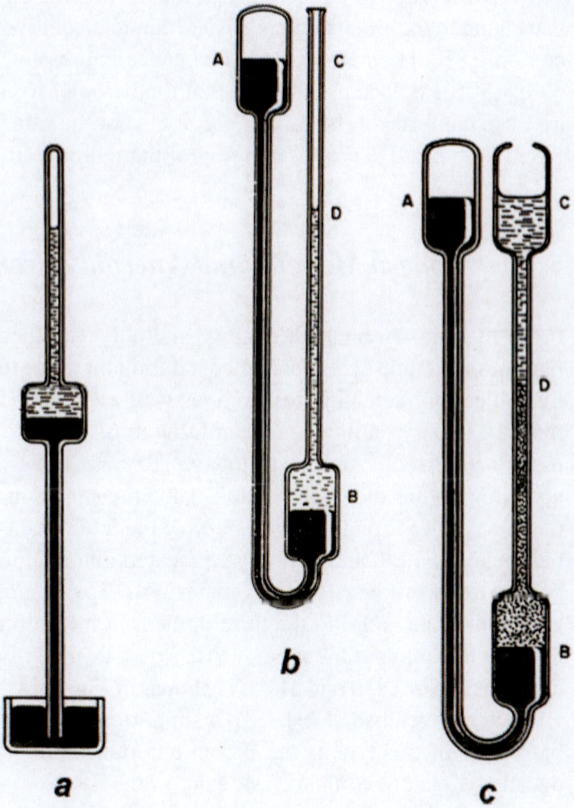

Hooke constructed a two-fluid barometer in 1668 and presented it to the Royal Society, Fig. 4.17b. He also uses water and mercury. It has advantages compared to Cartesian, but also a disadvantage because it quickly gets dirty through the tube through which the liquid moves. Hooke found the solution to the problem in a barometer with three liquids. He used mercury, turpentine oil, and alcohol. Atmospheric pressure is read in both cases by the position of point D.

Such barometers were popular and used in Western Europe for almost two centuries (Halley 1686). There were other improvements during that period. An "L"-shaped curved tube was used. In Bernoulli's barometer, one arm was narrower than the other, which increased sensitivity, but was impractical for measuring. In 1749, the Swiss meteorologist de Luc (1727–1817) analyzed the sensitivity of barometers to temperature. He introduced the correction of the read state to the temperature.

French scientists Cassini and Lemonier noticed in 1740 that the sensitivity of the barometer increases when boiled mercury is used, and the readings agree better with each other.

Hook constructed a ship's barometer in 1667, composed of an air and alcohol thermometer with a movable scale. Such a barometer was described much later by the Russian scientist Lomonosov. In 1759, he wrote: "When double thermometers show a difference in temperature, it means that the air is becoming heavier, and the barometric condition is higher." And Mendeleev introduced the "differential barometer" in 1878. He measured small changes in pressure.

In 1805, Laplace pointed out that the barometric condition should be corrected, because the force of the earth's gravity is not the same at different heights or at different latitudes. He proposed a way of that correction.

4.3.6 Modern Mercury and Aneroid Barometers

From the mid-seventeenth century, mercury barometers, without the complicated two- or three-fluid types, occupied a dominant place for household use, as a decorative object, for scientific research, as well as for standard meteorological measurements. Mercury barometers are still the most accurate instruments. There are many manufacturers of barometers today. They are mostly alike. There are two types, according to the shape of the tank. The tank can be in the form of a cistern (cup) or in the form of a siphon (in the shape of the letter "U"). De Luc constructed a siphon barometer in 1770 and strongly advocated that it should be used. One type of such barometer is the Fitz Roy barometer. Fitz Roy is a famous English meteorologist who contributed a lot to the development of meteorology around the middle of the nineteenth century. One type of Fitz Roy barometer with a siphon, which was produced between 1870 and 1885, is shown in Fig. 4.18.

Henry Cavendish (1731–1810) supported cistern barometers. The most famous barometer of this type is the Forten barometer. Nicolas Forten (1750–1821) introduced one change to the classic tank in the form of a cup so that the lower part of the tank could be raised to a spike, Fig. 4.19. This made it possible to accurately determine the reference level of mercury in the tank.

There is a completely different type of instrument for measuring atmospheric pressure, namely aneroids (***rain guage***). The idea for the aneroid barometer was given around 1700 by the famous German mathematician Leibniz Strangeways (2002b). In a letter addressed to the well-known Swiss scientist Johann Bernoulli, he described a "small closed box" that could compress and expand under the influence of the increase and decrease in air weight. At that time, there was no good material or technique to produce such boxes.

Nicolas Jacques Conte had a similar idea in 1797. He proposed to make a box in the shape and size of a pocket watch. The bottom of the box would be made of solid steel, and the top of thinner steel, so that it could move under pressure changes. The air would be removed from it, and a spring would be placed inside that would return the lowered box. The movement of the top of the box would be transmitted by the pointer to the pressure scale. According to this idea, such a box was made, but it was very sensitive to temperature changes. The idea was finally realized and a practical instrument was produced by Lucian Vidi in 1843. He called the instrument aneroid (without liquid) and patented it for France, England, and the USA for the period 1844–1846.

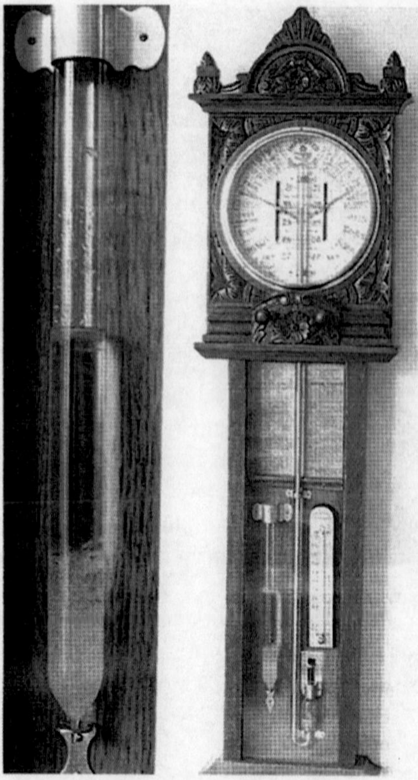

Fig. 4.18 Fitz Roy's siphon barometer (right) with "storm glass" in the lower left (left)

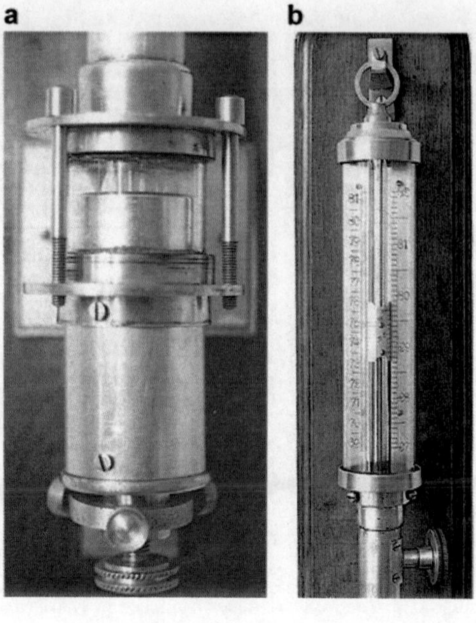

Fig. 4.19 Forten barometer (**a**) tank with a spike (left) (**b**) upper part with scale

A similar, though not identical, pressure gauge was patented in 1846 in Germany by the railway engineer Schintz. His instrument was in the form of a twisted tube with an elliptical cross-section. The curvature of the tube changed under the influence of internal pressure. Since 1848, this device has been installed in steam locomotives in Germany to measure water vapor pressure. Three months later, the Frenchman Bourdon patented a very similar instrument. It is quite possible that it was made independently of the others. Both Vidy's and Bourdon's pressure gauges were shown at the Great World Exhibition in 1851 in London.

Both were awarded medals. But then Vidi starts legal proceedings against Burdon. Vidi lost the case in the first two trials. It is a kind of natural justice because his "invention" is practically the one invented by Conte. However, Vidi complained a third time. Now he got a verdict in his favor. However, his patent expired and he was not granted an extension, so Burdon could continue his work without any problems. Bourdon became rich by selling instruments for various industrial needs. Vidi remained a poor businessman because his aneroid found application, not in the industry, but for meteorological measurements. The external and internal appearance of an aneroid, produced in 1880, is shown in Fig. 4.20.

Aneroid is used in an instrument for pressure registration, which is called a barograph, Fig. 4.21.

Before the invention of the aneroid barometer, mercury barometers with a siphon were used to register pressure (Strangeways 2002a). One such barograph from 1880 is shown in Fig. 4.22.

4.3.7 In-Door Barometers

When talking about barometers, attention is rightly paid to the method of operation, accuracy, sensitivity, etc. However, there are very widespread barometers, where the price is also their decorativeness. One of these, which combines both accuracy and decorativeness, is Hook's mercury barometer with a circular scale.

This barometer was described by Hooke in 1665 in the book "Micrography." It is a mercury barometer with a siphon (the lower part of the tube bent in the form of a

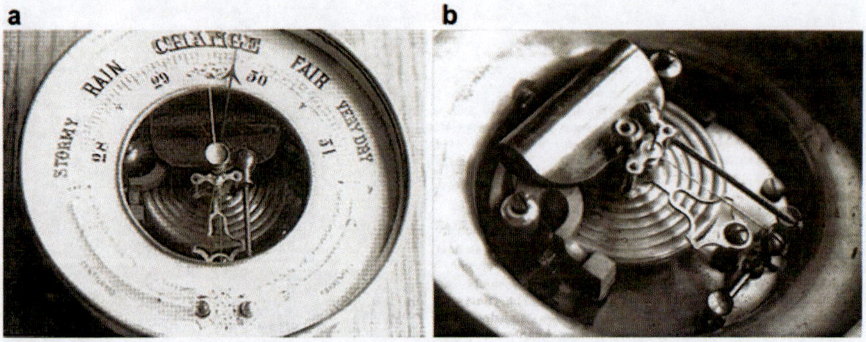

Fig. 4.20 Aneroid barometer produced in 1880. (**a**) External appearance; (**b**) inner part

Fig. 4.21 A barograph with a column of eight Vidy boxes

Fig. 4.22 "An 1880 barograph featuring the receiving part of a mercury barometer with a siphon"

Fig. 4.23 Barometer with circular scale: (**a**) front part with two scales and pointer; (**b**) the rear part with mercury tube, float, rope, swivel wheel, and a counterweight

large printed letter J). A float is placed on top of the mercury in the shorter part of the tube. A rope is attached to it, which is passed over the reel. A counterweight is hung at the end of the string. A pointer is attached to the reel. In front of the pointer is a circular scale, which is engraved in pressure units and/or with time markings in certain segments of the scale. As a unit of pressure, lengths (heights of the mercury column) expressed in English inches (25.40 mm) or in Paris inches (27.07 mm) were used. A full circle on the scale corresponds to a pressure change of 2 English inches. On the scale, a value of 29.5 inches is marked on the "weather time type scale" with "variable" (weather). Going towards lower values of pressure, every half inch there are labels in order: rainy, heavy rain, stormy. When going from the 29.5-inch mark to higher pressure, every half inch the marks are in order: clear, steady clear, very dry. It is not known by which rule these two scales were set. In 1688, the scientist Dalense added two more states to this descriptive scale of six weather conditions: very dry and very stormy. These descriptive scales are still used today with aneroid barometers. The front part of the house barometers was made with a lot of decorative details, which depended on the manufacturer. In Fig. 4.23. the front part of the barometer with a circular scale and the back part where there is a mercury barometer with a siphon, a float, a coil of string, and a counterweight are shown.

4.3.8 Change in Atmospheric Pressure with Height

One of the scientists who repeated Torricelli's experiment was the Frenchman Blaise Pascal (1623–1662). With the help of his father Etienne, he experimented with pipes of various shapes and lengths. He used mercury, water, and red (black) wine. Since wine is lighter than water, the pipes were about 12 m long. But, unlike the others, Pascal was not satisfied with repeating already-known experiments (Spiers 1937).

He was thinking about what this measured pressure represents. He saw that Torricelli's idea of the atmosphere, as a sea of air whose total weight we measure with a barometer, was correct. This meant that the pressure, going up, had to go down. To check this assumption, he needed to measure the pressure at the top and at the base of the mountain.

Blaise Pascal (1623–1662)

He was born on June 19 in Clermont-Ferrand, France. His mother died when he was 3 years old. He, his brother, and his sister were raised by his father, Etienne. Pascal's family moved to Paris in 1629. Blaise was immediately recognized as a child prodigy in mathematics. As early as 1642, he invented an addition and subtraction machine, but it was expensive to manufacture. His analysis of cycloids inspired others to formulate their study. He was also a pioneer in many areas of fluid mechanics. Once Pascal met with the professional gambler Chevalier de Mer. De Mer lost a large sum of money gambling. That is why he was interested in Pascal to find a method that would bring him success in gambling. It is not known how Merr fared in his professional work, but it is known that Pascal, solving his problem, was the originator of the theory of probability.

Pascal had a great accident in 1654. His racehorses almost killed him completely. He barely survived. Then he completely turned around in his spiritual life. Having experienced a mystical revelation, he turned to faith in God. Pascal joined an extremely strict sect within the Catholic Church. That sect was later condemned by the Pope himself. He presented his philosophical and religious ideas in "Provincial Letters." He lived a strict ascetic life in the monastery of Port-Royal des Champs, 15 km southwest of Paris. Pascal died on August 19. When his body was being prepared for burial, it was noticed that he had a belt around him with pointed nails pointing toward his skin.

Fig. 4.24 Pi de Dom is surrounded by layered clouds

Since he was seriously ill and lived in Paris, where there are no mountains, he could not make these measurements personally. That is why he asked his son-in-law, Florin Perrier (1605–1672), to carry them out. Perrier lived in the mountainous region of central France. The proposed measurements were made on September 19, 1648, along the mountainside of the volcanic mountain Pi de Dom (1464 m above sea level) Fig. 4.24. Now the famous Pi de Dom observatory is located on top of the mountain.

Perrier invited several prominent ecclesiastics and laymen from his town, Clermont-Ferrand, to join him. First, they set up a barometer at the foot of the mountain and took a measurement. The mercury column was exactly 28 Paris inches high (1 Paris inch = 27.07 mm). Then they climbed to the top of the mountain and took a measurement. The height of the column was less and was 24 and 2/3 inches. Perrier made an additional five measurements at various points on the top and got the same result. When descending from the top, they took measurements. At the same time, the barometric condition increased. They took the measurement again at the base and the height of the mercury was the same as at the beginning of the measurement, 28 inches. The results of the measurements confirmed Pascal's idea that the air is elastic and that the pressure decreases with height.

There is evidence that at the end of 1647, such measurements were carried out in Warsaw by V. Magni. So that was before Perrier's measurements. This fact has captured the attention of scientists. A few centuries later, in 1906, the historian F. Matisse accused Pascal of forging a letter written by Perrier on November 15, 1647. Investigating the facts, another historian, J. Messner, concluded that Perrier probably sent the letter to Pascal, which Pascal slightly altered before printing. This shows that there was a certain vanity in affairs of this kind.

It is worth noting here that Pascal then calculated the total mass of the atmosphere. He found it to be 8.28·1018 pounds (5.25·1018 kg). Pascal did not say the calculation method, but he emphasized that it could be calculated by a child who knows how to add and subtract.

4.3.9 Misconceptions When Interpreting the Barometric Condition

The invention of the barometer was a key turning point in understanding the nature of air. Torricelli's tube was used to measure all over Europe. One would think that all the problems related to that have been overcome, but they only appeared then. It was necessary to interpret why the pressure changes at some point in time, in various weather situations, etc. Practically all thinking people of that time were involved in that polemic. Today it just seems funny what people didn't know back then. For this reason, only a small part of the understanding of the nature of pressure in the period of about 50 years after the invention of the barometer will be listed here.

In February 1645, less than a year after Torricelli's experiment, Cardinal Giovanni Carlo de' Medici had several of Torricelli's tubes in Rome, on which he read the condition for a long time. He found that the mercury level did not match the observed change in temperature and humidity. It was concluded, therefore, that the pressure does not change due to the properties of the air from the immediate environment, but due to the air from a wide area. Thus began long attempts to find a scientific basis for weather forecasting, and to find a reasonable explanation for irregular changes in the barometric state (Frisinger 1974).

Observations at the very beginning imposed a great unknown about that "corner" of nature that makes up the atmosphere. Namely, it was observed that the mercury usually reaches the highest level when the weather is nice and the lowest during ugly, rainy weather, when the air is full of "water vapor and vapor," and, apparently, "the heaviest." That knowledge was enormously surprising and even disappointing. Nature does not work fairly, many thought. Thus, Pierre Gassendi (1592–1655, philosopher who reactivated Epicurus in the West) wrote in 1654:

> … When the sky is clear and northerly winds are blowing, the mercury in the tube always remains higher than when the sky is covered (with clouds) and when southerly winds are blowing. But, when the sky is clearer, the weight of its air should be much less than when it is gloomy, so that it puts much less pressure on the mercury in the cistern, and consequently the mercury comes out of the tube and should have a lower height.

Gassendi immediately, as a thinking man, explains the disappointing fact: "The reason for this may be that in good weather, as well as in cold weather, vapors do not come from the ground, as they do in warmer and cloudier weather." There are some particles that cause these vapors to press upwards instead of downwards.

Cloud particles are of this type. "Indeed, you have not seen them suspended unless some force carries them up and sustains them, just as an owl rises and flies through the air".

According to this, it was thought that there were particles that pushed upwards, thus lightening the air. This idea is similar to when Boyle performed 29 air pump experiments. Once he opened the tank from which he pumped the air. The smoke particles are drawn into the tank. Of course, we know today that it happened because of the so-called pressure gradient forces, and what kind of force facilitated the "cloudy air," had yet to be figured out.

John Wallis (1616–1703) a professor from Oxford, found that quicksilver (mercury) rises when there is a thick fog. He attributed this to the weight of water vapor in the air. It also rises in sunny weather, partly because "water vapor rises with the sun, and partly because the heat increases the elasticity of the air." In cloudy and rainy weather, the height decreases "because the air is lighter due to the precipitated precipitation." In windy weather, "I found that the height generally drops and that this is more pronounced than in rain." I explain it by the fact that the wind moves the air and does not allow it to push down much, similar to a swimmer. "I have never found it to be lower than in a strong wind".

This shows how much pressure measurement has brought freshness to the search for the causes of events in the atmosphere. Wallis discovers four causes: (1) the weight of water vapor; (2) that the elasticity of air is independent of pressures, due to its weight; (3) the effect of discharged rain, and (4) the effect of wind.

Giovanni Alfonso Borelli (1608–1679) was one of the most active members of the "Academia del Cimento" and the first meteorologist who made systematic wind measurements with a thermometer and barometer and observed other weather phenomena. Borrelli accepted two of Wallis's ideas and had his own specific explanations. Thus, Borelli points out that the particles of humidity in the cloud somehow lie on top of each other and thus support part of the weight of the air above the cloud. This is why the pressure is lower in cloudy weather. The existence of this idea can be found in the letter that Prince Leopold (a smart and scientific man) sent on December 15, 1657, to his brother Ferdinand II. In that letter, Leopold rejects such an explanation. He points out: "I doubt that such minute particles of moisture, which are in the cloud, and descend little by little, either as rain or fog or in some other form reach the earth, can produce the same effect of retaining the upper layer as a solid body, if the cloud is like a brush."

Borelli sent Prince Leopold a report from Pisa on March 5, 1660, in which he pointed out: "The mercury in Torricelli's tube was extremely high that morning, higher than it had ever been observed in the past three years; practically '20 degrees' higher." From the second letter, it can be seen that the usual height of mercury is about 460° of that scale. Borelli points out: "This monster of nature shows that the air lying above Pisa is excessively and extremely larger than it has been on other occasions, due to the mixture of other materials, water vapor, liquid or ground with it. We will see if this will be followed by extremely heavy rains. Or, if the material is not watery, and the wind cannot disperse it, we shall see that it is not something that precedes the comets... If nothing of this happens, I suppose that at different moments of time, the air may become more or less heavy, due to the rising of particles from the ground."

The correspondence between the prince and Borelli was frequent. In a letter dated March 16, 1660, Borelli writes: "... You point out that the duration of a strong wind can accumulate a large amount of air above Pisa and its surroundings, and that because of this the weight of the air can increase the height of the mercury in the pipe... I predicted heavy rainfall or something else that precedes a comet."

Borelli was probably the first prognosticator to ever give a bad prognosis on scientific grounds. Borrelli expressed doubt that a strong wind could produce such a large effect. Ten years later, Borelli accepted Wallis's ideas. Borrelli believed that the mercury is usually high before a long and continuous rain, but descends when the rain begins to fall.

In those years, the Royal Society also dealt with the problem of changing the barometric state. Discussions were lively in January 1677. Robert Hooke opined that a high reading on the barometer means that various vapors have been added to the air. With this, he explained the great heaviness of the air during the long blowing of eastern winds and its lightening during the blowing of southern winds. In the first case, the air passes over the vast land and thus collects in itself a large amount of vapors, which remain suspended in it. In the second case, the air passes over a large ocean surface, which gives a smaller amount of the parts that make up the air.

The great scientist of that time was Halley (1656–1742). He also did not fail to observe the barometric condition. Based on observation and theoretical experience, he wrote the first formula for pressure change with height. He also dealt with thoughts (speculations) about the connection between the barometric state and the type of weather. In the work, he published in 1686, Halley presents the theory of the hydrostatic balance between air and water vapor. He also explains why the low value on the barometer in calm weather before rain: "Air, being light, can no longer hold suspended water vapor because it is specifically heavier than the environment in which it swims so that it descends towards the earth, and as it falls, it meets other liquid particles, they unite and thus create small drops of rain." He introduces a lot of terms, some of which are unacceptable today.

In the record from 1690, it is clear that they practically continuously monitored the barometric condition, so changes that happened quickly could be noticed. "On the evening of the 28th of November, Mr. de Hara observed that the mercury in the barometer, which was 28 inches high, descended in a very short time to 26 inches and 10 lines. The wind was extremely strong at the time. Mr. Varigon says it could be because the wind splits up and goes around the air column". Obviously, the interpretation that the pressure comes from the weight of the column of air above that place is taken literally here. That pillar is now seen as a solid obstacle.

There are also very witty comparisons, which clearly highlight someone's wrong attitude about the cause of pressure changes. So Bernadino Ramazini (1633–1714) from Modena writes in his book: "It is difficult to explain how the air can be lighter when it contains rain (or when it is falling) and heavier when the weather is clear. It would be like if someone claimed that a woman is heavier after the birth of her baby than when she was pregnant." He tried to explain the change in the barometric state by saying that there are "earthy, salty and nitrogen vapors" in the air. Because they absorb moisture and are heavier fall to the ground in the form of rain. They lighten the atmosphere. When the weather is nice, these particles dry out and return to the atmosphere. This happens especially with the north wind.

From Ramazini's book, one can find out about various discussions and positions on barometric conditions and weather. This is stated by de Franco Tarigi, a professor from Modena, who claims that "when water vapor begins to condense and fall, it no longer has weight and is not part of the air." The "weightless" state of raindrops was also discussed by the Royal Society in 1678 and 1679. It was only around 1695 that it took the position that it did not support Torti's theory.

In order for the views on this to be clarified, Ramazini wrote to Leibnitz around 1700 and presented Torti's problem to him. After some time, Leibniz writes him a letter from Hanover in which he describes an experiment that could answer the question. The experiment is as follows: Water is poured into a deep vessel and a blocked hollow ball made of some heavy material is placed in it so that it floats. When the hole in the ball is opened, the ball will fill with water and sink to the bottom. As the ball descends, Leibniz claims, the water is not in equilibrium and will lose some of its weight during that time. This Leibniz experiment was later judged to be worthless. Leibnitz probably thought that the movement of the sphere mechanically causes a compensatory upward movement of the water and that because of this it exerts less pressure, and loses some of its weight. According to him, this would have the following analog in atmospheric conditions: "When the weather is good, water droplets are so small and scattered in the air that they cannot descend, like pieces of butter (fat) in milk before it is heated. The mercury begins to descend a little before they reach us". Here one can see how much confusion reigned at that time about these fundamental questions, even among such men of great scientific reputation as Leibniz.

Something more should be said about misconceptions about how strong wind affects the barometric condition. This was discussed by the Paris Academy in 1690. After that, there were many different opinions on this matter. The first is what was advocated by Pierre Varigon (1654–1722). He believed that the strong wind somehow prevented the air pressure from fully acting on the mercury in the barometer tank. Another thought was that fast horizontal movement reduces the weight of the moving body. It was also widely believed that opposite winds blowing in an area tend to collide more often, thus making the air denser. There were also opposing opinions, that these collisions of winds create a moderate vacuum. Finally, there were those who took into account that air can have both an ascending and a descending component of movement. In this regard, it is instructive to mention Leibniz, who wrote a letter from Hanover on February 26, 1700, to someone from the Paris Academy, regarding the influence of wind on changes in barometric conditions. He emphasizes: "The air will be supported by a strong wind, and especially by a wind that moves away from the ground and tends to move upwards." Obviously, he is not referring here to convective movements, which were not known at that time, but to the vertical component of strong air currents.

Then they were impressed by the observed fact that in Western Europe, with easterly and northerly winds, higher atmospheric pressure prevails than with westerly and southerly winds. We find a clear and well-argued reason for this in the doctoral dissertation written in Jena by Johann Grugerus in 1701. The idea is clear. He starts from the already-known fact that westerly currents prevail in moderate latitudes. Therefore, he says: "If the wind blows upon our area from the north or the east, or from the area between these, it will be opposed by a constant wind coming from the opposite direction. Therefore, the air around us must accumulate, contract, and become heavier, so that its weight acts on the mercury to rise. In the case of southerly and westerly winds, or from the area between them, a constant wind is added to them, so it will pull the air and thin it. The air becomes lighter, and the mercury descends."

It is interesting to mention the opinion of the famous experimenter of that time, Francis Hoxby the Elder (circa 1713) about the oscillations of the barometric state in strong mossy winds, which we now often call "pumping," caused by the dynamic effect of the wind on the pressure in the closed rooms in which it is located barometer. This is an artificial phenomenon that has nothing to do with anything that happens in the free atmosphere. However, on days with strong winds, the barometric condition is lowered. Hawkesby placed the tank of the barometer in a closed box and allowed a strong current of air to flow around it. The pressure began to drop. It could have been the opposite effect if the dimensions of the tubes through which the air flowed were different. This was noticed by Christian von Wolff (1679–1754). Wolff has great merit in the development of meteorology because he claimed for the first time, in 1709, that the wind blows from an area of higher pressure to an area of lower pressure.

The man who tried the hardest to explain the movement of the mercury column under the influence of the wind was de Marjan (1678–1771). In his prize question from 1715, he described in detail how moving bodies, including air, exert less pressure (and are lighter) on the ground beneath them than when they are at rest. It can be seen that he was influenced by ancient mystical attitudes. He also connected it with the well-known fact that you can cross thin ice if you move quickly, without the ice breaking. His views were later refuted by the Swiss meteorologist de Luc (1727–1817) who calculated that a wind of 100 km/h would change the barometric condition by only 1/244th of a line.

In order not to get the wrong idea that no one from that period understood the nature of pressure changes, let us quote the opinion of the great Swiss mountain explorer, Benedict Saussure (1740–1779). Analyzing the small pressure changes near the Equator, he pointed out … that the pressure changes mainly due to the air temperature. "These changes occur in part due to the movement of air from one place to another. The increase in pressure in Europe in the northerly winds, which are dry, occurs because the cold air pushes the one that was above our heads. That's why we have good weather and an increase in the barometric condition."

4.3.10 Atmospheric Pressure Force

The first to clearly show that the atmosphere causes enormous pressure was Otto von Gerik (1602–1686). To demonstrate the existence of a vacuum, in 1650, Gerik pumped air out of a wooden water barrel with a pump he constructed. The wooden barrel was not strong enough and broke. He then repeated the experiment with the copper ball, and it deformed when the pumping began. To maintain the created vacuum, he had to use a sphere with thicker walls.

Otto von Gerik (1602–1686)

He was born on November 12 in Germany. He attended the University of Leipzig from 1617 to 1620, the University of Hellest in 1620, Jena from 1621 to 1622, and Leiden. He studied law, mathematics, and construction. In 1626, he became an alderman of the city of Magdeburg. From 1618 to 1648, Germany went through great violence. A 30-year war was fought over who would be king of Bohemia, a Protestant or a Catholic. Although France, Spain, Sweden, and Denmark participated in the war, the war was mainly fought on the territory of Germany. Gerik supported the Protestants and became quartermaster general of the Swedish king Gustavus Adolphus I. He performed a similar duty in 1635, with the ruler of Saxony. He became the mayor of Magdeburg in 1646.

Fig. 4.25 Gerik's experiment with the air pressure force

To measure atmospheric pressure, Gerrick made two bronze hemispheres. He placed a valve on one part, which could be closed and opened. He pumped the air out of the sphere formed by joining the two hemispheres. He attached one hemisphere to a holder fixed to a wooden post, Fig. 4.25.

He pulled the lower hemisphere down with weights. The spheres did not separate. He continued the load until they broke apart. To demonstrate the power of atmospheric pressure to the German Emperor Ferdinand II, Gerik repeated the experiment with hemispheres in Magdeburg in 1657. He used several horse carts to separate the hemispheres, Fig. 4.26.

He amazed those present with the power of atmospheric pressure. In fact, the hemispheres were held by pressure gradient forces (the difference between the surrounding pressure and the vacuum in the sphere). In modern conditions, this power is manifested if a passenger plane door is accidentally opened during flight. All passengers who are not properly restrained will be sucked out (because the pressure in the plane is higher than the atmospheric pressure at flight height).

4.3.11 Laws of Pressure

The first law of atmospheric pressure was formulated by Robert Boyle. He and his assistant Robert Hooke performed a series of experiments right after Torricelli's experiment. They found that in a vacuum, a feather and a metal object fall at the same speed and that sound is not transmitted through it. They also discovered that

Fig. 4.26 Magdeburg experiment

objects cannot burn in a vacuum and that animals cannot live in it. Despite these facts, Boyle never publicly declared that the vacuum existed.

They experimented with an air pump. They discovered the "air spring" effect. Namely, when the compressed air was allowed to expand, they noticed that it returned the pump piston, like when a spring is compressed. Then they took a glass tube in the shape of the letter "J," which was closed on the short side. They poured mercury through the long arm. The air in the shorter arm was compressed to a smaller volume as the mercury was higher in the longer arm. Repeating the experiment for 44 positions, they came to the law that the product of pressure and volume is constant at unchanged temperature, $pV = \text{const}$. This is known as "Boyle's Law." Boyle published those results for the first time in 1662.

The same law was independently discovered by the French priest Mariot (1620–1684) so that law is known under the name "Boyle – Mariott's Law." Thomas Andrew (1813–1885) was the first to discover that below a certain temperature (called the "critical temperature"), carbon dioxide does not behave according to Boyle's law. After that, Amjet studied this problem, using a long vertical tube placed in the shafts of a coal mine (to conduct experiments at pressures greater than atmospheric). He concluded that because each gas molecule has its own individual volume, the space available for the movement of molecules must be smaller than the space occupied by air. Thus, the Boyle-Mariot law was corrected, which now reads $p(V - a) = \text{const}$, where a is a positive parameter.

In 1801, the English scientist John Dalton (1766–1844) formulated a law that states that each component of a mixture of gases causes the same pressure as if it alone occupied the entire space of the mixture at the same temperature. It follows that the total pressure of the gas mixture is equal to the sum of the pressures originating from each component individually.

The French scientist Denis Papin (1647–1714) discovered that when the pressure above a liquid is lowered, its boiling point is lowered, and when the pressure is increased, its boiling point is raised. The boiling point of water at an atmospheric pressure of 1000 mb is 100 °C, while at a pressure of 850 mb (which corresponds to a height of about 1500 m above sea level), the boiling point is 96 °C, at 750 mb (which corresponds to a height of about 3000 m) around 92 °C, etc.

Papin was a scientist, mechanic, and inventor of safety valves and steam engines. He proposed to the French king, Louis XIV (1643–1715), to use his water pump, with which he would feed the fountains of Versailles. French finance minister Colbert refused. However, Papin used his machine to propel ships via a paddle wheel. Papin also found the first pressure cooker in 1681, which he called the "digestor" (digestor), Fig. 4.27.

The boiling point increases in the digester, so it cooks much faster. When he visited England in 1679, Papin prepared food for members of the Royal Society in his pressure cooker. Then the English king Charles II (1630–1685), the founder of the famous Cavendish Observatory, ordered a digester for himself in 1675. Papin moved to London in 1684, where he worked for 3 years as the temporary head of the experiments of the Royal Society.

Papin apparently nicely applied the law he discovered about the dependence of the boiling point on atmospheric pressure (and in general on pressure, produced in

Fig. 4.27 Digester

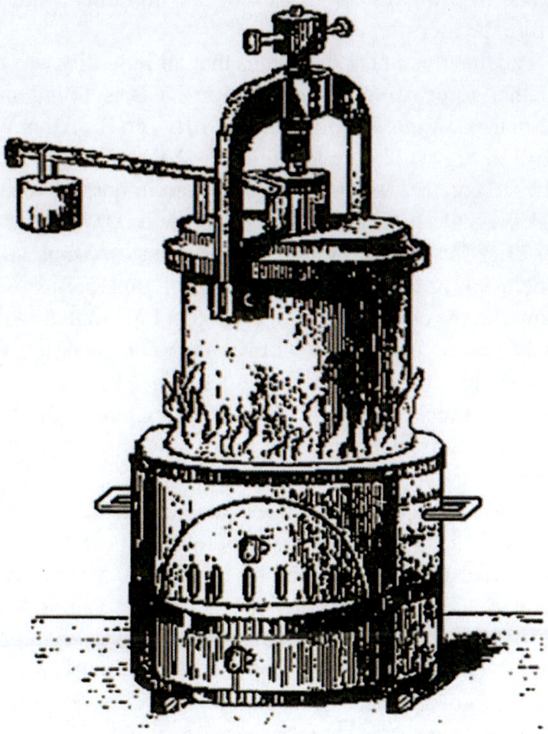

any way). However, this idea is not his, and it is not so "recent." Namely, the great Venetian travel writer Marco Polo described in 1280 his experiences from a trip through the Vaška Valley (Central Asia). "The mountains are so high … that everything you see is extraordinary. The air is so thin, and a fire that burns does not give the same warmth as a fire in the lower regions. It is not effective when cooking food." This is, as far as is known, the first record showing that boiling occurs at lower temperatures (fire is not efficient for cooking!) when the atmospheric pressure is lower (the air is so rarefied!).

4.4 Higrometer

4.4.1 The Invention of the Hygrometer

The origin of water vapor, and its properties, has long been disputed among scientists. There was no real concept of what air humidity actually is. However, this did not prevent the invention of instruments for measuring air humidity well before the concept of humidity was clarified. Ignorance of the nature of air humidity automatically meant that the measurement method could not be accurate, but was entirely relative. Because the sequence was not respected: first a clear concept, then instruments, instruments for measuring air humidity were found over a hundred years before the thermometer.

The first document that states that air humidity was measured for the purposes of weather forecasting on a scientific basis is found in the work "Works" by the German, Cardinal Nicolas Cusa (1401–1464). In that work, he wrote: "… if we put a ball of dry wool on one end of the balance scale, and stones on the other end, in order to create a balance between them in normal weather (in loco ed aere temperato), we will see that the wool will be heavier when the air is more humid than when it is drier. He who takes this difference into account will have a much more accurate conclusion about the future change of time." The imbalance in weight could be shown on the circular scale of the scale, Fig. 4.28. Such hygrometers were used until the beginning of the eighteenth century (De Saussure 1783).

It should be noted here that such hygrometers were known much, much earlier. Two thousand years before Christ, the Chinese used charcoal as an absorbent of moisture from the air. From the difference in the weight of the coal, they calculated the moisture content.

The balance hygrometer was also described by Leonardo da Vinci in his diary from 1483 to 1486. Instead of wool, he suggested the use of cotton. Since that time, there have been many attempts to exploit the hygroscopic properties of various materials. The Venetian physician Santorio Santorini (1561–1636) used a twisted thread made of animal intestines (sheep or cat) to indicate air humidity. The thread (rope, twine) would relax at higher air humidity. A talented experimenter, Robert Hooke, proved successful in this area. He constructed a hair hygrometer. He described it in his *Micrography* from 1665. He used chamois beard hair as a

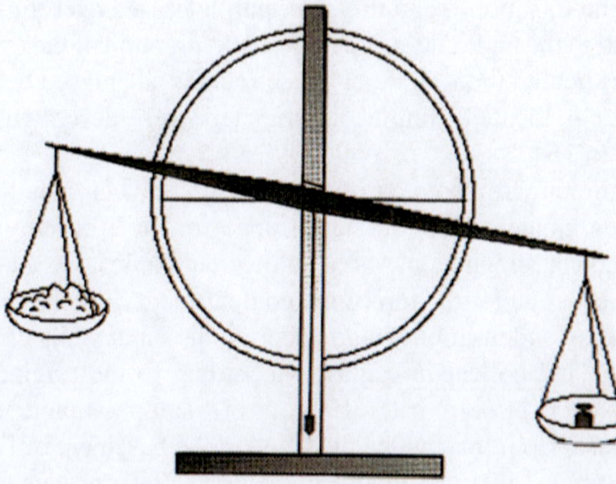

Fig. 4.28 Balance hygrometer

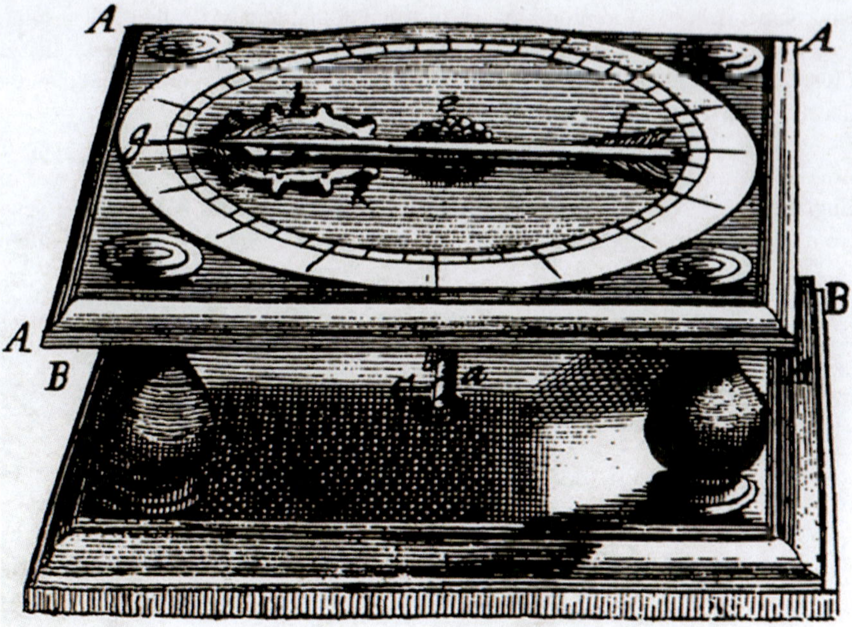

Fig. 4.29 Hooke's hygrometer

receiving part. A set of hairs were twisted and straightened depending on the humidity of the air. That twisting was transmitted via the pointer to the circular pointer, Fig. 4.29.

The Academia del Cimento used a hygrometer with a string, made of twisted narrow strips of sheep intestines. One end of the rope, about five meters long, is attached at one end to support, and the other part is passed over the moving reel. A weight is attached to the end. The stretching and contraction of the string was shown on a vertical scale behind the weight, or on a circular scale placed behind the reel on which the pointer is located. Similar hygrometers were described by the French scientist Mersen in 1644.

The condensation hydrometer, constructed in 1655 by Ferdinand II, Grand Duke of Tuscany, holds greater scientific significance for the Academy del Cimento's requirements. The instrument consists of an inverted conical vessel placed on a tripod. A glass is placed under the top of the compartment, Fig. 4.30. Ice was placed in a conical vessel. Condensation would occur on the outer walls of the vessel. The drops would flow and collect in a glass. According to the original record dated August 27, 1655, it can be seen that 9–10 drops of water per minute were condensed (poured into a glass) from this device, installed in the basement of Ferdinand's palace. Another instrument that was placed in the Great Hall "produced" 11–13 drops per minute. It is unlikely that the basement is drier than the Great Hall. Obviously, in addition to humidity, condensation is significantly affected by the temperature of the environment (that is, the difference between the temperature of the ice and the environment). It is a forerunner of the instrument that today is called an "absolute hygrometer." With this "mostra umidaria" (humidity indicator), it is shown that in Florence, the south and west winds (coming from the sea) are more humid than the north and east.

There is some evidence that Christopher Wren (1632–1723), designer of London's famous St. Paul's Cathedral, used the condensation hygrometer in England. In 1687, Amonton constructed a hygrometer consisting of a ball sewn from leather (like today's soccer ball) and a tube inserted into the ball. The volume

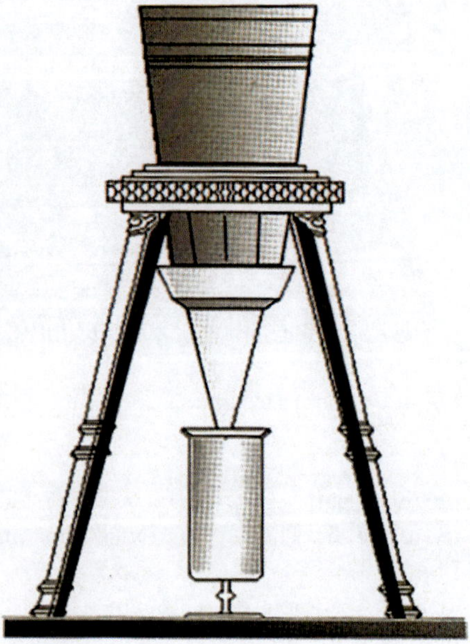

Fig. 4.30 Condensation hygrometer

of the ball changed depending on the humidity of the air, so the liquid from it rose or fell down the tube.

All these instruments were indicators of moisture in the air. There was not much scientific accuracy in all of this, until the famous German mathematician Johann Heinrich Lambert (1728–1777), Fig. 4.31.

He intensively studied the properties of the hygrometer and was the first to give the instrument the name hygrometer. The word is a coin, composed of the Greek words "hygros" (moisture) and "metron" (to measure). He published his research on humidity measurement in 1774, in the work *Suite de L' Essai d' Hygrome-trie*.

Lambert made a lot of measurements and always tried to present the results mathematically. He found an empirical law of changes in humidity during the year, as well as during the day. He determined the changes for three places: Berlin, Sagan, and Wittenberg. He also found a relationship between the daily and annual course of humidity and temperature. He expressed the movements with appropriate equations as a function of the height of the Sun. He was the first scientist who presented meteorological data in the form of a graph, rather than in the form of a table. This represented a fundamental advance in the way of analyzing meteorological data because it is easier to see how they are measured. Lambert also dealt with measurements of other meteorological quantities, such as pressure, wind, etc.

The Swiss Deluc (or de Luc) was active in the development of the hygrometer. Between 1773 and 1791, he published several works in the magazine *Philosophical Transactions*. In it, he points out that a good hygrometer should meet three requirements: that it has fixed points on the hygrometer scale; to use a standardized scale; that equal intervals on the scale correspond to equal changes in humidity. He suggests that the fixed points on the scale should be determined according to the highest humidity. In a work published in 1773, he gives a description of an instrument

Fig. 4.31 Johann Heinrich Lambert

consisting of a mercury thermometer with a cylinder-shaped reservoir made of ivory. The cylinder is about 2.5 cm high and about 0.7 cm wide. The walls are quite thin. The volume of the cylinder changes due to humidity, which causes the mercury in the hygrometer tube to rise and fall. A similar hygrometer was used by the Maynham Society. These hygrometers have not survived to the present day. In 1781, Deluk constructed a hygrometer that used whalebone as a reactive material.

Horace Benedict Saussure (1740–1799)

He was born in Geneva, where he became a professor of philosophy at the age of 22. In the beginning, he was interested in biology, and immediately devoted himself to the geology and meteorology of the Alps. The Alps particularly attracted his attention. He was the first great mountaineer scientist. The observations he made while moving around the Alps gave him a strong incentive to construct numerous instruments or to improve existing ones. The most obvious example of this is the hygrometer with human hair, which he described in the paper "Essais sur la hygrométrie" (Discourse on the Hygrometer). His most extensive book is *Voyage dans les Alpes* (Journey in the Alps), in which he describes most of his scientific discoveries in the Alps. There is no doubt that the enormous effort involved shortened his life. He died at the age of 59, after 8 years of poor health. He can serve as an example of a selfless scientific worker.

4.4.2 Hair Hygrometers

Between 1780 and 1830, many hygrometers were constructed. One of them has found the widest application to this day. It's a hygrometer with human hair. Lo and behold, it was later determined that it is best to use natural blonde hair! The inventor of this hygrometer is the Swiss naturalist Benedict Saussure (1740–1799).

By measuring, he determined that human hair, which is free of grease and other impurities, stretches (lengthens) when it absorbs moisture and shrinks (shortens) when it dries. The change in length, from the state when it is the wettest (when saturation is reached) to the state when it is completely dry, is 2.4%. He found that hair that had been boiled in a soda solution was much more brittle and more stable as an indicator of moisture. He introduced the division of the hygrometer scale into one hundred equal parts between the state of complete saturation (the hygrometer is placed in a space where water is boiling and covered with a wet cloth) and completely dry air. Figure 4.32 shows two forms of his hygrometer.

Saussure conducted experiments very carefully. He determined temperature corrections based on the readings of the hygrometer. He analyzed the effect of air expansion and contraction. Knowing the opinion of Volta, that the atmosphere in the upper part is mainly composed of hydrogen, and in the lower part of carbon dioxide, Saussure examined the behavior of the hygrometer in such experimental conditions.

Regarding the discussions that took place about the influence of water vapor on the change of the barometric state, Saussure wrote the following in his *Discourse on Hygrometry*: "I took a large glass sphere in which the air was saturated with 16 °R (Reomir's scale). I removed almost all the water vapor from the ball. Because of this, the pressure in the ball decreased by only 1/84. This means that if the entire atmosphere were to pass from being completely saturated to being completely dry, which of course does not happen, the lowering of the mercury column would be only 1/2 Paris inch". When we compare this attitude with the current knowledge about the cause of pressure changes, then we can only admire his contributions.

Along with sound reason, the precision of experimentation, breadth of reasoning, and great commitment to work, one more of his character traits should be highlighted here. Saussure was extremely moderate in expressing his views. He did not arrogantly, haughtily, and ignorantly put anyone's opposing position under his feet. Here is an example of how he presented his views: "Without pretensions to find a complete solution to such a difficult problem, which such great scientists have tried to solve, I will content myself with proposing some general views...."

At that time, the concept of saturated water vapor was known, but the concept of relative humidity was not yet known. Saussure knew that the change in hair length is not only proportional to the change in water vapor content in the air but also depends on the temperature. These circumstances later led Gay-Lisak and August to find a new (psychrometric) way to measure air humidity.

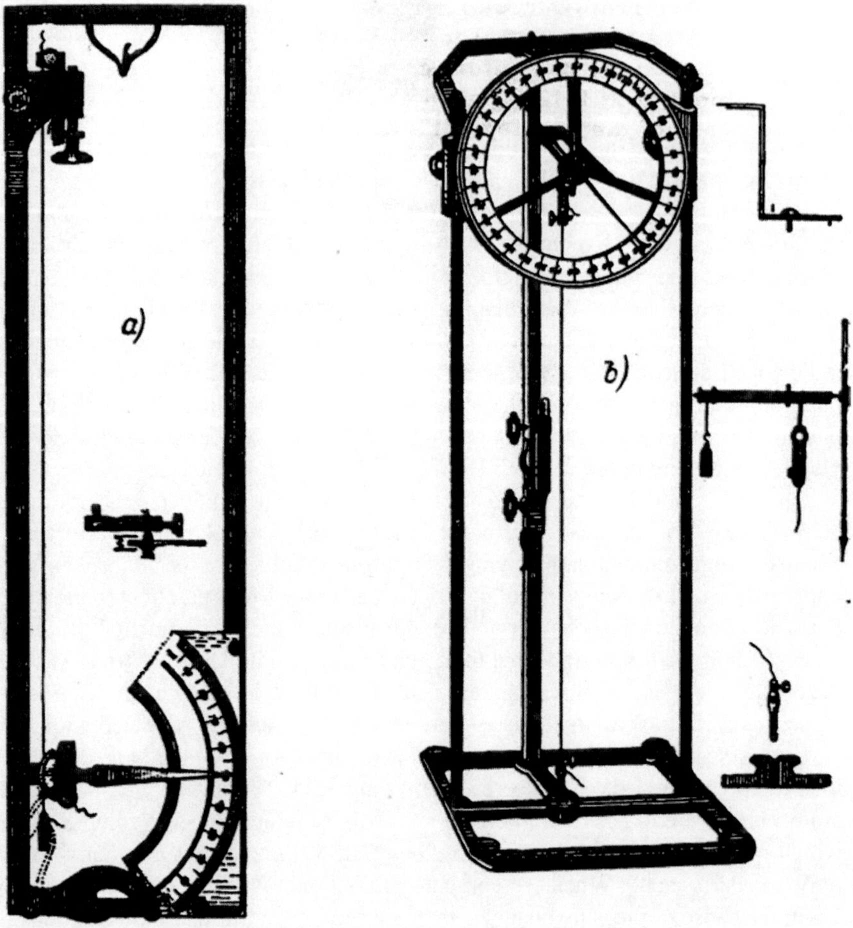

Fig. 4.32 Two models of Saussure's hair hygrometer: (**a**) with arc scale; (**b**) with a circular scale

4.4.3 Psychrometers

William Cullen (1710–1790), a professor of medicine in Edinburgh, noticed that a thermometer that had previously been dipped in alcohol (this was done by medics to disinfect it after use) showed a lower temperature. Cullen was the first to explain that this cooling occurs due to evaporation from it. After this, this principle was used as a new method for measuring air humidity, the dry and wet bulb method. In August 1822, James Ivory (1765–1842) showed that relative humidity can be measured using two identical thermometers, one of which has a wet tank. These are the so-called dry and wet thermometers. He also derived an approximate formula for relative humidity, expressed using dry and wet thermometer temperatures and

Fig. 4.33 August's psychrometer with minimum and maximum thermometer

atmospheric pressure. Three years later, the German scientist Ernst Ferdinand August (1795–1842) obtained a similar result. He called a pair of dry and wet thermometers a psychrometer (from the Greek word "psychos" (cold-bottom), and "metron" – (measurement)), Fig. 4.33.

Later, ventilation was introduced around the tank of the wet thermometer, in order to promote the evaporation of water from the cloth with which the tank was wrapped.

The first to introduce wet bulb ventilation was Welsh, director of the famous Kew Observatory in England. He did it in 1852. Unfortunately, meteorologists did not see the importance of the principle of ventilation, so that method was not widely accepted. It was only around 1880, thanks to the famous aerologist Asman (1846–1928), that ventilation was introduced into widespread use. That pair of thermometers is called a "ventilation hygrometer." Around 1891, Asman used a centrifugal hair dryer with a spring. Sprung later calculated the psychrometric constant. The ventilation hygrometer was widely accepted and is used throughout the world to this day.

In 1854, a professor of the college in Paris, Henri Reznoll (1810–1878) introduced the dew point hygrometer. That instrument consisted of a thinly polished silver thimble (cup) into which the reservoir of the thermometer was placed. Next to the thermometer, a narrow tube is lowered into the cup, which is connected to the fan at the upper end. Some volatile liquid is poured into the cup. When the fan is turned on, bubbles pass through the liquid. The liquid evaporates and cools. Very quickly the temperature drops low enough for dew to form. The temperature that should then be read is called the dew point temperature. The relative humidity can be calculated from this temperature and other values.

4.5 Rain Gauge

The rain gauge is the simplest meteorological instrument, and its history goes back much further than the thermometer or barometer. The first measurements of the amount of precipitation were made in the Orient, where for thousands of years life and progress depended on timely and sufficient amounts of precipitation (Reynolds 1965). The oldest records of it come from India and are found in the work *Arthasastra* (Scientific Politics). It was written by Chanakya, the minister of Chandragupta, a ruler who succeeded in uniting India from 321 to 296 BC by his authority. He established an administrative system. In a chapter entitled 'Agricultural Management', Chanakya writes: "The amount of rain that falls in the Jāṇgala area is 16 drones, half as much as in the Anupanam area; 13.1/2 drones in the Aśmaka area; 23 drones in Avanti and a huge amount in the Aparantan area… When one-third of the collected rainfall falls in the beginning and end months of the rainy season, and two-thirds in the middle, then it is considered that the rainfall is very uniform." The names of the provinces are from that period. Jāṇgala is a dry region, Aśmaka is in southern India, and Avanti and Aparantan are in the west. Drona is a measure of volume in ancient India. From all this, it can be seen that about 2300 years ago there were constant measurements of rainfall in the whole of India. Çankoya also gives a lot of other interesting information in that part. Depending on whether the amount of precipitation is higher or lower, he points out which plants are suitable for growing in certain areas.

Another area where rainfall was measured in ancient times was Palestine. Written documents say that in that area in the first century AD, priests measured the amount of precipitation using special vessels. The measured quantities were expressed in "tofahs" or "tefahs," the ancient measure of length, which was about 9 cm. Therefore, the height of the water layer was measured, as it is done now. According to the data, it can be seen that in the spring months in Palestine, 6 tofas of precipitation (about 54 cm) fell. It is something more than nowadays. Now, from January to April, 40–43 cm of rain falls. These measurements were used by the Jewish priests to predict the fertility of the year, and what the harvest would be like.

Precipitation measurements have been carried out in China for a long time. In their "Mathematical treatises" from the twelfth and thirteenth centuries (during the

Sung dynasty), they talk about the calculations of the amount of precipitation, measured by rain gauges. During the Ming Dynasty, in 1424, each region had to send reports on the amount of rainfall. The rain gauges used to measure precipitation were in the form of cylinders with a height of 48 cm and a diameter of 22.5 cm. Rainfall measurements in China continued and were carried out until the beginning of the eighteenth century. For now, it is not known where these data are located. The depth of snow in the mountainous regions of China was measured using bamboo sticks for several centuries BC.

Around the same time, precipitation measurements began in Korea. It is recorded in the historical annals that around 1443 Emperor Sejo had a device for measuring the amount of precipitation. The device consisted of bronze vessels 30 cm deep and 14 cm in diameter. The vessel was placed on a pillar. Every time it rained, the person in charge had to take a measurement and inform the emperor about it. Unfortunately, no data on the performed measurements has been saved. There are records of rain gauges that were used in Korea in the middle of the eighteenth century. The rain gauge was 32 cm high and about 25 cm in diameter.

The modern rain gauge comes from Galileo's student Benedetto Castelli (1577–1643). Castelli began measuring rainfall in Perugia in 1639. It is said that Castelli constructed a rain gauge after seeing the sudden rise in the level of Lake Trasimeno after a rain.

In the French city of Dijon, rainfall was measured using a rain gauge in the 1760s and 1770s. The name of the observer is unknown. The upper part of the instrument was in the shape of a square. The water was collected in a cylindrical vessel. The rain gauge was placed on a specially placed holder 1 m high in front of the window. One would say that it was not a representative place to measure. These measurements were used by Mariott when he analyzed the movement of surface waters.

The oldest rain gauge in England comes from Sir Christopher Wren. In 1662, he used an empty water container to measure rain. Fifteen years later (1677) Richard Glounley constructed a rain gauge in the form of a funnel, 14 inches in diameter, which was connected by a hose to a container where the weight of the collected rainwater was measured. It was a balance rain gauge. He made measurements from 1677 to 1703. Robert Hooke constructed a similar rain gauge (ombrometer) in 1695. That type of rain gauge was used to measure rainfall at Gresham College in London. He later converted it into an ombrograph, an instrument that records precipitation data.

There were a lot of rain gauges and rainfall measuring points. An attempt was made to find some regularity in precipitation from the data. Measurements have shown that the amount of precipitation increases when going towards the top of the mountain. This has led many to the wrong view of how rain occurs. This will be discussed in one of the next chapters on precipitation and clouds. To prove that over flat terrain, precipitation decreases with height, William Heberden the Elder (1710–1801) took three identical rain gauges. He placed one in the courtyard of the house, another above the highest chimney of the house, and the third on the roof of Westminster Abbey. They were placed so that the tower did not obscure the others. He carried out the measurement for the whole year, from July 1766 to June 1767.

The data showed that there was a marked increase in the amount of precipitation from the top of the house to the ground. The total measured rainfall during that period on the roof of the tower was 12,099 inches, on the roof of the house, was 18,139, and in the yard of the house 22,608 inches. They took this as evidence that raindrops are formed by the merging (combination) of several smaller drops on their way to the ground.

Detailed comparisons of precipitation amounts measured in Petrograd at the Main Geophysical Observatory were made by Wild. He made measurements at 1, 2, 3, 4, 5, and 25 m high. He found that the change in altitude is greater in winter than in summer.

Such a change could be explained by the attitude of Benjamin Franklin (1706–1790) and later Wilhelm Dove (1803–1879). They believed that when falling cold drops encourage water vapor to condense on them, and therefore grow as they fall towards the ground. This explanation was insufficiently precise. According to this, half of the precipitation would be generated below the clouds, near the ground. They overestimated the contribution of condensation to raindrop growth and neglected droplet enlargement due to collection. In summer, even below the base of the cloud, condensation does not occur on the drops, but the drops evaporate.

Others have also made similar measurements of precipitation with height. The results were similar, but the interpretations were different. To demonstrate why precipitation changes with altitude, Howard conducted a seemingly absurd experiment in London in 1812. He placed two rain gauges at the same height on the roof of a building but placed one on the west and the other on the east side of the roof. He found that the air vortices that occur significantly affect the reception of precipitation in the rain gauge. And Stevenson verified it in 1842 (Fig. 4.34). He found that in strong winds, the amount of precipitation at the same height varies a lot, while in calm weather these differences are not noticeable.

From all the analysis of the measurements, it became clear that the receiving part of the rain gauge must be changed, to protect from the drift of raindrops and snowflakes with the wind. This drift from the receiving part is more pronounced in the case of snowflakes. Wild, conducting experiments between 1878 and 1879, suggested that cross partitions be placed in the receiving part of the rain gauge, which would prevent particles of precipitation from being carried away. The attempt was not particularly successful.

At the same time as Wilde, in 1879 Francis Naifer, a professor at the University of St. Louis, tried to solve the observed deficiency in the rain gauge. He placed a concentric protective ring around the receiving part of the rain gauge, which was 0.5 inches away from the rim of the rain gauge. At the same time, measurements were made with ordinary rain gauges at various heights. The experiment showed that with the new rain gauge, precipitation was higher by 6%, at a measurement height of 2 m, while the increase was 18–50% at a height of 300 m.

This protective ring is installed especially on special rain gauges, which measure precipitation over a longer period (season), in more inaccessible mountain areas. It is called a totalizer.

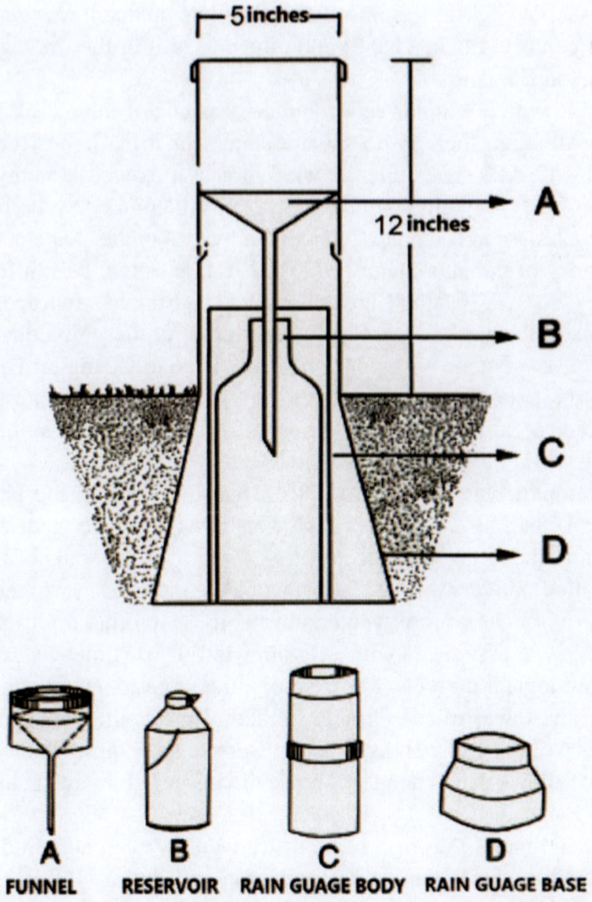

Fig. 4.34 Components of English rain gauges with beaker. Lengths are in inches (1 inch ≈ 2.5 cm)

4.6 Wind Measurement

4.6.1 The First Weather Vanes

We have seen that even in ancient Greece and during the Roman Empire, wind roses were introduced (D'Alembert 1747). The first observations, which were made within the Florentine measurement network, use the wind rose, but the winds from certain directions were given popular names: tranmontana, grecale, levante, shiroko, libezio, etc. They also distinguished the wind at ground level (calling it "lower wind") and the wind at height (the so-called "upper wind"). We find the method of identifying the wind through the direction from which it blows, with the learned monk Alcumnus (735–804). His wind rose, similar to the Greek one, has twelve basic directions (D'Arcy 1918). It is difficult to determine when and how the eight-way rose, now in use, was introduced. It is known that it was widely used until the sixteenth century. Locke (1666) and others used it in the first systematic wind measurements in England.

A wind vane with a wind direction indicator is one of the oldest meteorological instruments. Although there is no specific name for it in Greco-Roman literature, Livy (59 BC – 17 AD) states that the wind indicator existed in many Italian cities. They were probably brought from the east, from Anatolia or Syria. In the very distant past, the Chinese and Japanese made wind vanes in the shape of dragons.

In the history of the Han Dynasty (23–220), it is stated: "… In the first year of the Yang-cha (AD 132), Chang Chien distributed instruments for recording the wind and vibrations of the earth." There is no description of these first devices. In *Shang Fu Huan Tu*, it is written: "… south of the entrance to Chang'an Palace there is a tower… and on it stand copper birds that turn in the wind." In Europe, it was customary to place weathervanes on the towers of tall buildings, some of which, of extraordinary workmanship, date back to the early Middle Ages.

Bishop Rampert was the first to place a weathervane on the church tower in Bricken (Tyrol), in 820. Sometimes such weather vanes were made in the shape of a rooster, symbolizing the vigilance of the church over the souls of the dead. They were often called "singers for time," or, it would be nicer to say, singers of the times. Because it shows what kind of weather awaits us, given that traditionally a certain type of weather was associated with a certain wind. Those timers were then, and still are, real technological marvels. For example, the largest weathervane in the world is located on the tower of the Giralda Cathedral in Seville, made in the fifteenth century. He is 6 m tall and weighs 1250 kg. Despite its bulkiness, it is very sensitive and precisely follows wind changes. The Sevillians gave him the nickname "faithful Giraldini."

The weather vane in the form of an instrument for accurate wind measurement was constructed by the Italian astronomer Ignacio Danti († 1586). He called it an "anemoscope." He described it in 1578. The direction wing was in the form of a flag and was connected to the vertical shaft. It seems that it had a lot of friction, so the measurement accuracy was not satisfactory.

4.6.2 Measuring Wind Speed – History of the Beaufort Scale

The wing for the direction of the wind generally no longer required any special modifications, but the scientists devoted themselves to finding methods for measuring the wind speed. One of the oldest estimates of wind speed comes from the famous mathematician Girolamo Cardano (1501–1576). He assumed that during strong storms the wind blows at a speed of about 45 m/s (a distance of 50 "dogs" covered in the time of one pulse – heartbeat). Marriott estimated a much lower speed, at about 10 m/s. In 1705, Deržan, observing the movement of snowflakes, concluded that wind speeds during a storm range from 22 to 27 m/s.

Approximate estimates of wind speed were also made during the first organized meteorological measurements. The first speed scale with four-speed categories was introduced by the English Royal Society around 1655. The speed scale was described by Locke in 1666 as follows: "The force of the wind is marked with 0, 1, 2, 3 or 4. Zero means that the wind is so weak that it does not move the leaves on the trees; 1 indicates a gentle breeze that moves only the leaves on the plants; 4 means very strong wind and 2 and 3 represent speeds between 1 and 4. These speeds are not accurately determined, I think, but it's better than nothing."

A similar speed scale of four intervals was introduced by the Mannheim Society in 1781. The Beaufort scale, which is still used today, was introduced in 1838 (Fry 1967; Kinsman 1969; Friendly 1977).

The introduction of a unique precise scale for wind speed became a necessity. Seafarers especially felt the need for it. At the end of the eighteenth and the beginning of the nineteenth century, the navy was already very developed. Merchant ships sailed the seas, looking for new routes. Seafarers wanted to avoid weather uncertainties on the old routes as well. Logbooks from earlier voyages were of little use to them. Everything about time is imprecisely recorded in them. In particular, a precise scale for speed was missing, because there are no objects at sea like on land: grass, leaves, trees, etc.

In 1703, the work *Historical Review of Great Windstorms* was published in England. It contains a descriptive scale of speeds by an unknown author, of 13 characteristics of speeds, starting from complete silence, calm weather, and then to stormy weather and storm. The author himself states that the assessment of wind power is relative, and to some extent depends on the strength and construction of the ship.

A more precise assessment of wind power, without the use of instruments, was introduced by the English admiral, Sir Francis Beaufort (1774–1857). At the age of 13, he was a cabin worker in the English Navy, and in 1805 he became the commander of the ship "Woolwich." With it, he carried out hydrographic observations of the eastern Mediterranean from 1809 to 1812. In 1812, Beaufort was seriously wounded in an encounter with pirates. He was then retired and withdrawn from active service at sea. From 1814, he was a member of the Royal Society, and between 1829 and 1859, he worked as a naval hydrographer.

Bofor je 1807. godine uneo neke izmene, tako je pored nule imala još 12 stepeni. Beaufort kept meteorological diaries from 1790, first in the form of short comments and descriptions of general characteristics of the weather, and later regular two-hour meteorological observations. Weather forecasts were also written in his diaries. Even after leaving the ship, Beaufort made regular observations and kept a diary of the weather, until his death. On a trip in January 1806. Beaufort gave his scale of wind and weather for the first time, noting: "From now on I will estimate the strength of the wind according to the following scale because nothing is more uncertain about wind and weather than statements like *moderate wind*, or *cloudy weather*." His scale started from 0 (calm) and ended with 13 (storm), (Fig. 4.35). Beaufort made some changes in 1807, so it had 12° in addition to zero.

Until 1838, the Beaufort scale was used only by him in his diary. Even he himself did not use it in the official ship logs. In 1838, the English Admiralty introduced the

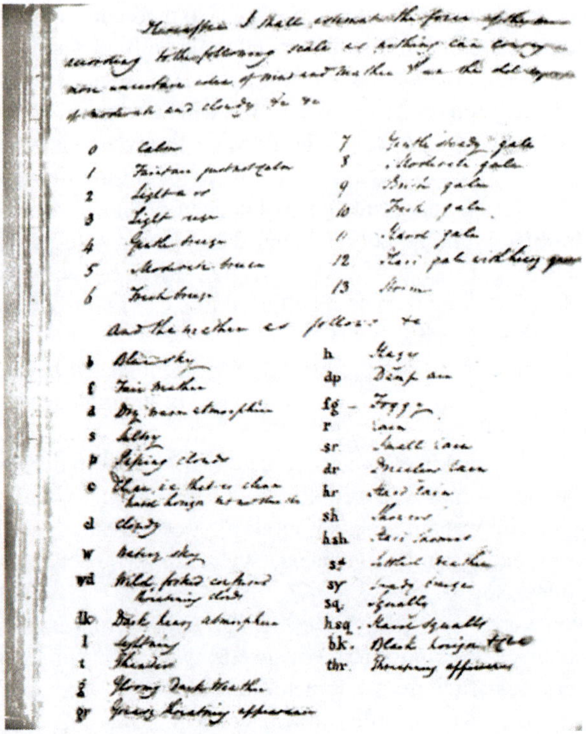

Hereafter I shall estimate the force of the wind according to the following scale as nothing can convey a more uncertain idea of wind and weather than the old expressions of moderate and cloudy &c &c

0	Calm	7	Gentle steady gale
1	Faint air just not calm	8	Moderate gale
2	Light air	9	Brisk gale
3	Light breeze	10	Fresh gale
4	Gentle breeze	11	Hard gale
5	Moderate breeze	12	Hard gale with heavy gusts
6	Fresh breeze	13	Storm

And the weather as follows &c

b	Blue sky	h	Hazy
c	Fair weather	dp	Damp air
d	Dry warm atmosphere	fg	Foggy
s	Sultry	r	Rain
p	Passing clouds	sr	Small rain
e	[illegible]	dr	Drizzling rain
d	Cloudy	hr	Hard rain
w	Watery sky	sh	Showers
wd	Wild [illegible] clouds	hsh	Hard showers
dk	Dark heavy atmosphere	st	Settled weather
l	Lightning	sy	[illegible]
t	Thunder	sq	Squally
g	Gloomy dark weather	hsq	Hard squalls
gr	Greasy threatening appearance	bk	Black horizon
		thr	Threatening appearance

Fig. 4.35 The original Beaufort scale from 1806

Beaufort scale into regular use in the Navy. It represented a big step in the development of meteorological research. Because now a huge amount of data collected from various ships could be used for analysis in a unique and precise way. Over the next two decades, different scales were used around the world. At the First International Meteorological Conference held in Brussels in 1853, it was decided to introduce a unique system in this area. The Standing Committee at the First International Meteorological Congress, held in Utrecht in 1874, approved the use of the Beaufort scale in international meteorological dispatches (telegrams). With minor changes, that scale was approved by the World Meteorological Organization, founded in Geneva, in 1951. The scale has been converted into speeds expressed in m/s or km/h. This can be seen in Table 4.1.

The first instrument for measuring wind speed was described around 1450 by the Italian mathematician Leon Battista Alberti. He described it in his treatise *On the Pleasures of Mathematics* as a weather vane "having a small plate suspended from the end of the vane shaft." The vane moves the board so that it is always exposed to the wind. Below the moving part of the plate, there is an arc with a scale that shows the deviations of the plate from the vertical. Apparently, Leonardo da Vinci described a similar instrument around 1500. He is often credited with being the inventor of the anemometer. But, as you can see, that is not true.

Table 4.1 Beaufort scale with descriptions

Beaufort number	Wind name according to WMO from 1964.	m/s	km/h	Description of phenomena on land	Description of phenomena at sea
0	Calm	0–0.2	<1	The smoke rises vertically	The sea as a mirror
1	Light breeze	0.3–1.5	1–5	The direction of the wind is observed by the movement of the smoke, not by the weather vane	Creases are formed but without any foam
2	A breeze	1.6–3.3	6–11	You can feel the wind on your face	Small waves, even shorter but more pronounced
3	Light wind	3.4–5.4	12–19	Leaves and twigs are constantly swaying; light flags develop	Very small waves; crests begin to break; foam with a slatted appearance; sometimes scattered "white kittens"
4	Moderate wind	5.5–7.9	20–28	The wind raises dust and sheets of paper; it starts small branches	Small waves that become longer, "white kittens" frequent
5	Moderately strong wind	8.0–10.7	29–38	Thinner leafy trees begin to sway; small waves are formed behind crests on land waters	Moderate waves that take on a more clearly elongated shape
6	Strong wind	10.8–13.8	39–49	Big branches are moving; the buzzing of telegraph wires is heard; the difficult use of an umbrella	Large waves begin to form; crests of white foam are spread everywhere
7	Very strong wind	13.9–17.1	50–61	All the trees are swaying; walking in the wind difficult	The sea is rising; the white foam that forms from the breaking waves begins to elongate into streaks downwind
8	Stormwind	17.2–20.7	62–74	The wind breaks the branches on the trees; walking against the wind is, in general, impossible	Moderately high waves of great length; eddies of sea droplets begin to break off from the edges of the cross
9	Storm	20.8–24.4	75–88	Light damage to buildings occurs (gutters tearing off, chimneys collapsing and tiles being removed)	High waves; thick streaks of foam down the wind; the crests of the waves begin to rock and break into rollers

(continued)

Table 4.1 (continued)

Beaufort number	Wind name according to WMO from 1964.	m/s	km/h	Description of phenomena on land	Description of phenomena at sea
10	Violent storm	24.5–28.4	89–102	It rarely occurs inland; uproots trees from the ground; there is great damage to buildings	Very high waves with long crests in the form of plumes; the foam produced in large pieces elongates into thick white furrows downwind; visibility reduced
11	Hurricane storm	28.5–32.6	103–117	A very rare occurrence, accompanied by large-scale destruction	Extremely high waves (small- and medium-sized ships can occasionally be lost from sight)
12	Hurricane	32.7–36.9	118–133	–	The air is full of foam and mist from the drops of the sea; the sea is completely white due to the fog from sea droplets

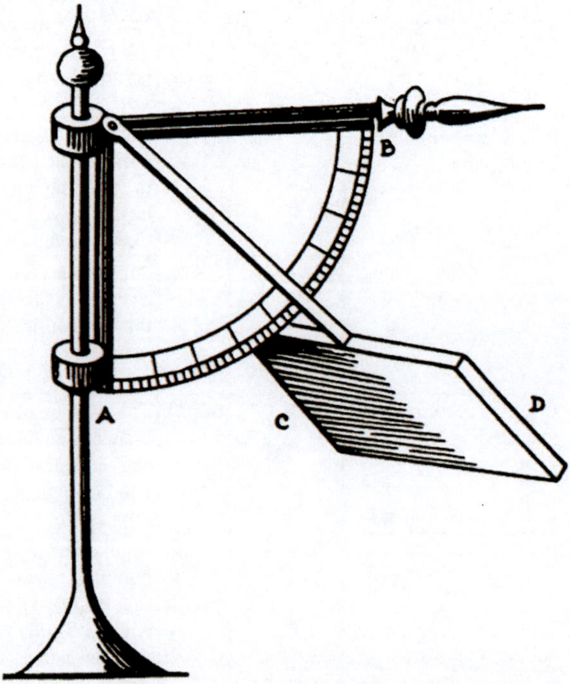

Fig. 4.36 Hooke's anemometer

An anemometer of this type was constructed by Robert Hooke in 1667, Fig. 4.36. The instrument is very practical, which is why it was used for over a hundred years until in the nineteenth century it was replaced by an anemometer with hemispheres

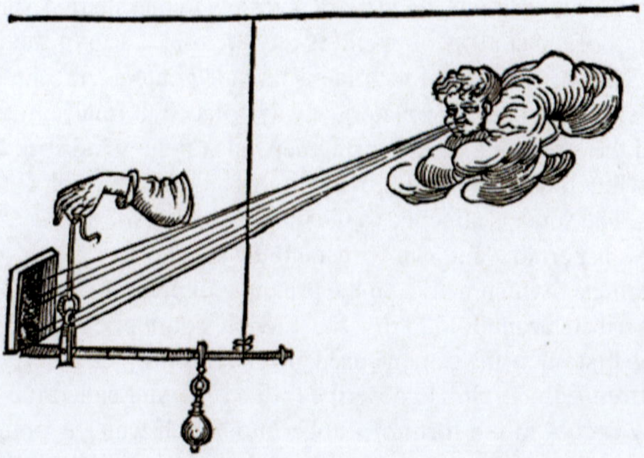

Fig. 4.37 Schematic of the Santorini anemometer

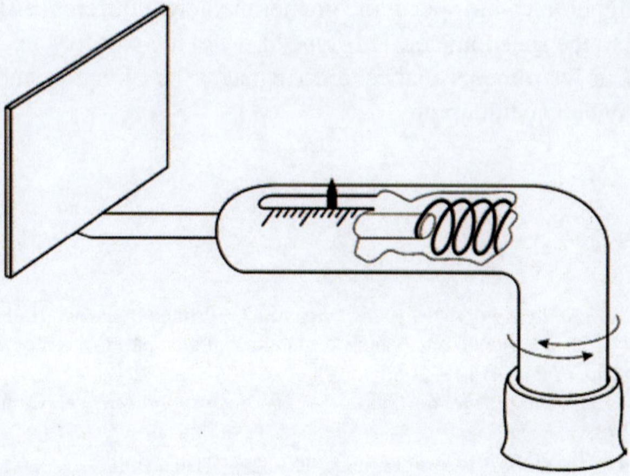

Fig. 4.38 Boujer's anemometer

and tubes that work on the principle of changing pressure depending on the wind speed (Laughton 1882).

Professor of medicine in Padua, Santorio Santorini, in 1625 constructed an instrument that can be called a balance anemometer, Fig. 4.37.

Wind pressure on a vertical plate attached to the end of the bucket is balanced by moving the weight along a standard weight scale. With a connection attached to the other arm of the scale, the entire device is balanced, until a balance is established between the strength of the wind and the weight. Obviously, it was not easy to measure with this instrument, especially if the wind is, as it usually is, mossy.

Santorini's concept was used by Pierre Boujer during the construction of an anemometer for measuring speed at sea, Fig. 4.38.

It consists of a square plate attached transversely to one shaft. A spring is attached to the other end of the shaft, which enters the horizontal arm of the pipe bent at an angle of 90°. The vertical part of the pipe is threaded onto an attached vertical shaft, around which it can rotate. The plate is always placed normally in the direction of the wind, and the stretching of the spring depends on the wind speed. On the upper side of the spring, there is a longitudinal slit on the tube, through which the pointer (spike) slides. The spike is attached to the moving end of the spring. This instrument is obviously transportable and can withstand the severe storms that occur at sea.

The anemometer, which works on the principle of pressure dependence on speed, was first described around 1721 by the French court priest, Pierre Daniel Xet. However, the first such instrument used in meteorology came from the scientist James Lind from Edinburgh. He described it in 1775 and called it a "wind gauge." The receiving part is in the form of a tube into which water is poured. The water level is the same on both legs of the pipe, and that is the zero state. When one leg of the pipe is connected to a horizontal pipe that passes through the shaft for the direction of the wind, then a difference in the height of the water in the U-pipe will appear. The higher the wind speed, the greater the height difference. This difference is transmitted to the speedometer. This type of receiving part for measuring speed is also installed in instruments that record (register) the direction and speed of the wind and are called anemographs.

References

Boyle, R. 1969. *Continuation of new experiments*, 68–73. Printed by Henry Hall.

Celsius, A. 1742. Observationer om tvenne bestandiga grader pa en thermometer. *Vetenskaps-Akademiens Handlingar* 4: 197–205.

Crombie, A.C. 1953. *Augustine to Galileo – The history of science*. Cambridge: Harvard University Press.

D'Alembert, J. 1747. *Reflexions sur la cause generale des vents*. Berlin.

D'Arcy, T. 1918. The Greek winds. *Classical Review* 32: 49.

Day, J.A., and F.H. Ludlam. 1972. Boyle. *Weather* 27: 449.

De Luc, J. 1784. Recherches sur les modifications de l'atmosphere, 2nd ed. *Paris* 2: 3–5.

De Saussure, H.B. 1783. *Essais sur l'hygrometrie*. Neuchatel.

Fahrenheit, D.G. 1724. Experimenta et observations de congelation aquae in vacuo factae. *Philosophical Transactions of the Royal Society* 33: 78–89.

Friendly, A. 1977. *Beaufort of the admiralty: The life of sir Francis Beaufort 1774–1857*. London: Hutchinson.

Frisinger, H.H. 1974. Mathematicians in the history of meteorology: The pressure-height problem. *Historia Mathematica* 1: 263–286.

Fry, H.T. 1967, 1968. *The emergence of Beaufort scale*. Mariner's Mirror.

Halley, E. 1686. On the height of the mercury in the barometer at different elevations above the surface of the Earth; and on the rising and falling of the mercury on the change of weather. *Philosophical Transactions of the Royal Society of London*: 104–116.

Hooke, R. 1665. *Micrographia*. London.

Kinsman, B. 1969. *Historical notes on the original Beaufort scale*. Vol. 39. Marine Observer.

Laughton, J.K. 1882. Historical sketch of anemometry and anemometers. *The Quarterly Journal of the Royal Meteorological Society* 8: 162–164.

Middleton, W.E.K. 1964. *The history of the barometer*. Baltimore: The Johns Hopkins Press.

———. 1966. *A history of the thermometer and its uses in meteorology*. Baltimore: The Johns Hopkins Press.

Pascal, B., trans. I.H.B. and A.G.H. Spiers. 1937. *The physical treatise of Pascal*. New York: Columbia University Press.

Réaumur, R. 1731. Second memoire sur la construction des thermometres, dont les degres sont comparable.... *Mémoires de l'Académie Royale des Sciences*: 250–296.

Reynolds, G. 1965. A history of rain gauges. *Weather* 20: 106.

Strangeways, I. 2002a. Barometric pressure, aneroid barometers. *Weather* 57: 204–209.

———. 2002b. Barometric pressure, mercury barometers. *Weather* 57: 132–143.

Chapter 5
Beginnings of Meteorological Measurements and Observations

5.1 Introduction

With the invention of basic meteorological instruments, quantitative measurements began. This contributed to significant progress in the development of meteorology in the seventeenth century compared to the previous period in which only visual meteorological observations were made. In that instrumental period, we distinguish two phases of development. The first phase is related to measurements carried out individually, in certain places, without any mutual communication. The second phase is related to coordinated measurements in several places. This is the phase of establishing a meteorological measuring network of stations.

However, the first step in collecting meteorological data is the data obtained by observing the weather. The observations undoubtedly date back to the very beginnings of civilization. But those systematic and carefully conducted observations come from the ancient Greeks, from the fifth century BC. Those observations had to be carried out systematically because, for example, Theophrastus formulated the empirical rules of weather forecasting. Ancient weather observations were mainly related to the description of the sky (cloudy, rainy, clear, etc.), wind direction, warm, cold, etc. Precipitation was undoubtedly the earliest observed.

Meteorological observations in the past never stopped completely, but there were periods when they were interrupted. Roman historians devoted considerable space to atmospheric phenomena in their yearbooks. Those chronicles were more complete throughout the Middle Ages, and observations were especially improved at the beginning of the thirteenth century. According to Hellmann, the first man in the West to make daily meteorological observations was the Englishman William Merle, a member of Merton College. He kept a daily meteorological diary. These observations were made in Oxford, from 1337 to 1344. Other examples of meteorological reports were found in manuscripts in the University Library in Basel, for the period 1399–1406. The form of such reports is as we will state for September 5,

M. Ćurić, V. Spiridonov, *History of Meteorology*,
https://doi.org/10.1007/978-3-031-45032-7_5

1399: "AM (before noon) beautiful and clear; PM (afternoon) cloudy and hot; after sunset, dark clouds with glimmerings and faint thunder; heavy rain during the night."

In the sixteenth and seventeenth centuries, European scientists paid great attention to meteorological observations. Thus, the German mathematician and astronomer Johann Werner (1468–1528) made continuous meteorological observations from 1513 to 1520. Tycho Brahe (1546–1601) made regular weather observations in Scandinavia between 1582 and 1598. The German astronomer and mathematician Johann Kepler (1571–1630) started keeping his weather diary in 1600 and did not stop it for the rest of his life. He also started making regular meteorological observations in Prague in 1604.

5.2 Beginnings of Meteorological Measurements

Most of the leading scientists of the seventeenth century, in addition to observations, began to make regular meteorological measurements, with instruments that had just been invented (Forster, 1815, 1816; Kington, 1974; Pearce and Smith, 1984). So Galileo and his friends started measuring the air temperature in 1613. René Descartes, despite his obstinacy regarding the concept of the functioning of the atmosphere, contributed to the progress of knowledge about certain phenomena and efforts to find and improve meteorological instruments.

The first meteorological measurements in Europe, with instruments including a barometer, began in Paris, Clermont-Ferrand, and Stockholm, between 1649 and 1651. Descartes made these measurements in Stockholm with the French ambassador to Sweden, Chanute. Descartes wrote: "An extremely high pressure of 28 inches and 7 lines was recorded at Stockholm on December 8, 1649." It was very cold that day. Or: "6. of May (George's Day) 1650, there were 26 inches and 4.3/4 lines, when the weather was fair and cloudy with a south-west wind."

In Paris, the first regular measurements began at the suggestion of Pascal. This can be seen from the notes of his son-in-law Perrier: "... having made measurements on the Pi de Dome, ... Monsieur Pascal informed me at Clermont-Ferrand, where I lived, that the pressure not only varies from place to place but that in changes in the same place. A change in weather changes the level of mercury in the tube. The rise or fall of the mercury will depend on whether it is more or less cold or warm, dry or wet..." On Pascal's recommendation, Perrier began daily measurements of the mercury level from the beginning of 1649, and he carried them out until the end of March 1651 years. One of his friends started measuring pressure in Paris on August 1, 1649, and continued to measure it until March 1651.

The first long-term measurements in Paris continued a little later. The first observer was Ismail Bouglio, a famous astronomer who supported the Copernican theory. Bulgio kept a meteorological diary from May 1658 to September 1660. His observations, which he made in Serpent Street, were as in Table 5.1.

After Bouglio, observations and measurements were continued by the famous mathematician and astronomer de la Haer (1640–1718). To familiarize himself with

Table 5.1 A part from Bullio's diary of meteorological measurements and observations

	Dated in May	Weather	Degrees	
♄*	25	Noon	25	Sun, clouds, moderately warm
☉	26	Morning	22½	At 8 AM rain, clouds, sun, east wind
☽	27	–	25	Warm rain, southerly wind
♂	28	–	21½	Cold rain, north wind
☿	29	Noon	25	Clear, west wind
		Night	21½	
♃	30	Pm	22	Clear, sunny
		Afternoon	26	
♀	31	Noon	27	Sun, clouds, south wind, warm, rain at 9 PM

The signs in the first row of the table denote the following: (*) sign for Saturn, (♄) Saturday, (☉) Sun, (☽) Moon, etc.

temperature measurements, he traveled to Florence. He made observations and measurements at the Paris Astronomical Observatory from 1664 to 1672. After his death, the temperature measurement was continued by Jacques Cassini and his son Cesar. From 1732 to 1740, systematic temperature measurements were continued by Reomir (1683–1757). He then emphasized the problems that arise when measuring temperature. Thus, he states that it is necessary for thermometers to have the same scale. In addition to this, the measurement should be made "in a place where the thermometer is exposed to the surrounding air. The place for the thermometer should be on the north side so that it is never directly heated by the Sun. There should also be no wall nearby that can reflect the sun's rays. To compare the temperature, one must know whether the measurement is made on the first, second, or third floor."

The first accurate measurements of precipitation also began in France, but they were not carried out at the Paris Observatory. The measurements were made by Pierre Perrot, brother of the famous architect Claude and writer Charles Perrot. He started measurements in Paris in October 1668 and carried them out until January 1674. He found the mean annual rainfall in Paris to be 19 inches (about 514 mm, the current mean is about 575 mm).

Later, these rainfall measurements were continued from 1688 to 1717, in order to ensure the supply of water reservoirs from which the Palace of Versailles was supplied. It is interesting to compare those rainfall amounts with those measured later. The data are shown in Table 5.2.

The Royal Society of London played a significant role in the development of meteorology in the seventeenth century. Its members Hooke and Boyle made a particularly big contribution. These scientists argued that meteorological measurements and observations should be made because of the psychogeographical descriptions of an area, or as Boyle wrote "because of the natural history of a given area." So, even then, they were aware that meteorological data is a basic feature of nature and man. In the seventeenth century, there were many articles on meteorological measurements and observations, carried out by Weil, Wallis, and Huygens in the Society's journal "Philosophical Transactions."

Table 5.2 Average rainfall in Paris (in mm) in two periods

Period	I	II	III	IV	V	VI	VII	VIII	IX	X	XI	XII	Per year
1668–1717	31	31	34	31	40	45	55	66	50	50	44	34	511
1806–1885	42	39	33	36	42	51	54	51	49	53	52	50	552

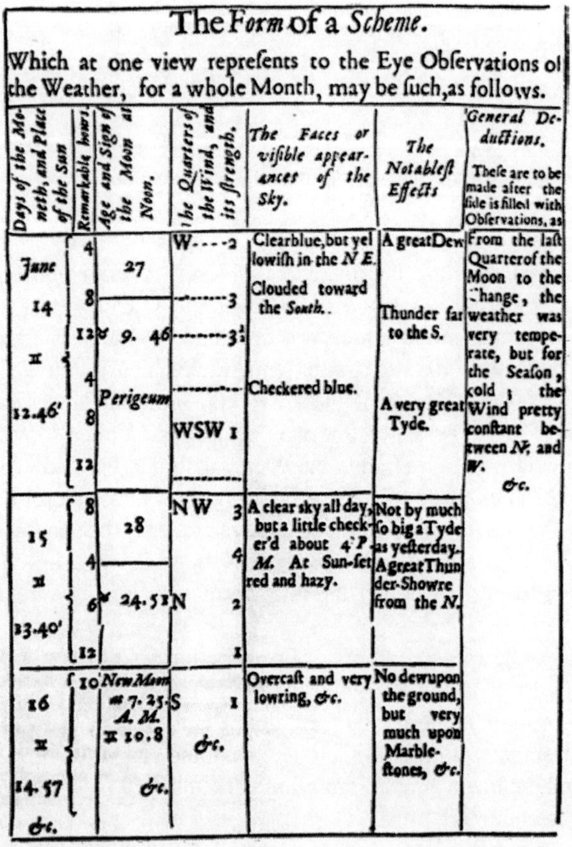
The Form of a *Scheme*.

Which at one view repreſents to the Eye Obſervations of the Weather, for a whole Month, may be ſuch, as follows.

Days of the Moneth, and Place of the Sun	*Remarkable hours*	*Age and Sign of the Moon at Noon.*	*The Quarters of the Wind, and its ſtrength.*	*The Faces or viſible appearances of the Sky.*	*The Notableſt Effects*	*General Deductions.* These are to be made after the ſide is filled with Obſervations, as
June 14 ♊ 12.46′	4, 8, 12, 4, 8, 12	27; ♉ 9. 46; *Perigeum*	W----2; --------3; --------3½; WSW 1	Clearblue, but yellowiſh in the *N E.* Clouded toward the *South.* Checkered blue.	A greatDew. Thunder far to the S. A very great Tyde.	From the laſt Quarter of the Moon to the Change, the weather was very temperate, but for the Seaſon, cold; the Wind pretty conſtant between *N.* and *W.* &c.
15 ♊ 13.40′	8, 4, 6, 12	28; ♈ 24.51	NW 3; 4; N 2; 1	A clear sky all day, but a little checker'd about 4 *P. M.* At Sun-ſet red and hazy.	Not by much ſo big a Tyde as yeſterday. A great Thunder-Showre from the *N.*	
16 ♊ 14.57 &c.	10	New Moon ♈ 7. 25. *A. M.* ♊ 10.8 &c.	S 1 &c.	Overcaſt and very lowring, &c.	No dew upon the ground, but very much upon Marble-ſtones, &c.	

Fig. 5.1 Hooke's time observation pattern

Hooke wrote a special manual for meteorological observations. In Fig. 5.1, an example of observations made according to his instructions is shown. In the second column, there are astrological signs (appearances of the moon, signs of the zodiac, etc.). The strength of the wind was estimated on the basis of a scale of four intensities. They recorded the state of the sky and other most significant phenomena very carefully. The last column showed a detailed monthly weather overview.

It can be seen from the form that observations were not scheduled at certain times but when something significant happened during the day. Copies of this form were sent to people who were interested in observing. This represented a turning point in the sense of introducing order into the way of recording 8-hour phenomena. In this sense, Hooke can be called the father of modern meteorology.

The first instrumental measurements in Germany were made at the Jesuit College in Osnabrück, around 1654, when they received a thermometer from Florence for parallel measurements. At the beginning of 1679, Rayer (1635–1714) began, at the suggestion of Leibniz, with regular measurements of temperature and pressure in Kiel. Measurement results are not saved. However, the data of the measurements made a little later by Camerari (1665–1721), a professor of botany in Tibigen, have been preserved. He started measuring at the beginning of 1691. The data were published in the "Tübingen Meteorological Ephemeris."

Instrumental measurements in Russia began between 1720 and 1730 (Tikhomirov, 1931). The first measurements were made in the command of secret affairs, during the time of Aleksey Mihailovich. The measurements were taken by guards in the Kremlin. However, it is known that earlier, around 1689, there were three mercury barometers with a copper scale, in the house of V. Golitsyn (whose family member was the famous Sarevna Sofiya Alekseevna). They seem to have served only as decorative objects of the house. In Petrograd, measurements were made in 1722 by Admiral K. Kruis, on the personal order of Peter I. In the order dated February 28, 1722, it is stated: "His Majesty orders your Excellency to keep an accurate diary of the weather and wind and to send here weekly reports." "Here" refers to Moscow, where, at that time, Peter I lived. Some copies of those reports have been preserved. The content was of the following type: "21. April, Sunday. In the morning, strong wind from the southeast and rain; in the middle of the day the same; in the evening southeast and south-southeast wind; during the night westerly wind and strong westerly-northwesterly; water overnight two feet above normal. April 22, Monday. North-west wind in the morning; water as previously stated. Cloudy and cold..., with light northwest wind at noon, and rain in the afternoon. Nice and calm weather until before dawn." These reports refer to the year 1722. Such reports exist from April to August 1722. It is not known whether such reports are still conducted.

The first instrumental measurements in Russia, for a short time, from December 24, 1724, to June 23, 1725, were conducted by an English priest in Petrograd, Thomas Consett. Part of the measurement was presented in the English magazine "Philosophical Transaction" (vol. XXXVIII, 1733).

In the main geophysical observatory, the reports of an unknown author who started measurements on December 1, 1725, are stored. Regular measurements of air temperature, pressure, wind direction and speed, cloudiness, hydrometeors, flooding, and freezing of rivers were recorded in the diaries. Until March, measurements were taken twice a day and after that three times a day. The reports were first written in German, and from 1728, they were kept in Latin. There are many linguistic errors in them.

Incidental measurements in many places of Russia were carried out from 1731 to 1780 by Johann Lerke, a military doctor. Lerke carried a Fahrenheit thermometer and a mercury barometer with him on his many trips to Russia, Persia, and Bessarabia. Wherever he stayed, he made measurements and observations three times a day. In addition to measuring temperature, pressure, humidity, and other 8-hour phenomena, he recorded the phenological stages of plant development, the freezing, and melting of ice on rivers, the occurrence of various epidemics, etc. The wind was recorded according to a four-level strength scale, to which a fifth was added, for destructive storms. His complete diaries for a period of 31 years (until 1761) have been preserved. From a later period, some data have been lost. Lerke, in addition to many places in Russia, made the first instrumental measurements in Moscow, from September 13, 1731, to February 15, 1732. Longer measurements in Moscow (from October 1, 1779, to the end of 1784) were made by Engel, a corresponding member of the Academy of Sciences. His observations from 1783 to 1784 were published in the collection "Ephemerides," the journal of the Mannheim Society.

It is interesting to list the measurements made by probably the most qualified people ever. Namely, for the needs of the Russian Academy of Sciences, measurements were made from 1726 to 1743 by academician Kraft, from 1743 to 1768 by academician Braun, and from 1769 to 1800 by Euler, the son of a famous mathematician. Unfortunately, the data from those measurements were not saved, and the observation logs were lost. All that remains are some conclusions, summaries, and average values.

In the far east of Russia, the first instrumental measurements were made in Ohock, in the period from 1785 to 1795. Measurement logs are kept at the Hydrographic Institute. Data on temperature, cloud cover, hydrometeors, storms, aurora borealis, and the date of melting and freezing of the Ohock River were recorded.

In America, the first measurements began in Brazil (location unknown). The measurements were made by the Dutch doctor Margraf between 1640 and 1642. A little later, measurements begin in New Sweden, near today's Wilmington. The measurements were made by Minister Campanis, between 1644 and 1645. An extract from Margraf's measurements was published in 1658 in "On the Nature of India." A brief description of measurements in New Sweden was published by Campanis' nephew in Stockholm in 1702.

John Leaning, born in Scotland, began his measurements in Charleston, South Carolina, where he lived. He started measuring air temperature in 1730 and pressure in 1738. He stated the reasons for which he started the measurements in a letter he sent to the Royal Society in London. In a letter dated June 22, 1741, he writes: "... The first reason I set out on this path was that I wanted to experimentally determine how different seasons affect the human organism, and to establish some knowledge about the origin of epidemics." He further gives a detailed description of the instruments he used to measure and the places where they were placed. He took measurements in the morning, at three o'clock in the afternoon, and when the weather worsened. He measured temperature and pressure, air humidity (with a hygroscope), precipitation, and cloudiness. Lining carried out measurements until 1750. After him, measurements were made by his colleague, doctor Chalmers, until 1759.

At Harvard College, mathematician Winthrop began to perform measurements in 1742. He took measurements in the morning and in the evening, until 1763. He measured both maximum and minimum temperatures using thermometers that were placed on the north side of the house. Average temperatures during that period were slightly lower than they are today. Between 1770 and 1780, meteorological measurements began in many places in the USA (Bradford, New Haven, Salem, and New York). The first measurements in Quebec (Canada) began in 1765.

5.3 The First Networks of Meteorological Stations

We have seen that great scientific names were involved in the invention of meteorological instruments. The most famous people of that time were the bearers of meteorological measurements. Despite the fact that all mental forces were engaged in measuring and interpreting those measurements, or thanks to their engagement, it was noticed that something qualitatively new should be done. It was necessary to make parallel measurements in many places over a wide area. Individual measurements, no matter how good they were, could not contribute to the further progress of meteorology. That is why the most influential people tried to establish organized measurements in several points, i.e., to organize networks of meteorological statıons.

5.3.1 The Measuring Network of the Academy del Cimento

In the seventeenth century, efforts were made to establish some kind of international network of places where meteorological measurements are made. The first such network was established in 1654 in central Italy, within the society known as "Academia del Cimento." The Duke of Tuscany, Ferdinand II, and his brother Leopold took over the patronage of this great undertaking. Regular meteorological observations and measurements were organized by the secretary of the Society of Jesuits, Antinori.

The network of stations (also called the Florentine network) consisted of stations in Florence, Valombrosa, Cutigiano, Paris, etc. 6.2. Measurements in this network were carried out daily, with great care, several times a day. Thus, measurements were taken five times a day in Florence. Temperatures were measured with two thermometers, one facing north and the other facing south. The barometric condition is read, unfortunately, still on different scales. Air humidity was measured. The state of the sky, wind direction, etc. were recorded. All observations and measurements were recorded in the same forms. Those reports were sent to the company for data comparison (Fig. 5.2).

Unfortunately, on July 14, 1667, after 10 years of existence, the Society ceased its work. The reasons were not of a professional nature. The Academy stopped working when Leopold became a cardinal of the Catholic Church. It seems that Leopold became a cardinal on the condition that he dissolves the Society, which,

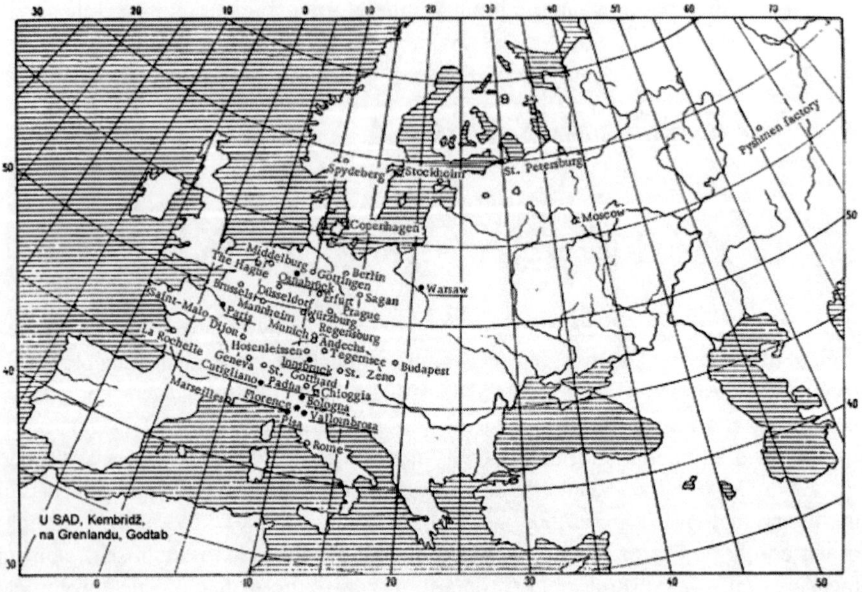

Fig. 5.2 Measuring networks of the Accademia del Cimento (underlined) and the Mannheim Society

with its proper understanding of the place of science, was very unpleasant to the church. One member of this Academy, Antonio Oliva, fell into the "claws" of the terrible Inquisition, so he committed suicide to save himself from torture.

5.3.2 *The First Attempts to Establish a Measurement Network in France and Germany*

The French Academy of Sciences, immediately after its foundation in 1666, began with meteorological measurements at its Observatory in Paris. Since 1688, the Academy has assigned one of its members to carry out systematic measurements, including pressure, temperature, precipitation, and phenomena. The observed data are published in the annual reports of the Academy. The Academy failed to organize measurements in several places (there were some measurements in Dijon and elsewhere), but the importance of that institution was great, due to the attention to the value of the measurements.

In Germany, the doctor Johann Kanold tried to organize an international network of measurements in 1717. He managed to collect measurements from several places in Germany as well as from abroad. He systematized and printed the collected reports in the magazine "Breslauer Sammlung." The magazine was published every 4 months, from 1717, for the next 10 years. Then the efforts to establish an international measurement network moved to England.

5.3.3 *Measurement Network of the Royal Society of England*

We have seen that in 1663, Robert Hooke wrote instructions on how meteorological measurements should be made. Much later, in 1723, the secretary of the Royal Society, James Jurin, invited all observers who had the equipment and were willing to prepare and send to the Society annual reports of their daily observations and measurements. In the invitation, Djurin attached instructions on how the measurements should be performed. According to the instructions, measurements were to be taken at least once a day and to include temperature, pressure, wind direction, and speed; the amount of rain or snow between two measurements; and the state of the sky. The barometric condition was especially to be read during strong storms, indicating the condition before the beginning of the storm and after its end. Observations and measurements were recorded in a diary that had six columns in which to write: date and time of observation, barometric condition, temperature, direction, and wind strength (with a scale from 0 (calm) to 4—the strongest wind, detailed but brief description weather and amount of rain or snow (in inches or 1/10 inch)). Average monthly and annual values should have been reported. One copy of their diaries was to be sent once a year to the Secretary of the Society. The secretary compared the data. The collected data are published once a year in the Society's journal "Philosophical Transaction."

The society enjoyed a great reputation in the scientific world, so many responded to Djurin's invitation. It really became the first real international data collection center. In the next few years, starting in 1724, copies of the report were sent to London, from such famous places as Uppsala from Sweden, Abo from Finland, Naples, Rome, from India, North America, etc.

Among those who responded to the call was the American mathematician **Isaac Greenwood** (1702–1745). In 1728, he suggested to Djurin that the land network of stations should be expanded but also that observations on numerous ships should be included. He saw in this a great benefit, both for meteorology and for maritime navigation. From the reports thus collected, knowledge could be obtained of the prevailing winds at various latitudes and in various months of the year, including the season when hurricanes are very active (Fig. 5.3).

Fourteen years after Greenwood's proposal, Roger Pickering prepared for the Royal Society a proposal for what should be a diary of meteorological measurements and what instruments should be used for measurement (Fig. 5.4).

As was the case with Djurin's proposal, the Society accepted Pickering's proposal. But the initial enthusiasm quickly waned, and the proposal was forgotten. The problem was not a misunderstanding of the Meteorological Society but that the Society faced many other scientific problems that needed to be solved. It became clear that it was necessary to establish a special society dedicated only to meteorology.

Honoured Sir, Boston N England. May 1. 1727

The Occasion of this Letter to You from One so little known by You, & of so obscure a Name is ye Indisposition of ye Reverend Dr Cotton Mather who would otherwise have wrote Himself.

The Doctor has been for a long Time endeavouring to engage his Friends in making Meteorological Observations that Our Country might not be ye most backward in following Your ingenious *Invitatio*, but ye unskillfulness of Some, & ye Business of others have I think hitherto prevented his Design, excepting only that He has so far prevailed upon an Ingenious Tradesman of this Town Mr Feveryear by Name, by presenting him with a Barometer, & some other things as to obtain ye Inclosed Observations.

My Charge from ye Doctor at present is, to acquaint You that ye Character of ye Author of these Observations is such that You may depend upon ye accuracy of them, and tho' there are Some few things admitted as for instance ye Degree or Strength of ye Wind what is inserted especially ye Barometrical Observations are performed with a great deal of Exactness; and I think have One thing peculiar to Them wch is that They are ye first Set of Such Sort of Observations that was ever made in New England.

I should have examined ye Tube, and Mercury with which these Observations were made but my Opportunity is so Short at present that I have not Time.

I shall add no more than just to observe to you that I sent last Fall to ye Rev. Mr Derham a large Collection of Observations on ye Weather, being those of Mr Robie's wch were sent to Me during my Residence at London, but having departed thence before They arrived, they were again returned to Me in N England, these were Observations for 10, or 12 Years successively wch may perhaps be of Some Service to your Design.

Also, That if it would be acceptable to ye R. Society to have an Annual Account of ye State of ye Weather in these parts of ye World, especially at Cambridge, that I shall have all ye Opportunities imaginable thereunto, being chosen by ye College their Hollisian Professor of ye Mathematicks & Experimental Philosophy wch place has some peculiar Advantages for Observation, above most of ye same Nature in ye world, being accommodated wth a very large Apparatus of Glasses, and other Instruments, besides by its Institution, furnish't wth 10 Pensionary Scholars of ye 2 upper Classes who will always be ready to continue on ye Observations in case of Sickness, absence, or any other Accident. If Likewise it may be of any Service to send out such astronomical Observations as we shall make, you may expect a constant account of such Eclipses &c. as may occur, being furnished with very good Instruments for that Design, Virt. ye same Quadrant that Dr. Halley had to observe on ye Southern Constellations at St Hallena [sic] besides several good Telescopes.

I am,

Sir,

Your most Obedient Humb^{le} Servt.

Isaac Greenwood

Dr. Jurin

Dated May 1 1727 To Dr. Jurin

Fig. 5.3 Letter from Isaac Greenwood to Jurin

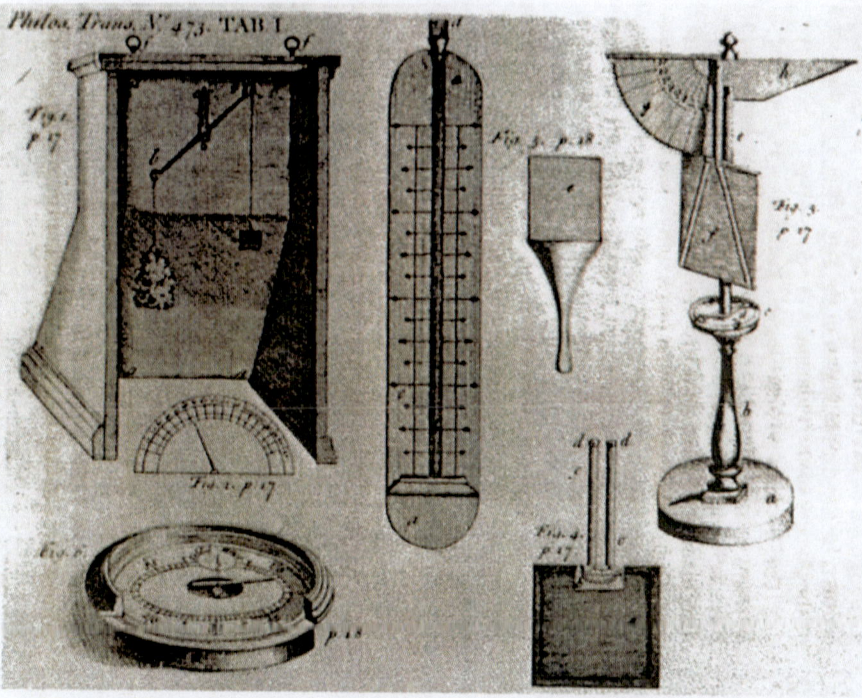

Fig. 5.4 Pickering's meteorological instruments

5.3.4 Siberian Meteorological Network

For the needs of the "Great Northern Expedition," a network of meteorological stations was organized in Siberia in 1730 (Fig. 5.5).

The expedition established measurements in Kazan in 1733. The eighth observer was Kunjicin. Next, in 1734, other stations with observers were established: Ekaterinburg (Patishchev), Tobolsk (Mirovich), Yamishev (Kal-berk), Yeniseysk (Charoshnikov), Tomsk (Salomatov), Turukansk (Sorokin), Irkutsk (Kanaev), Yakurtsk (Pekov), Selenjinsk (Tretyakov), Nerchinsk (Kovr-gin), and in the Argun (Kovirin) silver mine.

Instructions for measuring at these stations were written in 1732 by D. Bernoulli. The instructions refer to "barometric," "thermometric," and "hygrometric" measurements, determining the density of water from precipitation and in the seas, magnetic observations, etc., experiments that must be done to determine the movement of the pendulum, as well as other phenomena that must be observed on the ground. Later, measurements of the speed of sound, depending on temperature and pressure, warm and cold winds, etc., were included in the program. Detailed instructions are given for observing various phenomena, such as lightning, hail, rain, snow, dew, salt, fog, rainbow, shooting stars, "arcs of fire," and other light phenomena, aurora

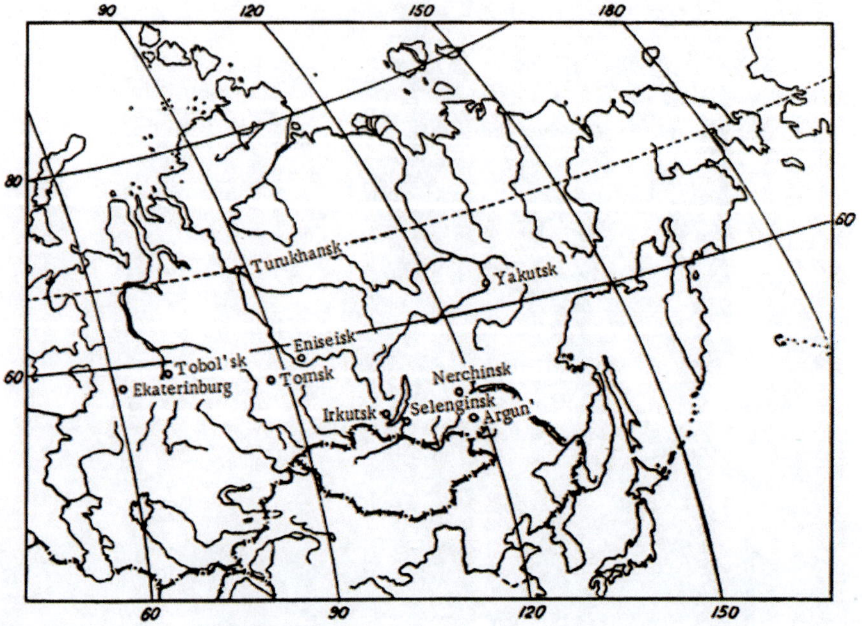

Fig. 5.5 Siberian network of stations

borealis, corona around the Sun and Months... There is also a suggestion on how to measure the "size and number of raindrops," the amount and appearance of snow (whether the flakes are big or small), etc. The strength of the wind was determined according to a four-level scale. Observations of the freezing and melting of ice on Lake Baikal were determined for the Selenjinsk station.

It is interesting to note that an observer in Tomsk, a Cossack named Petar Salomatov, noticed an interesting phenomenon. On November 21, 1734, during a severe frost, the mercury froze in the form of lumps. This possibility was examined 50 years later by the famous academics Lomonosov and Braun. They, experimenting on December 25, 1759, confirmed that it could happen.

Observations and measurements in the network established for the needs of the "Great Northern Expedition" continued for a longer period. Thus, it was measured in Yekaterinburg until 1746, in Irkutsk until 1747, in Yakutsk until 1749, etc. Unfortunately, only a small part of this data has been preserved, so the climate of Siberia in the eighteenth century could not be done. From this, it can be seen that this was a truly grandiose undertaking of measurement and observation. Today, such extensive and precise measurements could only be carried out at a small number of observatories.

5.3.5 *Lambert's Proposal for a Worldwide Network of Stations*

Johann Lambert, a famous constructor of many meteorological instruments, proposed a worldwide network of meteorological stations in 1771. He chose the places of observation and measurement, so that they could represent the environment. He divided the entire globe into triangles with one point in the center. In addition, he made a denser network for Europe (Fig. 5.6).

Lambert predicted that the observations are made simultaneously according to the moment when it is noon in London. Using the symbols used by Muschenbrock (1692–1761) when making meteorological observations in Utrecht in 1728, Lambert suggested what symbols should be used to record phenomena in the new world network. Thus, he suggested the symbol for snow (++) for the hail (· ·), etc. He suggested how to find the mean wind when the fraction of wind blowing from certain directions is known. Of course, he used vector addition. This method has been used for a long period of time. Lambert's network remained only one good suggestion. It was only much later that the conditions for the establishment of a global world measurement network matured.

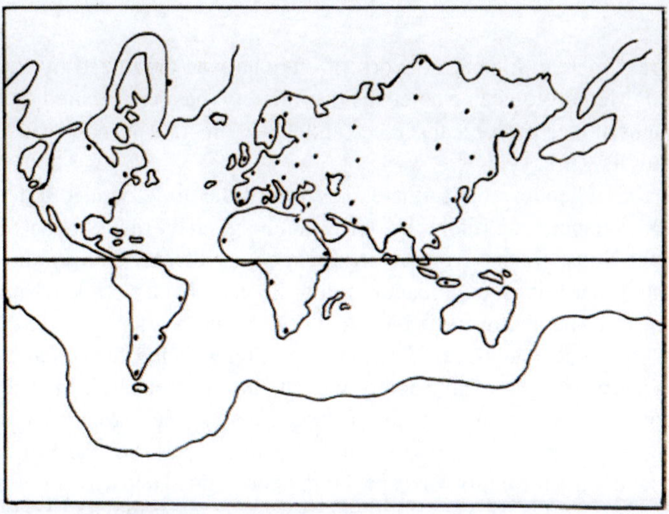

Fig. 5.6 Lambert's proposal for a worldwide network of meteorological stations

5.3.6 Meteorological Network of the Royal Society of Medicine of Paris

Louis Côté (1740–1815), a meteorologist and priest from Montmorency near Paris, collected and published meteorological measurements and observations from two continents, Europe and America. Côté compiled a form that the Royal Medical Association from Paris sent to various places in the world with the aim of collecting meteorological data that allows monitoring the spread of various diseases. Kote published these data in a reduced volume every year, in the form of mean and extreme values for each month, the number of days with precipitation, and other data. He published data from 1776 to 1786 in the Society's journal "Histoire de la Sociétè Royale de Médicine." At its beginning in 1776, the network consisted of 36 stations, and by 1786, the number had risen to 65. The stations were scattered over a vast area, from Petrograd to San Domingo, Haiti.

Kote knew how important it is to properly mount instruments and work according to specific instructions. That's why he compiled detailed instructions for each station. And Kote, like Djurin in England earlier, when summarizing the results of the measurements, pointed out a number of shortcomings, among them: the data come from randomly distributed stations and the observers were randomly selected; the observation time is different, and measurements are made with different instruments; the observation method is not the same. Because of all this, it was difficult to compare the data with each other.

5.3.7 Mannheim Meteorological Network

The first real meteorological network of stations was organized by the so-called Mannheim Meteorological Society. This Society has contributed a lot to the advancement of meteorological science and, therefore, it is necessary to know some details from its work.

Prince Karl Theodor of Palatinate, a patron of the arts, founded the "Privileged Society for Science and Literature" in Mannheim. It is thought that the idea of founding the Society came from Georg Stengel, a professor from Mannheim, who was, among other things, an amateur meteorologist. He carried out meteorological observations in Mannheim from 1758 to 1777. At the beginning of 1778, he transferred meteorological measurements to his son Stefan. Thus, the actual organizer of the Privileged Meteorological Society was Stefan Stengel. With his friend Hemer (the author of several minor works on meteorology), he presented the Society's work plan to Prince Carl Theodore. On September 15, 1780, the Prince signed a letter approving the formation of a special meteorological department of the Society under the name "Societas Meteorologica Palati-na"—a Privileged Meteorological Society. Hemer was the secretary of the Society. It was generally known as the Mannheim Meteorological Society, after the place where it was founded.

Among the objectives of the Society were "... finding trained observers in selected places and establishing contact with them; finding new instruments and working to improve existing ones; checking everything done; making copies; answering and asking questions; finding new methods of observation; collecting observations, translating them into Latin for wider use; attach remarks to the observations after comparing them; preparation of results for publication; do everything to advance the development of this still imperfect science." These summary tasks of the Society were printed in the first number of their magazine "Ephemerides"—Passages.

The society prepared a circular letter and sent it on February 19, 1781, to 30 places, academies, scientific societies, and observatories. These institutions were offered to participate in organizing the observation, with a note that the necessary instruments, instead of buying them, will be provided by Prince Carlo Theodore. It is interesting to note that they did not receive answers from the Royal Society of London, the Irish Academy of Edinburgh, and the Astronomical Observatory of Vienna. Apparently, the Islanders wanted to prevent anyone else from taking primacy in this area as well. They noticed the need for this kind of work, but they do not support it, because they are not organizers. This principle of theirs is recognizable even in recent history.

Many responded to the company that they wanted to participate in the planned observations. The number of those selected in 1781 was 14, which later rose to 39. Therefore, the Society managed to establish wide international cooperation among scientists. They created the "Monitum ad Observatores" Instructions for the observers. Observation times have been established (7, 11, 14, and 21 h). Each station was provided with a thermometer, barometer, hygrometer, and compass for measuring the magnetic inclination.

All instruments were calibrated and reliable. By calibrating the Manhaj thermometer in 1840, which was used in Prague, after a long period of use, it was shown that the measurement error was less than 0.1°. All measurements were recorded in one copy. Unique symbols were used for occurrences. The observations were, following the prescribed procedure, printed in the Society's magazine "Ephemerides," from 1781 to 1792. The last issue of "Ephemerides" was printed in 1795. The society stopped working in 1799. It was a great loss for the science of that era.

The Mannheim Meteorological Society did not live long, but its activity had an enormous contribution. It paved the way for the development of modern meteorology, which is impossible without a well-organized measuring network of stations. The company was far ahead of the times in which it worked. His example had to be followed, but only after 50 years, when the first meteorological institutes with their measuring networks were founded. Not only the organization of the Society was an example to meteorologists from the later period. The quality of the observed data was such that later climatological studies by A. Humboldt and L. Bush, and the first synoptic studies by Brandes, came from the data printed in "Ephemerides."

The Mannheim Society also had other scientific activities. In issue 2 of their journal, Stiglener published a paper on pressure changes, based on a 1776 comparison of observations in London, Regensburg, and Petrograd. It was shown that the low pressure "travels" from west to east, which was later confirmed by Brandes and Dove. The society also analyzed disturbances in the atmosphere during the summer of 1783, which they linked to the eruption of the Hekla volcano in Iceland.

In accordance with the custom introduced several centuries before the founding of the Mannheim Meteorological Society, they printed the "Ephemerides" in Latin, which had been the international language of scientists for almost a thousand years. In the eighteenth century, Latin was used less and less. In the nineteenth century, only a few selected, real scientists used Latin (e.g., Brandes). With the termination of the work of the Mannheim Society, the long period of development of meteorology, in which real science was born, ends. The first period was a period of speculation.

One of the few stations that was in the network of the Privileged Society from Mannheim, and continued to operate to this day, is the Clementinum in Prague. As an example of good, continuous work, it deserves to be said a little more about it in this history of meteorology. Monitoring a long series of data is of exceptional value for any climatological analysis. The meteorological station Clementinum in Prague has a continuous series of observations and measurements from 1775 until today (Belohlavek, 1977). Interest in meteorology in Prague dates back to the Renaissance. The Bohemian King and Roman Emperor Charles IV founded the Charles University in Prague in 1348. The fame of the University was magnified by the astronomers Tycho Brahe (1546–1601) and Johann Kepler (1571–1630). Both are well known for the meteorological observations they made in Prague (Brahe in 1597, and Kepler in 1605). Kepler's Journal of Observations has been preserved. Meteorological observations in Clementinum were started by another astronomer and meteorologist, Joseph Stepling. The building—Klemetinum Tower—was built in the heart of the old town of Prague, in the classic baroque style, between 1650 and 1756. It was built by the Jesuits. The complex consisted of three churches, a monastery, 32 rooms, a school, a library, and an astronomical observatory. It is located on the right bank of the Vltava (Moldava) River, Fig. 5.7.

Joseph Stepling was born in 1716 in Regensburg, Germany. He studied at the Jesuit Seminary in Olomouc and Prague. After serving as a priest, he devoted himself to making various instruments of his own. He made a Reomir-type thermometer for meteorological measurements. That thermometer was very accurate and long-lasting, as evidenced by the calibration in 1829. He started measuring the temperature in Clementinum in January 1752. The thermometer was hung on the window on the north side and was read at sunrise and sunset. He started measuring precipitation in February of the same year. The barometer was also read twice a day. He also recorded the cloudiness. After, unfortunately, there is only partial information about the measurements, especially after 1773, when the Jesuits stopped this activity. Fortunately, Stepling's student Antony Strand replaced him in 1775 as director of the Astronomical and Meteorological Observatory in Clementinum. From that period, observations were made daily, without interruption, even during two world wars. That station has the longest continuous series of observations in Central Europe. It represented the backbone of the meteorological network of stations in Bohemia and, later, the Austro-Hungarian Empire.

Fig. 5.7 Clementinum tower in Prague

In 1799, Strand was replaced as director by Martin Adam David. He was Strand's longtime assistant. He also replaced him as the chair of astronomy in Prague. David was also rector of the university in 1816. He retired in 1824 when he was replaced as director of the Observatory by the prominent meteorologist Carl Kreil. He contributed to the improvement of the design of meteorological instruments. The Prague Observatory continued to gain importance when Fritz (who was born in Prague, was attached to the Observatory, and wrote treatises on the weather in Prague) became the responsible person for meteorological and phenological observations in the newly founded "Central Agency for Meteorology and geodynamics of the Austro-Hungarian Empire" in Vienna. Another Prague native, K. Jelinek (1822–1876), also contributed to the observatory. He was the director of the Central Agency in Vienna and contributed to the founding of the International Meteorological Organization.

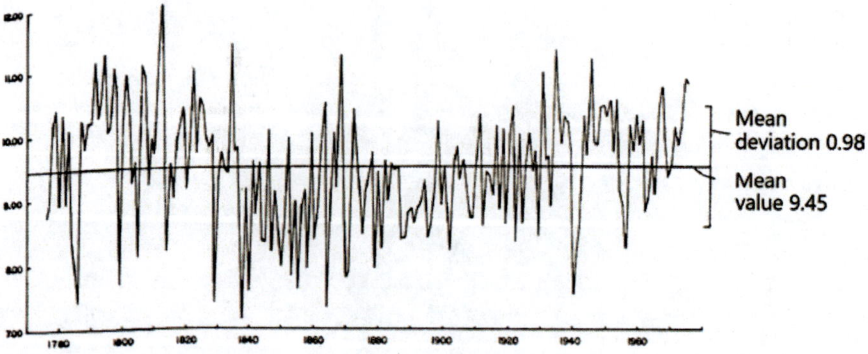

Fig. 5.8 Annual temperatures measured in Clementinum, Prague, 1775–1975

From the Clementinum station, from 1775 until today, there are data on pressure; average, maximum, and minimum temperature; wind direction and strength; and cloudiness and precipitation. Since 1845, the relative humidity of the air has also been measured. From that precious series of data, in Fig. 5.8, the trend of average annual temperatures in the 200-year period, 1775–1975, is shown. There is a global decrease in temperature from 1790 to 1850, followed by an increase in the next 25 years and an equal decrease in the next 25 years. From 1900 to around 1940, there was an increase, and from 1940 to 1970, there was a global decrease in temperature.

References

Belohlavek, V. 1977. The meteorological tradition of Prague. *Bulletin of the American Meteorological Society* 58: 1056–1057.

Forster, T. 1815. *Researches about atmospheric phaenomena*. 2d ed. London: Baldwin, Gradock and Joy.

———. 1816. Specimen of a new nomenclature for meteorological science. *Gentleman's Magazine 86* part I: 131–132.

Kington, J.A. 1974. The Societas Meteorologica Palatina. *An Eighteenth-Century Meteorological Society, Weather* 29: 416–426.

Pearce, E.A., and C.G. Smith. 1984. *The times book world weather guide*. Times Books.

Tikhomirov, E.I. 1931. *Instruktsiya russkim meteorologicheskim stantsiyam XVIII v*, A handbook for the Russian meteorological stations of the 18th century. Izvestiya Glavnoi Geofizicheskoi Observatorii.

Chapter 6
Establishment of Meteorological Institutes (Services)

At the end of the eighteenth century, it was quite clear that an international meteorological network of stations had to be established. The initial practical efforts in this direction were made by the "Accademia del Cimento" society. Later, there were tries by Hook, Judin, and Dalamber. The Privileged Meteorological Society of Mannheim has the furthest and best-realized idea of a world meteorological network. Unfortunately, despite the perfect organization, that project did not last more than 10 years either. It was as if there was some invisible wall that was diverting from that so visible and clearly traced path. Perhaps it was vanity and the desire for domination in the case of certain influential countries. The only way to meet the need to establish a worldwide network of meteorological stations was to establish a network within individual countries. Their unification will occur only in the middle of the nineteenth century.

The establishment of meteorological networks was entrusted to scientific institutions and services. And not just measurement, by the nature of their work, they were responsible for developing knowledge in meteorology. It will be discussed in this chapter of the book. The presentation is easiest to follow through the description of activities in some of the most important countries. It would be more difficult to follow the chronology of events without taking into account the administrative division of the world.

6.1 Establishment of an Institute in Russia

At the beginning of the nineteenth century, meteorological observations began at the University of Moscow (Khrgian 1948; Budyko 1957; Rykachev 1899). Professor Strahov organized that work in 1808. The observation log from that period is shown in Fig. 6.1.

M. Ćurić, V. Spiridonov, *History of Meteorology*,
https://doi.org/10.1007/978-3-031-45032-7_6

Метеорологическія наблюденія въ Августѣ 1812.

Дни наблюденія	Часы наблюденія	Термометръ Реомюровъ Внѣш.	Термометръ Реомюровъ Внутр	Барометръ, Англійск. размѣра	Направленіе вѣтра	Состояніе Атмосферы
17. Суббота	Въ 8 часовъ ут.	+ $9\frac{1}{2}$	+ 15	29, 65	Сѣверо-Западный	Небо покрытое.
	Въ 2 часа п. п.	+ 17	+ $15\frac{1}{2}$	29, 68		Попеременно солнце сіяло и облаками было закрываемо
	Въ 9 часовъ веч.	+ $11\frac{1}{2}$	+ 16	29, 69	Сѣверный	Большею частію безоблачное небо
18. Воскресенье	Въ 8 часовъ ут.	+ $9\frac{1}{2}$	+ $16\frac{1}{2}$	29, 66	Южно-Восточный	Сіяніе солнца
	Въ 2 часа п. п.	+ $16\frac{1}{2}$	+ $17\frac{1}{2}$	29, 61	Сѣверо-Западный	
	Въ 9 часовъ веч.	+ 13	+ 17	29, 53	Южный	Большею частію безоблачное небо
19. Понедѣльникъ	Въ 8 часовъ ут.	+ 12	+ $16\frac{1}{2}$	29, 46	Южно-Западный	Небо покрытое
	Въ 2 часа п. п.	+ 17	+ 18	29, 42	Западный	Сіяніе солнца
	Въ 9 часовъ веч.	+ $13\frac{1}{2}$	+ 17	29, 36		Безоблачное небо
20 Вторникъ	Въ 8 часовъ ут.	+ $13\frac{1}{2}$	+ $16\frac{1}{2}$	29, 24		Попеременно солнце сіяло и облаками было закрываемо
	Въ 2 часа п. п.	+ $14\frac{1}{2}$	+ 17	29, 20	Южно-Западный	Небо закрытое. Но а к 5 часу п. п. дождь на $2\frac{1}{2}$ линіи
	Въ 9 часовъ веч	+ $12\frac{1}{2}$	+ 16	29, 26	Сѣверо-Западный	Безоблачное небо

Fig. 6.1 Observation notebook of Prof. Strahova in Moscow for August 1812

Observations were interrupted during the Patriotic War of 1812. The observation continued under the leadership of Prof. Dvigubski, in 1820.

After the foundation of the University of Kazan in 1804, meteorological measurement began. Universities have tried to hire observers for this work. Thus, the University of Moscow published in 1814 the observations made in Volodya by teacher Fortunator from 1808 to 1812. It has been shown that no institute or higher education institution can organize a wider system of meteorological observations.

The famous Russian scientist Karazin (1773–1842) proposed the establishment of a meteorological service that would work under the control of Russian educational institutes. In his letter sent to the Moscow Naturalists' Union on March 15, 1810, he proposed setting up a network of stations from Kola (in the north) to Tiflis (in the south) and from Libava (in the west) to Nizhne-Kamchatka (in the far east). The stations would be organized within local faculties. All stations in the network would work in a unique way. He also proposed the establishment of a "Society of Scientists" that would take care of these affairs. A little later, in 1818, Karazin revived the idea of establishing a "State Meteorological Committee" to supervise the affairs of the network. This was supported by the Department of Mines, which was then developing rapidly, so they were interested in meteorological and magnetism measurements.

Professor Cooper (1799–1865), head of the Department of Chemistry of the University of Kazan, was in charge of taking over the tasks related to this. He organized some measurements within the "International Days." From 1829, he organized magnetic measurements in Petrograd, Peking, and in Sitka, Alaska. Then, on November 16, 1826, A. Humboldt visited Russia. In a speech at the Academy of Sciences, he emphasized that magnetic measurements must be significantly expanded in Russia. Young academician Cooper proposed to establish a special observatory to monitor magnetic and meteorological phenomena. Within that, he suggested what equipment must be procured. "The project of establishing meteorological and magnetic observations" was presented by Cooper to the Ministry of Finance in 1833. Cooper laid out very clearly what the state would gain by adopting the proposed project. He points out: "... Russia would undoubtedly discover new national resources... if they knew its climate much more precisely. The government could then develop new branches of agriculture or industry. It would be known in advance which industry or agricultural activity we can plan in which areas."

The Ministry of Mining supported the establishment of magnetic measurements in several places (Nerchinsk 1832; Yekaterinburg 1836; Lugansk 1836; Tiflis 1837). Still, Cooper didn't give up. He asked Humboldt to write a letter to Emperor Tsar Nicholas I. And he did so in 1239. The authority of the famous scientist Humboldt contributed to Emperor Nicholas I approving the construction of a central magnetic and meteorological station. In 1843, he ordered the drawing up of a plan for the observatory building. In December 1848, the construction was completed. The observatory was opened on April 1, 1849. In the declaration read at the time, it is stated: "In order to make measurements and carry out large-scale physical experiments, the Main Physical Observatory in Petrograd is being established for the whole of Russia...".

It was one of the first central meteorological institutions in Europe and America. In the beginning, the staff of the observatory was small: Cooper (director), two older, and three younger observers. With this, a very large task of unifying meteorological research and practical work in all of Russia began. At that time, the observatory maintained 50 observation stations, which were scattered over the vast expanses of Russia. A great deal of work has been done to introduce unique instruments and measurement methods. Control of measurements by stations could not be carried out due to a lack of travel funds.

The observatory published the collected data in "Review of observations carried out outside the Main Physical Observatory and those carried out under its control." There were many errors in the reports. In the beginning, there was little interest in these data, primarily because the Observatory was within engineering and mining, and there was no contact with farmers and sailors. The development of science also could not be realized, because practically only Cooper worked at the Observatory as a scientist.

After Cooper's death (in 1865), Kamic (1801–1867) became the new director of the Observatory. He was the author of a well-known textbook on meteorology. He was the director for only 2 years, and he could not improve the work of the Observatory significantly. Kamica was succeeded by Rikachev and then Wild. The observatory was connected to them for the next 50 years.

M. A. Rikachev (1840–1919)

Rikachev graduated from the Naval Academy in 1865. After that, in 1866, he was sent abroad to familiarize himself with the meteorological service in other countries. His assignment was related to the Russian Navy's interest in meteorology. In the second half of the nineteenth century, the Russian fleet developed, and they made many long voyages. The Crimean War (1854–1855) caused the stimulus for the development of a meteorological service for the needs of the navy. Rikachev stayed in England until 1867. He worked at the Greenwich Observatory. When he returned to Russia, he worked in the hydrographic department of the Main Physical Observatory. He also held the position of assistant director of the Observatory from 1867 to 1896. From 1896, after the retirement of Wild, he was the director of the Observatory until 1913. Visiting the stations, he once traveled a distance of 9000 km, of which 2800 km he walked or rode on horseback. He was persistent in spotting problems and improving the functioning of the measurement system.

Heinrich Wild (1833–1902)

He was born in the canton of Zurich in Switzerland. He finished his studies in Zurich in 1854. He became an assistant professor of physics in 1858 at the University of Zurich. Afterward, he transferred to the University of Berlin, where he was sent by the Swiss Government to study the possibility of establishing a network of meteorological stations in Switzerland. During this time, Wild built several meteorological instruments, including the famous Wild wind vane. Moving to the Main Physical Observatory represents a new challenge for Wild. He was the director of the Observatory from 1867 to 1896. He proposed the establishment of a commission of the highest authority, which was supposed to formulate the reorganization of the Observatory and the entire observation network in Russia.

Rikachev and Wild were a good team to achieve success. During 1871 and 1872, they established 29 new stations. At that time, there were a total of 73 of them. The main job was to control the operation of the stations. Wild also took part in the control. From June to August 1872, he visited a large number of stations. The main problem was the impossibility of paying observers for their work. At all stations, even those in the most remote places, which have harsh climatic conditions, the work was completely voluntary. The observers were mostly professors, doctors, agronomists, pharmacists, military personnel, and sometimes even priests. The only satisfaction they had, along with scientific enthusiasm, was when they saw that their data was in the "Annals of the Main Physical Observatory" (Annals of the GFO). For the observers, a beautifully written diploma with the signature of the President of the Academy of Sciences was a special honor.

Reviewing and correcting of reports were organized in the Observatory. The observers were informed about the mistakes, and this obliged them to work more carefully. The reports were printed in the "Annals," instead of the earlier printing in the "Review." That work was apparently done well, and in 1874, at the Utrecht Conference, the Permanent International Committee on Meteorology adopted the "Annals" as an example of how it should be done in all other countries.

Under Rikachev's leadership in 1872 leadership in the "Time Service" department was established within the Observatory. The task was to apply meteorological observations. That department could not have been established without the help of the Ministry of the Navy. This assistance continued when daily weather bulletins began to be issued. The number of employees in the Observatory itself also increased (in 1899 there were 70 of them). Numerous young scientists who contributed to Russian and world science worked at the Observatory. The Academy of Sciences provided assistance during the printing of the Annals. By 1894, the number of stations was greatly increased. There were 650 stations of the II order and 1450 that measured precipitation.

It was quite monotonous work at the Observatory. Young scientists were not encouraged by new ideas. They could only print their weather analyses in the "Meteorological Bulletin Journal" supplement. Due to dissatisfaction with Wild's work, many scientists left Petrograd, such as Keppen, Braunov, and Sreznevsky.

Around 1860, it was clear that due to the urbanization of Petrograd, the measurement at the Main Physical Observatory had become unrepresentative. That is why a new building for magnetic and meteorological observations was built in Pavlovsk (near Petrograd). The building began operation in 1878 (Fig. 6.2). Its first director was Milberg. The measurements were representative there. Many instruments were constructed there. Toward the end of the nineteenth century, aerological measurements began at the Pavlovsk observatory. In 1902, a special department for sounding with kites was established. Continuous radiation measurements have been introduced since 1912. It is the first observatory in Europe with such measurements.

In 1886, the Irkutsk Observatory became the meteorological center for Eastern Siberia. The meteorological network in Russia has become too bulky to be controlled from one place. Indiscipline in measurement was very great. That is why the opinion that measurements must be reorganized at the local level has matured. Almost simultaneously, local networks emerged around 1885–1892.

In the Main Physical Observatory, they believed that the decentralization of Russian meteorology introduces irregularity in measurements. That is why they demanded that the Observatory be designated as the supreme controller. On December 24, 1912, the main physical observatory was declared by law not to be dependent on the Academy of Sciences, due to Rikachev's efforts. Academician Golitsyn became the new director of the Observatory on May 6, 1913. He strengthened the position of the Observatory so that a large number of talented young scientists worked in it again. So Mulatovsky started working on long-term weather forecasts; Friedman dealt with theoretical meteorology, Obrensky with electricity, etc. As a result of such work, in 1914, scientific results began to be published in the "Geophysical Collection." He went on to play a major role in the development of Russian meteorology.

Fig. 6.2 Pavlov Observatory

6.2 Organization of Meteorology in France

At the end of the eighteenth century, France went through great social changes. The revolution took place in 1789. Some scientists also see the causes of the revolution in the storms that almost completely destroyed the harvest the previous year. Therefore, the new government was obliged to bring some justice to the wider strata of people who depended on agriculture. And agriculture depends on meteorological events. The new government, due to attaching great importance to the revolution, decided to introduce a completely new calendar. The first year (instead of the appearance of Christ) was counted as the year of the revolution. The year was divided into 12 months with 30 days each. The other 5 (or 6 leap-year days) were counted as festival days. The months were named after the main meteorological characteristics of that period. The first month started on September 22.

Because it is a unique example of how much attention was paid to the meteorological factor in life, we list the names of all the months of the year, according to that new calendar, with the meaning and period to which it refers according to the usual calendar:

Vendemiaire – grape harvest month (September 22 – October 21)
Brumaire – the month of fog (22.10 – 20.11)
Frimaire – the month of frost (21.11 – 20.12)
Nivose – snow month (21.12 – 19.01)
Pluviose – rainy month (20.01 – 18.02)
Ventose – the month of wind (19.02 – 20.03)
Germinal – budding month (21.03 – 19.04)
Floreal – the month of flowering (20.04 – 19.05)
Prairial – month of grass (20.05 – 18.06)
Messidor – harvest month (19.06 – 18.07)
Thermidor – the month of heat (19.07 – 17.08)
Fructidor – fruit month (18.08 – 21.09)

This new calendar was in official use in France from September 1792 to the end of December 1805. After that, they switched to the old calendar.

From the beginning of the nineteenth century, numerous meteorological stations in France worked within the framework of scientific societies, university schools, etc. They all worked independently of each other. That is why in Paris in 1864, the Minister of Education took measures to organize meteorological observations in regular schools. Thus, already in 1877, 58 stations were working within schools. The Paris observatory (Fig. 6.3) found it difficult to manage the entire network. The lack of discipline was great. That is why the stations were under the jurisdiction of a special commission. At that time, the Commission was unable to detect meteorological problems.

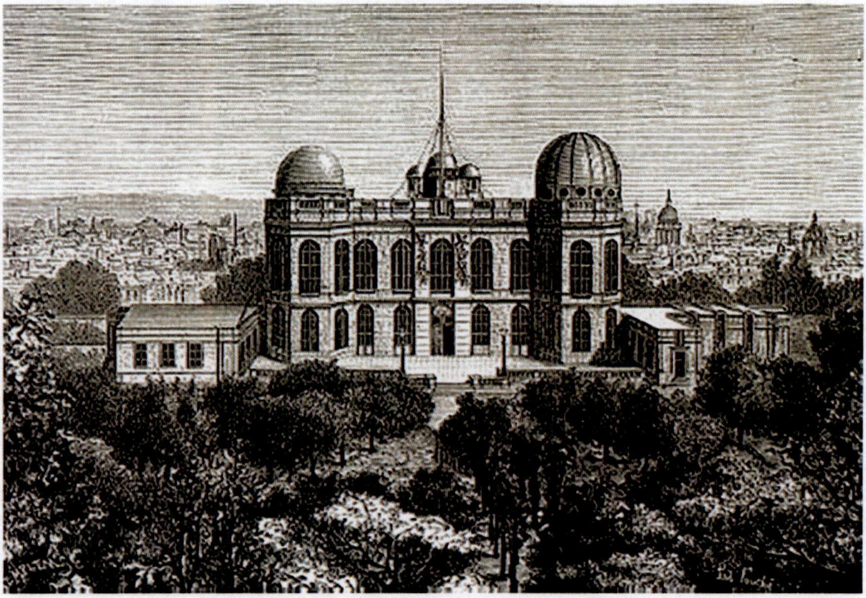

Fig. 6.3 Paris Observatory

After the death of Leverrier, the influential director of the Paris Observatory, the "Central Meteorological Bureau of France" was founded in Paris in 1878. The goal was to reorganize observations by regions and departments, issue warnings for ports and farmers, and organize studies of atmospheric movements. Maskart (1837–1908), a celebrated meteorologist, was appointed director. He appointed Fron as head of the warning service, Angot of the climatological service, Tisseran de Bora as a general meteorologist, etc. Already in 1903, there were 160 stations. Regional and mountain meteorological observatories were established in Bordeaux, Toulouse, Orleans, etc.

The inspector of general transport reorganized the completely independent rain gauge network, which started operating in 1854. The goal was to use this network to alarm the occurrence of floods, especially around the Seine River. Pluviometric measurements (precipitation measurement) together with data on the nature of the soil (whether and how much it lets water through) enabled such good analyses that the water level in the rivers could be predicted to an accuracy of 10–12 cm.

Later, the Commissions in the departments took care of the rain gauge stations as well. Commissions have achieved particular success in engaging local intellectuals to work in this area. They even published meteorological magazines in some departments.

In France, a special network for monitoring storm clouds has been organized. The reason was that the appearance of hail clouds is very frequent, and it causes great material damage, especially to grapes. The observation of such clouds began in 1865, under the supervision of the Paris Observatory. The idea was that even without instruments, such clouds can be monitored if a dense observation network is available. Observers were supposed to fill out a special form, a "bulletin d'orage," about each storm cloud and the damage it caused. France's storm monitoring network has served as a model for many other countries. The data collected in this network made it possible to analyze the movement of the "disaster line." It was the first Duran-Greville radio. He was the first to investigate the properties of cold and occluded fronts.

The Paris Observatory in Montsouris (southern part of Paris), which was founded in 1868, was the place where many useful works for meteorology were done. The composition of the atmosphere was analyzed. In 1870, the sanitary authorities installed several rain gauges around Paris in order to analyze the rainwater and draw conclusions about the chemicals in the air.

In general, it can be said that in that period, after the revolution in France, the spirit of practical, visibly useful research was developed. They sought to solve the problem of flooding from heavy rains, the problem of tracking and forecasting hail clouds, and the problem of pollution. Those efforts were felt in the meteorological science of the later era. Thus, for example, in the field of cloud physics and weather modification, French scientists made notable contributions (especially the Dessans family).

6.3 Meteorological Organization in England

The first meteorological observatories, both private and national, were founded in England before the eighteenth century (Radcliffe in Oxford, and Kew in Richmond). However, only in the nineteenth century did the efforts of individuals appear to coordinate their work into a single whole. As far as is known, Gleshir from the magneto-meteorological department of the Greenwich Observatory was the first to try to coordinate the work. It happened under really special circumstances. Namely, in 1840, the statistical department of the so-called "Main Registry" started a study on the connection between some social factors (primarily the number of deaths) and meteorological phenomena. For this need, Gleshir collected observation results from meteorological stations. In 1845, there were 9 stations, and in 1875, there were 50 stations. He published this data together with graphs of the number of deaths in England over a 20-year period.

The first official meteorological center in England was founded in 1855 under the name "Meteorological Department of the Ministry of Trade." Admiral Fitz Roy was appointed director. The main job of the Department was to collect, verify, and analyze meteorological observation data from ships and the coast. Such jobs were logical, considering the development of the navy.

After the death of Fitz Roy (suicide due to responsibility for bad weather forecasts), in 1866, the Meteorological Office was established, which was separated from the Ministry of Commerce. The service was administered by the Meteorological Committee of the Royal Society. At that time, the annual budget of the service was 4500 pounds sterling. The service consisted of departments for maritime meteorology, meteorological telegraphic reports, and agrometeorology.

As before, maritime meteorology played an important role. From 1867 to 1874, that department equipped Royal Navy ships and merchant ships with 1287 barometers, 1154 aneroids, and 6976 thermometers. All instruments were previously inspected at Kew Observatory. In 10 years, starting in 1876, 845 logbooks containing meteorological observations from all seas from Greenland to Australia were collected from just one merchant fleet. From year to year, the number of reports increased. In the period 1913–1914, that number was 279 per year.

In the first period, the English tried to find the best routes for navigation. At that time, they carried out particularly frequent meteorological measurements. Thus, during the year 1876, 125,000 observations were collected over a small area of the ocean (between 10°S and 20°N and between 10°W and 40°W). The collected data were processed in the form of a "climatological description of one square of the ocean." This work in the marine department was very demanding and prioritized. The affairs of "continental" meteorology were secondary.

On land since 1868, exemplary measurements have been made at seven first-class observatories: Kew, Aberdeen, Armagh, Stonyhurst, Glasgow, Wallendier, and Falmouth. All meteorological measurements from these stations were published regularly in the "Quarterly Weather Report" from September 1, 1868. These data represented a valuable scientific basis for the well-known study of the formation of the cold front in the "Bergen School." And here it should be noted that all meteorological observations in Britain were voluntary. Some were carried out within the Meteorological Society (later called the Royal Meteorological Society) and some within the Scottish Meteorological Society. The first mentioned society was founded on Glešir's initiative in London in 1850.

The Meteorological Society discussed the establishment of meteorological stations and the organization of phenological observations. They carried out a program of phenological observations in 1874, together with other related societies, and in 1876, they started observations at 22 stations. Since 1866, the Society has started publishing its magazine "The Quarterly Journal of the Royal Meteorological Society."

It should be noted that the measurement of precipitation was in a completely different organization, which was called the "British Rainfall Organization." It is a private organization organized in 1859 by Simmons (1838–1900). Later, the organization received support from the British Association for the Advancement of Science. The Rainfall Organization carried out rainfall measurements also on a voluntary basis. In the whole of Britain, measurements were made at 453 rainfall stations in 1862, 2100 in 1877, and 4200 stations in 1903. Different rain gauges were used for the measurements. The most common were those that are very similar in appearance to today's but had the receiving part only 1 foot (30.48 cm) above the ground (a significant part of the rain gauge was buried in the ground). The survey data were printed annually and were of great use to farmers. Since 1919, the organization of precipitation measurements has come under the jurisdiction of the British Meteorological Service.

Several well-known observatories operated in Britain. Greenwich and Kew are the most famous. At the Greenwich Observatory, there was a department for magnetism and meteorology. Kew Observatory was built on the outskirts of London in 1769, Fig. 6.4, as an astronomical observatory. Meteorological measurements have been taken on it since 1772. At the beginning, the measurements were irregular and of little value for scientific purposes. The situation improved when the Observatory came under the jurisdiction of the British Society for the Advancement of Science in 1842. From 1852, the director was Ronalds, and from 1852 to 1859, the well-known scientist Welsh. After 1871, Kew Observatory came under the responsibility of the Met Office.

A very important activity at the Observatory was the making of instruments. A wide variety of high-quality meteorological instruments were made, as well as for measuring magnetism. Registering instruments were especially well-known. The Kew observatory equipped all seven mentioned observatories with barographs, thermographs, and other instruments equipped with a photographic record of 8-h data. Beckley, one of the instrument makers, made an anemograph and pluviograph in 1869, with a mechanical recording mechanism. At the Observatory, they also calibrated the instruments. In 1876, they turned 4000, and in 1890, 20,000 instruments. Such work provided the Observatory with a large income.

Fig. 6.4 Kew Observatory taken in 1935

6.4 Organization of Meteorology in Italy

At the beginning of the nineteenth century, Italy was a predominantly agricultural country. Experts from agriculture and forestry were engaged in the analysis of meteorological conditions. They often established funds to carry out appropriate experimental research, including meteorological research. They were particularly interested in precipitation. Many scientists, amateur researchers, and others made meteorological measurements and made meteorological studies. All Italian astronomical observatories carried out meteorological measurements, in Milan (founded in 1760), Padua (1767), Rome (at the College Romano, in 1787 at the University of Turin (1820), etc. Many recognized astronomers, such as Secchi and Tachini, engaged in the construction of meteorological instruments and wrote meteorological studies.

As we pointed out earlier, the first attempt to organize a meteorological network of stations was in the seventeenth century in Florence. A century later, the director of the astronomical observatory in Padua, Taoldo (known for his analysis of the influence of the moon on the weather) tried again to establish a network of stations. He managed to get 60 observers from all over Italy interested in it. From them, he collected a lot of observational material from an earlier period as well. Based on these data, later (in 1839) the study "Description of climate and vegetation in Italy" was made. The study was done by Shaw, a famous Danish climatologist and botanist.

The current network of meteorological stations was founded in 1860 by Professor Francesco Denza from the Carlo Albert College in Moncaliera (near Turin). Denza organized a meteorological observation at his college in 1859. It was the best station at that time, which is why he received material compensation from the Government for observation. Denza organized a network of stations, especially in northern Italy, which was known as the “Alpine-Apennine Meteorological Association.” Many individuals have privately taken part in this project. The number of stations in the network increased. There were 28 stations in 1873 and 113 in 1879.

In 1872, the observatory in Moncelliera organized a network of rain gauges and began to publish measurement data. In Italy, the Central Meteorological Service “Ufficio Centrale di Meteorologia” was established in 1876, instead of the previously established Directorate for Meteorology at the Ministry of Agriculture, Industry, and Trade (founded in 1864). Measurements were taken at 9 am, 3 pm, and 9 pm. In the network of 38 stations since 1878, three altitude stations have been included, Stelvio (2543 m), Col di Valdobia, and small St. Bernard. In southern Italy, a private network was organized around 1877 by professors de Đorđi and Fitipandi.

In the meantime, a lot of experimental stations were organized for the needs of agriculture and forestry. At those stations, along with other instruments, radiometers were used to measure the total radiation energy. The device consisted of two connected spheres, colored black and white. There was a volatile liquid in the white one. The amount of evaporated liquid served as a measure of the total absorbed heat from the Sun.

Rainfall measurement is also organized within the Ministry of Agriculture. In Italy in 1894, there were a total of over 100 Class II stations and 550 stations that measured temperature, wind, and precipitation.

At the Observatory in Milan, there have been continuous meteorological measurements since 1763. In 1876, the director of the Schiaparelli Observatory organized a network of storm monitoring stations. He tried to gather as much detailed information about storm clouds as possible. He even used meters to measure the height and azimuth of the place where lightning appeared. Schiaparelli was not impressed with those measurements, but the basic data on rain, city, and storm onset time were detailed and accurate. In Italy too, the development of meteorology was aimed at solving practical problems, storms, and precipitation, which was very important for agriculture.

6.5 Development of Meteorology in Belgium

In Belgium, the main meteorological activities were at the Royal Astronomical Observatory in Brussels. Its director was Ketelet (1796–1974) Fig. 6.5.

In 1823, he created the Observatory’s work plan. The plan was accepted in 1826, and construction began in the following years. In the work plan, Ketelet stated: “... It is impossible to analyze (astronomical) observations without knowing the structure of the atmosphere and the earth.” That is why there was a department for magnetic and meteorological studies from the very beginning. Regular meteorological measurements began on January 1, 1833, and phenological observations began in 1839.

Fig. 6.5 J. Ketelet

Ketelet applied general statistical methods for processing meteorological data. He wrote an extensive monograph on the climate of Belgium (1845–1857). A revised edition was printed in 1867, entitled "The Meteorology of Belgium Compared with that of the World." Given that Belgium is a small country with a uniform background, there was no immediate need to perform meteorological measurements in several places. Ketelet provided instruments for some stations, but no attention was paid to their operation, control, collection, or publication of data.

The current measuring network was established by Ketelet's successor Huzo. On January 1, 1878, he founded three "international meteorological stations" and 30 climatological stations. Since that time, observations have been very regular and of high quality. Observations were made twice a day at unusual times, 8 am and 1 pm. Following the example of the French, they too, in 1870, established a network of over 100 points for monitoring storm clouds. The Meteorological Service of the Brussels Observatory started working as a separate institution in 1898, with director Lancaster.

6.6 The Development of Meteorology in the Netherlands

The Dutch Meteorological Institute in Utrecht began its work in 1854. Five years before this, the famous scientist Biss-Balot (1817–1890) built a small magnetic observatory in Sonnenberg. It was actually a small cellar in an earthen embankment. Bis-Balot's assistant Krek made meteorological observations there from the very beginning. The Dutch Royal Institute was founded on February 1, 1854, with these measurements. After that, Bis-Balot established a network of four stations (Utrecht, Groningen, Gelder, and Vlissingen). They were equipped with recorders, and observations were made at 8, 14, and 20 (or 22) hours. In 1905, there were a total of 15 such stations and 200 stations that measured precipitation, phenological, and other observations. Since 1849, data has been printed in the "Meteorological Yearbook of the Netherlands." Observations from other areas of Europe, and even further afield, from Nagasaki, Panamariboe, and Buenos Aires were printed in that yearbook. Apparently, Bis-Balot was trying to help international meteorological publications.

The Netherlands had many colonies, especially in Southeast Asia. That is why they paid great attention to meteorological observations on ships. A special center for collecting such data is established within the Meteorological Institute. Those who had the best ship logs of meteorological observations were awarded gold and silver medals.

The Netherlands was the first country to accept the resolution of the Brussels Meteorological Conference in 1853. They immediately began sending the ship's observation data to Washington. The Dutch Meteorological Institute was the first in Europe to start printing maritime meteorological charts in the 1950s, namely, Dutch charts of trade winds (trade winds) in 1856; North Atlantic Wind Charts, 1856; general wind charts of 1860; a collection of maps of stormy winds, rains, thunderstorms, and fogs, 1862; and many others. It can be seen that the Netherlands has adapted the meteorological service to the needs of a large naval power.

6.7 The Development of Meteorology in Germany

The meteorological service of Germany was for a long time influenced by the political divisions that ruled Germany until the 1870s. The small German states (not even the great Prussia) were not able to reorganize, firmly unite, and standardize the meteorological network, to start with research work and with the application of those results in practice.

The Prussian Meteorological Institute was founded in October 1847 as a scientific department of the Prussian Statistical Institute. Alexander Humboldt contributed the most to the organization of his work. In a letter to the Government, he describes the need for such an institution: "... Heat and moisture are the most important elements for the life of plants, and without numerical data on the variability of these two factors over many years, one cannot talk about yields from plants in the plains and mountains." Because of this, the main goal of the Institute was the development of agrometeorology.

35 stations worked within the Institute in 1848. The Director of the Institute, Malman (1811–1848), made a great effort during 1847–1848 to control the work of the stations and to supply them with instruments. Observations were made at 6, 14, and 22 (or 21) hours. He wrote the Observation Guide. He wrote an important work "On the mean distribution of heat on the Earth's Surface."

From 1849 to 1879, the Institute was managed by Dove (whose significant work will be discussed in the next chapter). After his death, the Institute was managed by Arnt, then Helman, and from 1885 to 1907, Bezold, known for his works on the thermodynamics of the atmosphere.

For many years, the Institute did not have its own observatory. An observatory was built in Podzdam between 1889 and 1891. Magnetic measurements on it began in 1890, and meteorological measurements began in 1893. The Director of the Observatory was A. Sprung (1848–1909). He constructed many instruments for the Observatory (balance barograph, cloud machine—cloud photography, etc.). It starts with actinometric measurements and registers the radiation of the Earth, the gradient of the electric field, and ions.

In other German states, the meteorological service started working later than in Prussia. In Bavaria, the main meteorological station began operating in Munich in 1878, under the direction of Bezold. During 1882, 45 first-, second-, and third-order stations and a large number of storm monitoring stations were operating in Bavaria. At that time, 16 stations were operating in the state of Baden, 24 in Wittemberg, etc.

6.8 Development of Meteorology in Norway

For the establishment of a meteorological service in Norway, the great need for meteorological weather reports for the huge fishing fleets and merchant ships played a decisive role. The director of the telegraph service, Nielsen, equipped five of his stations with meteorological instruments in 1860. From those stations, they sent reports by telegraph to Christiana (today's Oslo), and to the Paris Observatory, which at that time started making regular weather forecasts.

The weather observer at the Astronomical Observatory in Christiani in 1864 was Mon (1835–1917) (Fig. 6.6). He sent the Polytechnic Society a report entitled "Modern Meteorology and Weather Forecasting." This report caused many reactions. The society formed a group consisting of university professors (Firnley, Christine, and Sachs), Nielsen, and the director of the port Roll. The group created a study plan for the Department of Meteorology at the university and for special meteorological institutes.

The order establishing the Meteorological Institute was signed on July 28, 1866. Mon was appointed director, with two assistants. The institute started its work in

Fig. 6.6 H. Mon

December 1866. It had a total of seven stations (in addition to the previous five stations, a station was opened in Bergen and Christiani). Thanks to the significant help of the merchant and fishing fleet, the budget of the Institute was significantly increased, so Mon employed 9 more people at the Institute and increased the number of second-order stations to 80, and the number of rain gauge stations to 450.

The measured data from these stations enabled the creation of a large number of climatological studies. The most significant is "Climate – Atlas of Norway," written by Mon. That study was supplemented with data from 1874 to 1914 and printed for the jubilee of the Institute in 1917.

Scientific activity at the Institute was very fruitful in many areas of meteorology. The results of scientific work have been published in many studies. The first was the "Storm Atlas of the Norwegian Meteorological Institute," which was printed in 1870. Many theoretical works on the movement of the atmosphere followed. A particular contribution to science was the work of Moon and Guldberg (another university professor) on air friction. The Mon–Guldberg expression for friction is well known. The Institute actively participated in many meteorological expeditions, including observations for the needs of the International Polar Year during 1882–1883.

Within the framework of the international plan, the eight-throw of the Clouds was organized in Norway in 1895 and 1896. Data on this were published in 1900. These were the first attempts to systematically observe processes in the upper layers of the atmosphere. In order to analyze the state of the atmosphere by height, in 1909, a pilot-balloon measurement began in Christiani. The observatory in Oslo was opened in 1906, where, in addition to other measurements, measurements of atmospheric electricity were made.

6.9 The Development of Meteorology in Sweden

The first meteorological network in Sweden was organized in the period 1856 to 1858, within the Swedish Academy of Sciences, on the initiative of Edlund. The Academy founded the Central Meteorological Institute (Meteorologiska Centralanstalt) in Stockholm in 1873. Edlund was appointed director. In 1879, there were 32 second-order stations and in 1906, 38. In 1876, the Swedish Agricultural Society established a rain gauge network at more than 200 locations. Hildebrandson, director of the Observatory in Uppsala, contributed in particular to this. Meteorological measurements were made at that Observatory in 1722, but the data were not printed until 1855. The observatory organized a program of measurements for the needs of agriculture in 1869. Measurements and observations included: storm clouds, hail damage, nocturnal autumn, and spring frosts, as well as melting and freezing on many rivers and lakes in Sweden. In particular, for the first time, microclimate measurements of air temperature were performed at several levels above the ground. Movements of cirrus clouds were observed. Such cloud movements were organized by Hildebrandson at 21 stations, with the aim of studying airflow at altitude. A small department for nautical meteorology (Nautisk Meteorologiska Buran) was founded in Stockholm in 1878. Measurements were made on a small number of ships while sailing the Atlantic.

6.10 The Development of Meteorology in Spain

In 1737, Francisco Fernández Navarrete drew up a plan for establishing a network of meteorological observations. The proposal was sent to the Royal Academy of Medicine in Madrid. For 2–3 years, some members of the Academy sent weather reports, but soon that activity died down.

In 1784, the president of the Kampomani Royal Council issued an order to town chiefs and village leaders to send information about weather observations every 2 weeks. No measurements with instruments were made, only observations. This order was respected until the beginning of the nineteenth century. After the end of the Spanish War of Independence, in 1814, the request was renewed, and observations were made until about 1834. This information has obviously been collected over a long period of time, but from a scientific point of view, it is not particularly useful.

Similar to other countries, the Astronomical Observatory in Madrid served as the meteorological center for Spain. The observatory was built in 1847. In the meteorological department, the man in charge was Rico Sinobas. By the royal decree of October 8, 1850, the operation of 23 stations at universities and some schools (Institutos de segunda ensenanza) was ensured. In the beginning, the measurements were poorly organized, but in the period 1858–1865, the situation improved under the control of the "Statistical Commission" (Junta de Estadistica). Measurements were made at 9 am and 3 pm every day. In 1879, there were 22, and in 1900, 42 stations.

The best equipped were the stations within the maritime institute at the San Fernando Observatory, which had been operating since 1850. Organized measurements were also introduced on Spanish ships in 1874.

6.11 Development of Meteorology in Austria-Hungary

At the time of the founding of meteorological institutions, Austria, Hungary, and Czechoslovakia formed the Austro-Hungarian Empire. Meteorological observations in Austria began in the eighteenth century. In addition to those in Pilgram, Vienna (1762–1786), observations were also made within the famous Astronomical and Meteorological Observatory in the Benedictine Monastery in Kremsminster. Meteorological measurements began in 1763, and magnetic measurements in 1838. Meteorological observations in Austria were unified into one system in the nineteenth century, after the founding of the Academy of Sciences in Vienna in 1847. A year after that, Baumgartner, vice president of the Academy, contributed a large sum of money with which instruments were purchased for the stations. A commission was established within the Academy. That commission became the Central Institute for Meteorology and Earth Magnetism (Centralanstalt für Meteorologie und Erdmagnetismus). By the Government decree of June 23, 1851, Krel (1798–1865) was appointed director. Measurements began at the Institute in 1852, and in 1872, in a new building in Döbling, a northwestern suburb of Vienna. The institute was managed by Jelinek (1822–1876), a professor from Prague. Jelinek organized the printing of meteorological reports and established a weather forecast department, for which a new building was built.

After 1871, the Hungarian Institute was founded, and the entire network in Austria-Hungary was under the administration of the Institute in Vienna. The number of stations grew rapidly, from 177 in 1874 to 400 in 1908. There were more than 2200 rain gauge stations in the network. Here, too, the observations were on a voluntary basis. Despite this, disciplined work was done at the stations. Jelinek's "Instructions for Meteorological Observations" served as an example in other countries.

After Jelinek's death (1876), Han became the director of the Institute. He developed synoptic meteorology and climatology. Hahn and Jelinek helped found the Austrian Meteorological Society in 1863 (which was confirmed in 1865). The journal of the society "Zeitschrift Österreichischen Gesellschaft für Meteorologie" was later renamed to the well-known international journal "Meteorologische Zeitschrift."

After gaining some autonomy, the Hungarian Central Institute for Meteorology and Earth Magnetism was founded in Budapest on April 8, 1870.

6.12 Development of Meteorology in Serbia (Yugoslavia)

Vladimir Jakšić started his measurements on January 1, 1848. He first measured the air temperature using a "ten-part thermometer" and recorded the weather conditions in terms of rain, snow, and cloud cover. On December 1, 1850, he added to this program the measurement of daily precipitation, and from February 1855, the measurement of air humidity. They cover a period of 52 years, from January 1, 1848, to the end of 1899. During Jakšić's occasional absences from Belgrade, measurements were made by his friend Vuk Gavrilović. In the last years of Jakšić's life, measurements and observations were carried out by members of his immediate family, sisters, and daughter.

Jakšić did the most for Serbia's meteorology during his work at the Lyceum (University). As a Lyceum professor, he began intensive work on the establishment of a meteorological network in Serbia. The year 1856 can be considered the first year when a well-organized meteorological network worked in Serbia. That year, 20 weather stations were operating in Serbia. The following year, in 1857, 27 meteorological stations were operating in Serbia, which probably represents the densest meteorological network in the world at that time, if ordinary precipitation (rain gauge) stations are excluded. The instruments installed at the stations were unusually good for that time. Jakšić bought them from the Vienna Central Meteorological Institute, which at that time had the best meteorological instruments in Central Europe. Jakšić made other preparations in a timely manner in order to enable the most successful work in the elevated stations. He also wrote the Instructions for working at the stations. Jakšić's meteorological network functioned relatively well during the entire time he was a professor at the Lyceum. Thus, in 1862, it still counted ten stations and worked as a unit. After 1862, there was a sudden loss of stations, but observations were still made in some places.

The Department of Astronomy and Meteorology was founded at the Faculty of Philosophy of the Great School in Belgrade in January 1880. This came after an almost 20-year decline in meteorological activity in Serbia, excluding Jakšić's measurements in Belgrade (Dobrilović 1964). This event is a reflection of undoubtedly changed and new understandings of the importance of astronomy and meteorology.

The establishment of the Department and the election of Milan Nedeljković (1857–1950) as a professor of astronomy and meteorology represent one of the decisive moments in the development of meteorology in Serbia, and it can be said more broadly, in the entire former Yugoslavia. In the middle of 1879, Nedeljković, as a state scholarship recipient, arrived in Paris, where he spent almost 5 years completing his school education and special studies in astronomy and meteorology. He studied at the Sorbonne and the Collège d'France in the academic years 1879/1880 and 1880/1881. He spent from October 1881 to the end of July 1884 at the Astronomical School of the Paris Observatory, the Meteorological Institute of France, and the Observatory in Parc Saint-Maur, working on many astronomical and meteorological problems.

With unusually great persistence, Nedeljković founded a temporary (1887) and newly built Meteorological Observatory in Belgrade in 1891. He renovated and expanded the network of meteorological stations in Serbia. He once again started the restoration of the meteorological network after the First World War.

A significant event in the development of meteorology in Serbia was the election of Dr. Pavle Vujević (1881–1966) in March 1907, first as a temporary, and from 1910 as a permanent assistant professor of climatology and meteorology at the Faculty of Philosophy of the University of Belgrade. Vujević started his scientific work with unusual dedication. It has already been described in detail that since 1901 many valuable measurements and observations have been established and realized at the Observatory. By processing and theoretically analyzing these Nedeljković measurements, Vujević expertly presented and explained some unknown microclimatic phenomena in several of his notable works. Vujević's works were cited in a large number of textbooks and monographs on microclimate, in the journal *Meteorologische Zeitschrift* for 1911; Die Temperaturver-haltnisse der untersten Luftschichten, and also in Reports of the Vienna Academy of Sciences; Die Temperaturen verschidenartiger Bodenoberflachen—published in Meteorologische Zeitschrift 1912; and Ueber die Beschaffenheit der taglichen Temperturkurve—published 1914; and so on.

After the First World War, a new country was created—Yugoslavia. Therefore, the jurisdiction of the Observatory had to be changed, along with the other two similar institutions that existed in the new state. On that occasion, in 1921, a meeting of the three largest Yugoslav meteorological institutions, the Belgrade Observatory, the Geophysical Institute in Zagreb, and the Institute for Meteorology and Geodynamics in Ljubljana, was held in Zagreb. At this meeting, Nedeljković agreed with professors Mohorovicič and Gavaci to carry out the regional division of Yugoslavia in terms of work management.

According to this agreement, the Meteorological Observatory in Belgrade took over the management of the entire meteorological network in Serbia, Vojvodina, Macedonia, Montenegro, and a part of Dalmatia. This division remained in force, except for minor changes, until the beginning of the Second World War.

After a full 60 years, from May 1, 1887, to October 1, 1947, the Meteorological Observatory in Belgrade ceased to be a university institution. Within the Hydrometeorological Service of Serbia, the Institute, the Observatory continues to function as a special department of this institution, performing existing and newly introduced measurements. During the following years, a network of meteorological stations in the Belgrade area was organized within the Observatory, for the purpose of examining the city's climate.

6.13 Development of Meteorology in Other European Countries

In Portugal, meteorology developed later than most other countries. After the Meteorological Conference in Brussels in 1853, Pegliado drew up a plan for the construction of a meteorological observatory in Lisbon and for a network of meteorological stations. The construction of the observatory began in 1853. Meteorological measurements began immediately after its construction, on October 1, 1854, and magnetic measurements began in 1857. The observatory was under the administration of the Polytechnic School of the Ministry of the Navy. After Pegliado, the director was Silva, and from 1856 to 1874, Silveira. He contributed a lot to the development of meteorology. In 1861, the stations were under the administration of the Medical Academy. Stations were built on the Azores and Madeira in 1865. The number of stations increased from 11 in 1879 to 13 in 1905. The link between the observatory and meteorological stations in the Portuguese colonies of Angola and Macau was very weak. Good meteorological and magnetic measurements were made from 1863 to 1866 at the Faculty of Philosophy of the University of Coimbra. Measurements on ships were not numerous due to the relatively small number of shipping lines.

In Romania, the first measurements were made at the end of the eighteenth century (Hepites 1899). They are listed in Raičević's work "Review of Historical, natural and political conditions in Wallachia and Moldavia," which was printed in

Naples in 1788. The first systematic measurements were made in 1839–1840 by Pangrati and Stamati, instructors of the Lyceum in Iaša. At the same time, the optician Koc started measurements in Bucharest. A little later, observations begin in Ferestro, Sulina, Galata, and other places. The Romanian Meteorological Institute was founded in 1884, in Bucharest, with three meteorological stations and ten rain gauge stations. The network expanded rapidly so that in 1905, there were 66 stations, plus 350 rain gauge stations. The observatory in Bucharest was built in 1888 when seismological and meteorological measurements began. From 1885, the Institute started publishing "Annals," in which scientific works and observations from Bucharest were published. Since 1895, daily weather bulletins (without weather forecasts) have been issued.

In Bulgaria, there were occasional measurements in Sofia and Russia (Kaltchewa 1939). However, systematic measurements did not begin until 1887 in Sofia. They were carried out by Bachevarov (1859–1926), a university professor. The meteorological service was founded in 1890, under the leadership of Vasov (1856–1928), a famous Bulgarian pedagogue. He wrote a manual on the organization of meteorological measurements. In 1893, the Bulgarian Meteorological Institute had 8 stations of the second order and 55 rain gauges, and in 1926 there were 55 stations of the second order and 125 rain gauges.

6.14 Development of Meteorology in China and Japan

Numerous meteorological instruments were made in China, and precipitation and humidity measurements were carried out for several centuries BC. They recorded clear and cloudy days. However, as far as is known, measurements of pressure, wind, and other phenomena were carried out by an Englishman named Cunningham from October 1698 to January 1699 in Amoa. These data were published in "Philosophical Transactions" for the year 1699. Later, from 1757 to 1762, the Jesuit missionary Joseph Amiot measured pressure, cloudiness, humidity, and wind direction and recorded storms twice a day in Beijing. A short series of observations in Peking (from December 1830 to June 1831) was made by Fuss, who accompanied the Russian mission in Peking in 1830. From January 1, 1841, this Mission made regular measurements. From 1849, magnetic and meteorological measurements were made at the observatory, which was built with the approval of the Russian mission, until 1883. In 1873, the Jesuits established a Meteorological Observatory near Shanghai. The China Customs Service participated in the establishment of a network of stations in the Yangtze River Valley.

For many years, Japan was isolated from the technical and scientific achievements of the West. In the middle of the nineteenth century, they came into contact with European culture. The first meteorological observations were organized in 1862 in Yokohama, in 1872 in Hokkaido, and in 1875 in Tokyo. The first instructions for measurements were printed in 1880. The Japanese Meteorological Service was founded on August 3, 1887. By decree, the service was under the administration of the Tokyo Observatory, which was the country's central observatory. The network expanded rapidly so that in 1900 it had 80 stations, of which hourly measurements were made at 14. They even set up a weather station on top of Fujiyama (3700 m), the sacred mountain of Japan. Unlike the others, well-paid observers worked at the stations. They did very good observations. They recorded the storms in detail. The first climatological study of Japan was done between 1894 and 1895 by Nakamura. The first manual entitled "Modern Meteorology" was written by Okada in 1901.

6.15 The Development of Meteorology in the USA

The first meteorological measurements in the USA were made by Lining in Charleston in 1738 (Miller 1933; Brown 1940; Bates 1989; Fleming 1990, 1996). His interest in the influence of meteorological phenomena on the spread of disease was the impetus for the development of meteorology. The stimulus also came from migrations in the New World toward the West and South to new climatic conditions.

The establishment of the meteorological system in the USA is linked to the name of Joseph Lowell, an army general from 1818 to 1836. He wrote: "The influence of time and climate on diseases, especially on epidemic diseases, is well known... climate studies make it possible to discover the causes of these diseases." From 1819, observations began at many military locations. These data were collected in 1840 and analyzed by Espy, the first meteorologist in America.

At New York University, the measurement of temperature and precipitation was organized in 1825 and was carried out until 1850. Between 1837 and 1845, the Franklin Institute in Pennsylvania established a measuring network. Since its inception, the Smithsonian Institution has shown an interest in meteorology. This versatile scientific organization was founded in Washington by the bequest of John Smithson (an Englishman who died in 1829). The Smithsonian Institution sought to organize "a broad system of meteorological observations with a view to solving the problem of American storms." The secretary of the Institute, Henry (1799–1878), sent a circular letter in 1848 to all interested parties who could participate in this observation system. At the end of 1849, there were 150 interested volunteers. They were given instruments, standard barometers, and thermometers bought in London and Paris. Instructions were also written (printed in 1850). Numerous data were collected in this way. At that time, numerous works by the famous scientists Espy, Slott, and Coffin appeared. Coffin produced a collection of selected climatological papers entitled "Wind on the Globe." The Smithsonian Institute has provided significant support for actinometric (radiation) research. Thus, the Smithsonian Institute made a huge contribution to the development of American meteorology. The government organized another organization within the army called "Signal Service." At the beginning of 1870, all meteorological works in America were transferred to the Reporting Service, which consisted mostly of synoptic meteorology.

Mier played a major role in the establishment of the meteorological service in America. He joined the Reporting Service in 1860 when this service was given great importance during the Civil War (1861–1865). After the end of the war, the Reporting Service was significantly reduced, but Mier saw the importance of the Weather Service in peacetime conditions as well. Then the Time Service was founded, the founder of which was the army. Mier became a brigadier general in 1867. It should be noted that most of the employees of the Time Service were civilians. At its very beginning (1870), the Weather Service had close relations with the Smithsonian Institution.

The network of stations expanded rapidly. In 1904, there were 200 stations with paid observers and 3700 volunteer ones (which measured temperature, precipitation, and phenomena). The stations were set up to meet the needs of dynamic meteorology and weather forecasting, not climatological. Namely, they believed, as the great scientist, meteorologist **Abe** pointed out, in 1893, "... Only when we solve the basic problems of meteorology, it will be possible to solve the problems of ordinary climatology." Observations for climatological needs were made at 7, 14, and 21 h.

And the storm watch organization covered a large area. In the first year that the Reporting Service took over this work (1884), more than 13,000 reports were prepared. The reporting service did not have a special meteorological observatory for scientific research. They believed that such observatories should be on the mountain. Thus, in 1870, it was included in the work of the Observatory on Mount Washington in New Hampshire (altitude 1619 m). In October 1873, the meteorological station at the highest altitude in the world, Pankov Peak in Colorado (height 4311 m), began to operate (Fig. 6.7).

Meanwhile, there were those who thought that the Reporting Service should be a civilian organization. Her demeanor within the army was too artificial. Therefore, on July 1, 1891, the meteorological service was transferred to the Department of

Fig. 6.7 Observatory Pankov Peak

Agriculture and received the name "United States Weather Bureau." The director was Professor Harrington (1848–1926). Maritime meteorology (description of sea climate) developed completely independently. Its development is linked to **Maury** (1806–1873), a lieutenant in the American Navy. At the age of 19, he was a cadet on a ship. On the ship "Vincennes" from 1826 to 1830, he cruised around the world. Due to an accident in 1839, Maury became disabled and was unable to continue military service.

He moved to work for the American Hydrographic Service in 1841. Immediately afterward, he became the director of that service. He analyzed the meteorological information from the ship's logs. He came to the conclusion of which route is best when sailing from New York to Buenos Aires. When Maury's optimal direction determined by the prevailing wind is used, the journey from New York to the Equator is completed in 24 h, instead of the previous 41 h. He also analyzed other routes, knowing the climatology of the ocean. He began printing 1851 his famous "sea route charts," trade wind charts, storm and rain charts, thermal charts, etc. Using his maps, the trip from New York to San Francisco (around South America) was shortened from 180 to 135 days. The journey from England to Sydney and back was shortened from 250 to 130 days. Maury distributed 20,000 copies of these cards for free, all over the world. In return, he only asked that they send him the last observation from the ship's log.

Maurij published his book *Physical Geography of the Sea* in 1853. It was printed in more than 20 editions and translated into several languages. In it, he described sea currents in clear language (especially the Gulf Stream, sea salinity, evaporation, ocean depth, etc.). It describes the atmosphere, the climate of the oceans, the winds, the "cloud ring above the Equator" storms (clouds formed by the strongest convection), etc. He was still analyzing the ozone in the atmosphere. He linked ozone and wind direction. He wrote about the general circulation of the atmosphere, etc. His achievements in popularizing meteorology and oceanography are no less significant than his success in collecting a large amount of data on wind and sea currents.

6.16 Establishment of International Standards in Meteorology

By the middle of the nineteenth century, meteorological measurements were made in a large number of countries around the world (Fassig 1918; Khrgian 1970; Cressman 1983; Conover 1990). Networks of stations are established within individual countries. At that time, it was quite clear that the further development of meteorology cannot occur if uniform criteria are not established regarding the method of measurement, the type of instruments, and if the data is not exchanged on a global scale. In order to achieve all that, international institutions had to be established that would prescribe uniform norms.

On the initiative of the American researcher Maury, the British and American governments invited other countries to a conference in Brussels. The conference was held from August 23 to September 8, 1853. The conference was attended by representatives of Britain, Russia, France, the USA, Denmark, Belgium, the Netherlands, Norway, Sweden, and Portugal. The chairman of the Conference was the Belgian meteorologist **Ketelet. It was the first international meteorological conference.** The main goal of the Conference was to establish permanent international cooperation in the research of meteorological conditions above the sea. An agreement was reached on the form of observation logs to be used on military and merchant ships. Meteorological, astronomical, and magnetic measurements and observations are included in the diary. An agreement was reached that upon the arrival of the ship in the port, the logs must be sent to the appropriate services for analysis. The measurement dates were established at the Conference. The Fahrenheit and Celsius temperature scales are accepted, and ten cloud types are established that should be observed.

Immediately afterward, the conclusions of the Brussels Conference were accepted by the countries that did not attend it, Prussia, Spain, Sardinia, and Austria. As early as 1858, 19 countries accepted this measurement system on ships. Processing of meteorological observations was carried out in the USA, the British Meteorological Department, and the Netherlands Meteorological Institute. Thanks to this, many atlases, various maps, etc. were created.

After the establishment of international meteorological cooperation on measurements at sea, it was logical to establish such cooperation for measurements on the land. In this sense, Maury writes in a letter to Ketelet around 1860: "... It becomes very obvious that meteorologists on dry land should work together with sailors...". In order to compile weather maps, it was necessary to collect data from both land and sea. Undoubtedly, this was the main motive for the establishment of the International Meteorological Organization. Wild, Jelinek, and Bruns (1830–1881), director of the Astronomical Observatory in Leipzig, exchanged letters in this **regard. Their letter to the meteorologists of the world states: "If there is one field of science that benefits from a unified system, it is meteorology." This letter was sent** by Wild, Jelinek, and Bruns to the meteorologists of all countries in May 1872, inviting them to a conference in Leipzig. The letter also highlighted this: "Now, the interest in the development of meteorological analyzes requires the greatest uniformity of methods in different countries." The letter highlighted 25 content proposals that would be considered at the conference. Most of them are related to observation methods. The conference was held in Leipzig from August 14–16, 1872. In addition to a large number of German scientists, the conference was attended by delegations from 15 countries. At the Conference, there were big differences in the opinions of meteorologists, both regarding the best instruments, how to place them at the measurement site, and the choice of measurement units and observation time.

This was followed by the International Meteorological Congress, held from September 2 to 16, 1873, in Vienna. The Congress was attended by representatives of 20 governments and meteorological services from most European countries and America (Fig. 6.8).

It should be noted that the famous scientist **Dove**, who at that time was the director of the Prussian Meteorological Service, did not attend the Congress, at the

Fig. 6.8 Participants of the First International Meteorological Congress in Vienna in 1873, when the International Meteorological Organization was founded

Congress, the so-called Standing Committee. It was the first international meteorological organization. The Standing Committee then met in Utrecht (10–14 September 1874 and October 1878) and in London (18–23 April 1876). The second international meteorological congress was held in Rome in 1879. After that, the International Meteorological Committee met in Berne (1880), Copenhagen (1882), Paris (1885), Zurich (1889), Uppsala (1894), Petrograd (1899), Paris (1900), Southport (1903), Paris (1907), Berlin (1910), and Rome (1913). Meetings of the directors of meteorological services were also held in Munich (1891), Paris (1896), Innsbruck (1905), and, after World War I, in Paris (1919).

It was not easy to reach an agreement on uniform principles in meteorology. This was contributed by the political desires of individual states for dominance and inherited differences in measurement principles. The attitude in this regard of the British delegate Scott at the Leipzig Conference is typical. He came out strongly against the position of meteorologists to introduce the Celsius temperature scale. He also believed that it could not be decided by meteorological scientists but by state governments. This is why the Fahrenheit scale has remained in use to this day in Britain and the USA. Nevertheless, the desire for an agreement prevailed among the majority. Thus, in 1870, Russian meteorologists accepted the metric system (which was first introduced in France) and the Gregorian calendar. It was not easy to agree on the timing of the measurement either. For this reason, a decision was made in Vienna to accept the following variants of measurement terms: 6, 14, 22; 7, 14, 22; 7, 13, 21; 7, 14, 21; 8, 14, 20; 9, 15, 21; 10, 16, 22; 8, 20; 9, 21; and 10, 22 h.

In terms of instruments, it was agreed at the Congress of Vienna to use the hair hygrometer and the aneroid barometer as parallel instruments to the mercury barometer. Howard's cloud classification was adopted for cloud observation. Later, at the Munich Conference, it was somewhat modified by the proposals of Abercrombie and Hildebrandson. In Vienna, there was a proposal to establish an International Meteorological Institute. However, this was abandoned due to a lack of finance. Later, in Utrecht (1874), they concluded that: "... The establishment of a well-organized institute, which would be supported by an international fund, is currently unfeasible."

Within the framework of the International Meteorological Organization, there were agreements that had great scientific importance (Zigun 1955). Thus, the International Half-year was established in 1882–1883. The impetus for this came from the admiral of the Austrian Navy Weiprecht (1838–1881) in 1875. Namely, the Austro-Hungarian Monarchy organized the Arctic Expedition in 1872–1874. Then the Land of Francis Joseph was discovered. Experience was then gained on the basic principles of research in those areas in all fields of science. Weiprecht emphasized that the scientific success of that expedition is small because these are expensive projects. He emphasized that the goal of those expeditions should not be geographical discoveries but scientific research. The International Polar Commission was founded at the Conference in Hamburg in 1879. In Bern, it was agreed to establish eight polar stations as part of the Polar Expedition. Russia, Austria, Norway, and Denmark were responsible for that. Observations at those stations lasted from August 1, 1882, to September 1, 1883.

During the First International Polar, numerous measurements of meteorological and magnetic elements were carried out. These data were a good basis for important scientific studies. Half a century later (1932–1933), the Second International Polar Year was organized in order to further look at the many scientific problems of the functioning of nature in the polar regions. In that period, Soviet meteorologists contributed significantly to the success of that action. Other European and American countries did not make a significant contribution due to financial and political circumstances. But there were still strong scientific reasons to continue with this coordinated scientific action. Thus, the International Geophysical Year was held in 1957–1958. During that period, simultaneous measurements were carried out all over the world.

Thanks to the efforts of the International Meteorological Organization, uniform standards in meteorology were gradually introduced. Such standardized measurements are exchanged, which has made it possible to produce regular daily weather forecasts. In order, for such a system to function, it was realized that the states should stand behind it in their decisions. Thus, on March 23, 1951, the World Meteorological Organization was founded, which was recognized by the United Nations General Assembly in its resolution from 1951 as a specialized agency of the United Nations (Fig. 6.9). The governments of the countries did not directly stand behind the International Meteorological Organization. The World Meteorological Organization (WMO) was founded by the governments of the member countries of the United Nations. Thus, the WMO is part of the United Nations. The governments of the countries are now the guarantor of the constant income of this organization for its activities, while this was not the case before. The seat of the WMO is in Geneva (Fig. 6.10).

Fig. 6.9 Participants in the founding of the World Meteorological Organization, March 23, 1950

Fig. 6.10 Building of the World Meteorological Organization in Geneva (moved in in 2001)

References

Brown, R.H. 1940. The first century of meteorological data in America. *Monthly Weather Review* 68: 130–133.

Budyko, M.I. 1957. *Meteorologicheskie issledovaniya v SSSR* (Meteorological Research in the USSR). Meteorologiya i Gidrologiya, No. 11.

Conover, J.H. 1990. *The Blue Hill meteorological observatory: The First Hundred Years, 1885–1985*, 514. The American Meteor Society.

Dobrilović, B. 1964. On the development of meteorology in Serbia. *Papers* 5.

Fassig, O.L. 1918. A signal corps school of meteorology. *Monthly Weather Review* 46: 560–562.

Khrgian, A.Kh. 1970. *Meteorology – A historical survey*. Vol. 1. Springfield: U.S Department of Commerce.

Miller, E.R. 1933. American pioneers in meteorology. *Monthly Weather Review*.

Bates, C.C. 1989. The formative Rossby – Reichelderfer period in American meteorology, 1926–1940. *Weather and Forecasting* 4: 593–603.

Cressman, G.P. 1983. Francis W. Reichelderfer, 1895–1983. *Bulletin of the American Meteorological Society* 64: 398–400.

Fleming, J.R. 1990. *Meteorology in America, 1800–1870*, 264. Baltimore and London: Johns Hopkins University Press.

———. 1996. *Historical essays on meteorology, 1919–1995*. Boston: American Meteorological Society.

Hepites, Š. 1899. *Organisation du service météorologique de Roumanie*. Bucarest.

Kaltchewa, R. 1939. 50 Jahre meteorologischer beobachtungen in Sofia.

Khrgian, A.Kh. 1948. *Istoriya meteorologii v Rossii*, History of meteorology in Russia. Vol. II. Moskva: Trudy Instituta Istorii Estestvoznaniya.

Rykachev, M.A. 1899. *Istoricheskii ocherk Glavnoi geofizicheskoi observatorii*, A Historical Outline of the Main Geophysical Observatory. Sankt Peterburg.

Zigun, V.P. 1955. *Vtoroi kongress Vsemirnoi meteorologicheskoi organizatsii* (The Second Congress of the World Meteorological Organization). Meteorologiya i Gidrologiya, No. 6.

Chapter 7
Establishment of Weather Forecast Services

7.1 Initial Hints

Weather forecasting has been of interest to people in various historical periods (Kimball 1933; Barlow 1958; Frisinger 1967: Williams 1992). The need for a forecast based on scientific foundations was emphasized by two great scientists in the eighteenth century, Lomonosov and Lavoisier. Lomonosov, Fig. 7.1, in his work printed in Petrograd in 1759, "Discourses on the Accuracy of Maritime Routes," emphasized the following: "Predicting the weather is both necessary and useful, because the peasants need warnings since they would need good weather during sowing and harvesting, and they need rain to moderate the heat during the growing season (Gaigerov 1949; Crawford and Marsh 1989). At sea, sailors know how useful it would be to be informed about the direction of the wind and about the approach of a storm." He further points out: "All this can be achieved with the help of theory if frequent observations at sea and in the air are available... I have an idea of how to organize such observations and to provide various new instruments for this." Apparently, Lomonkov had in mind that a network of meteorological stations should be organized.

And Lavoisier (1743–1794), a versatile French scientist, was interested in meteorology in several works. In one he writes: "De Bohr was the first to propose simultaneous measurements in various places. That's why a few years ago, he organized several experimenters in various areas of France to measure with a barometer." Unfortunately, there is no information about the places or the date when the measurements were taken.

At the beginning of the nineteenth century, Brandes (1777–1834) drew the most important tool of modern meteorology, the synoptic map. He prepared a series of synoptic charts from 1816 to 1820, for each day starting from 1783. He used data

M. Ćurić, V. Spiridonov, *History of Meteorology*,
https://doi.org/10.1007/978-3-031-45032-7_7

Fig. 7.1 Mikhail V. Lomonosov

collected from the Mannheim Meteorological Society. On the maps, he drew lines of equal deviation of atmospheric pressure from normal. He wrote a paper about this in 1826, entitled "On sudden changes in pressure." It should be emphasized that these were not maps of the current state of the weather at the time but rather depicted the state of the past. At that time, the name "synoptic maps" was not yet known. He drew maps based on measured pressure in a large number of places in Europe. Even then, Brandes noticed that there is a connection between the distribution of pressure and airflow.

Another German meteorologist, Dove, drew air flow maps in 1828 for stormy days between 1821 and 1823. These maps were analyzed by Brandes. The maps were drawn without a geographical basis (without the basic features of rivers and mountains), so it is difficult to follow any regularity of the flow (Fig. 7.2).

The next to draw weather maps was Lums in the USA. In 1841, he analyzed the storm of December 20, 1836, using hourly observations from eight stations (New York, Baltimore, Albany, Fleming, New Haven, Gardner, Montreal, and Quebec). He drew a map of isolines of simultaneous occurrence of minimum pressure with wind direction and precipitation in several places. In 1846, Looms drew weather maps for several days from February 1842. He drew pressure and temperature deviations from normal and wind direction. It also entered data on cloud cover (clear or cloudy), rain, or snow. He marked all phenomena with different colors (dark blue, gray, yellow, and green).

The first weather maps in Russia were drawn by Smirnov, an assistant professor at Kazan University. In 1870, he published a series of maps of the European part of Russia, under the title "Prediction of weather and spring storms in Russia."

All of these weather maps showed weather that occurred well before the time they were analyzed. In order to plot the current weather, it was necessary to wait for the moment when the telegraph was found, with which meteorological data could

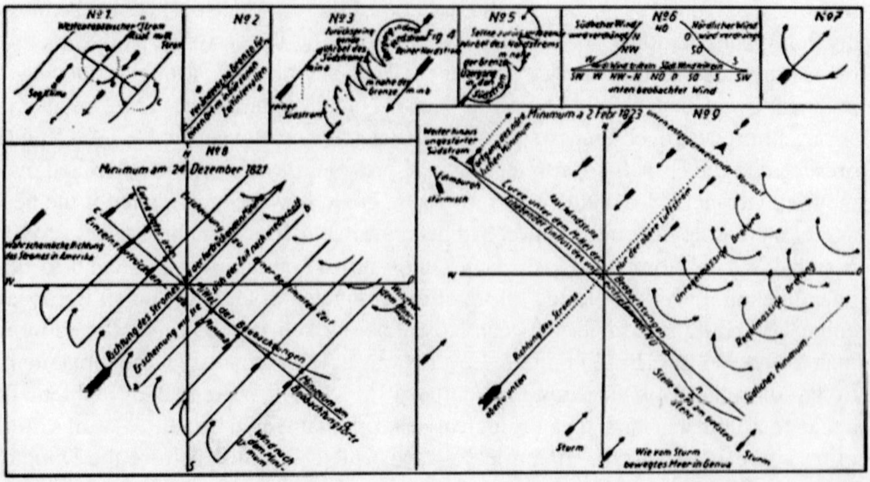

Fig. 7.2 Dove's charts show the movement of storms from 12/24/1821 to 2/2/1823

be collected quickly. That is why it is not unusual that in the nineteenth century, the field of meteorology dealing with weather analysis and forecasting was often called "telegraphic meteorology".

An assistant at the observatory in Prague, Krell (1798–1863), suggested in 1843 that telegraph poles be used to collect meteorological data and send weather forecasts. In America, in 1847, Redfield also suggested using telegraphs to study storms. This worked in practice in 1849 but in the simplest form. Namely, Henry, the secretary of the Smithsonian Institute, suggested that telegraph operators (telegraphers) type one word in the morning when turning on the telegraph: clear, cloudy, rain, and the like, instead of the usual word "okay." In Washington, based on these reports, simple maps of the current weather were drawn, which they used to inform the general public.

A similar collection of weather data in England was carried out in 1849 by Glešir. From June 14 of that year, he regularly printed a weather bulletin in "The daily news," based on 20 telegraphic reports. In 1851, in order to help Glacier, the Telegraph Company exhibited "maps of the current weather in some parts of Great Britain," during the first great International Exhibition (World's Fair) in London. Data on pressure and wind direction are plotted on the maps, and phenomena are marked with letters: o, cloudy; c, clear; and r, rainy. The cards were shown every day from 8.6. until 11.10.1851. years.

7.2 Leverier's Decisive Contribution

All these first attempts to display the current weather condition were of little use, due to the small amount of weather information and the fact that they were collected from a small area. The first attempt to overcome the initial shortcomings was made by Leverier (1811–1877), director of the Paris Observatory. He pushed for the introduction of international "telegraphic meteorology" in the early 1850s. He tried to reconstruct the weather service in France. That attempt was strongly aided by a practical event that befell the French Navy. Namely, during the Crimean War, a strong storm destroyed French ships in the Black Sea on November 14, 1854. It was presented to the French government that the previous day that storm had raged over the Mediterranean Sea and that its movement could have been predicted if the necessary meteorological information had been available. The French Minister of War, Marshal Whelan, ordered Leverrier to study all the circumstances related to this phenomenon. Leverje collected information from meteorologists from all European countries about the weather between November 12 and 16, 1854. It received more than 240 responses. Based on this, Leverje was able to reconstruct the movements of the storm that sank the ships. On February 16, 1855, he presented to Napoleon III a plan to establish a large meteorological network that would be able to warn sailors of an approaching storm. His project was accepted. With the help of the Director General of Posts and Telegraphs of France, on February 19, 1855, Leverier presented to the French Academy a weather map for the 10 hours of that day. It left a favorable impression that already from June 2, 1855, 13 stations in France regularly sent telegraphic reports to the Paris Observatory. It was more difficult to secure such reports from abroad. Nevertheless, such reports have been arriving since 1857 from 8 European cities (Petrograd, Brussels, Geneva, Rome, Turin, Madrid, Vienna, and Lisbon). Since 1858, the observatory has studied the daily "International Bulletin," which has become more and more complete (with more stations) from year to year. Leverrier asked the Minister of the Navy, Admiral Hamelin, to allow warnings of approaching storms to be sent to French ports. The admiral delayed issuing such permission.

At the same time, a weather forecast service was organized in Russia. In 1856, the main physical observatory in Petrograd began receiving telegraphic weather reports from 13 stations in Russia. Part of the report was sent to Paris, and in return, they received reports from Paris, Lyon, Madrid, and London. The Secretary of the Navy suggested that a service be established for accurate telegraphic weather reports. Therefore, the director of the observatory, Cooper, sent some people abroad in 1865 to be trained for such a service.

7.3 Bright Beginning and Tragic End of Fitzroy

In England, Admiral FitzRoy was active in the establishment of the weather service, within the Meteorological Department of the Ministry of Trade, of which he was the first head (FitzRoy 1863; Tikhomirov 1932). The British Society for the Advancement of Science addressed a letter to the Department of Trade in which it was pointed out: "... The importance of telegraphic reports is that they enable warnings to be sent of the possible approach of storms." After that, the Ministry assigned Admiral FitzRoy to receive meteorological telegrams from 11 British stations. These telegrams were paid for, unlike in other countries, where they were free. From September 5, 1860, summary weather reports were published every day in the evening edition of "The Times," and a little later in other newspapers.

Robert FitzRoy (1805–1865)

At the age of 14, he was a cabin crew on a British Navy ship. He quickly rose to the rank of captain of the Beagle. In that role, in 1828, he was assigned to patrol the coasts of Patagonia and "Tierra del Fuego." From 1831 to 1836, the ship Beagle, under the command of Fitz Roy, made the famous voyage around the world with Charles Darwin as a passenger. That trip had an epoch-making significance for biology and hydrography. Fitz Roy invested a significant amount of his money (about 6000 pounds sterling) in that trip, so that the trip would go as well as possible. On those trips, he proved to be an outstanding navigator. Despite this, strong and prolonged storms repeatedly caused minor damage to the ship. On his return from his voyage around the world, Fitz Roy was a highly respected man. He received the rank of admiral. He became a member of the British Parliament in 1843. Then, from 1843 to 1845, he was the governor of New Zealand. After the establishment of the meteorological department at the Ministry of Trade in 1854, he was appointed head of the department. This is where the most fruitful part of his life began.

Fitz Roy was a great practicing meteorologist. He acquired scientific knowledge about meteorology by solving various practical problems during navigation. He is an active observer of many stormy meteorological processes that he went through on the open sea. Thinking about them, he specialized in that field. His forecasts of approaching stormy weather were highly regarded among English, especially Scottish fishermen. He emphasized that "when forecasting, one should keep an eye on barometric and thermometer changes, but also on the appearance of the sky, especially at sunrise and sunset. Clouds

(continued)

(continued)
should be monitored, but also the behavior of sea birds." He adhered to the rule: "The more carefully we combine prognostic material obtained from different sources, the more successful the results will be." Fitz Roy saw that forecasts based on single-point observations (however accurate) could not be good. That is why he prepared synoptic maps. Sometimes the forecasts were not correct, not because of his negligence but because of objective circumstances. Journalists made a lot of noise about it. One could say that they are objective critics. However, journalists, then as now, were those who like to affirm the greatest ignoramuses and amateur bluffers. They did not realize that there was no better forecaster than Fitz Roy. With their criticisms, they made such an honorable, moral, and knowledgeable man decide to commit suicide. When she learned of his death, the wife of an Aberdeen fisherman exclaimed in despair: "Who will look after my husband now."

Fitz Roy began in February 1861 to study storm forecasts. He prepared small maps which he first called "synoptic maps". The word comes from the Greek word "synoptikos" which means showing a general view. These maps covered a small area (England, Ireland, and the northern coast of France). On these maps, he tried to show two currents, polar and equatorial. He emphasized: "... The weather in our country completely depends on the collision, interaction, mutual dominance of parts of these two opposing currents." Fitz Roy's analysis of these maps would always begin in the area of Ireland and Scotland, since that is where the currents most often meet. One example of such a map is shown in Fig. 7.3. His forecasts are highly valued by sailors. In the beginning, he sent warnings about stormy weather in 50 points, so that in August 1861, it would be 130 points on the British coast. Due to a lack of funds, Fitz Roy could no longer issue daily synoptic charts.

The preparation of storm forecasts in the Meteorological Department was almost entirely tied to Fitz Roy. The opinion of the English public about the success of these forecasts was tied only to his name. With the exception of seamen, the English public did not think well of these forecasts. Disbelief was expressed in the daily newspapers. In addition to the general public, skepticism was expressed by scientists and the Royal Society. One of his failed forecasts, when he had only a small amount of data from a small area, was severely criticized in the press. A veteran of meteorological science, an old sea wolf, who knew how to push his research ship through the biggest storms, he could not withstand the criticism. He shot himself in the head on April 30, 1865.

The tragic death of the victim of science is a dark stain on the entire English meteorology of the time. For years, scientists in England ignored the reasons for FitzRoy's death and his contributions to synoptic meteorology. After his death, a commission was formed to investigate the activities of the Meteorological Department. They concluded: "FitzRoy took the first step towards placing storm forecasting on a clear and accurate basis." He forced scientists, and society in general, to pay attention to this area of meteorology, and he gave his life for those

Fig. 7.3 One example of Fitz Roy's synoptic chart

efforts. "The commission further concluded: "... There are still no real scientific arguments for daily weather forecasts. The accuracy of those forecasts has not been generally demonstrated, and there is no data to indicate their validity." After the death of Fitz Roy, the publication of weather forecasts was discontinued. The entire Meteorological Department was closed on December 7, 1866.

At that time, only "storm" forecasts were talked about. Obviously, then the weather forecast had a narrower meaning than it does today. Back then, it was considered that the forecast was good if only the worst, stormy weather was announced. Therefore, it is not considered a success if good weather is correctly forecast. By those criteria, even today many meteorologists would deserve criticism like FitzRoy.

7.4 Weather Service in Various Countries

Netherlands

The establishment of a weather forecast service has also begun in the Netherlands. In 1859, the Dutch Government adopted a special law on telegraphic data collection and telegraphic transmission of weather information. Bis-Balot, the director of the Meteorological Institute in Utre-ht, particularly stood out in this job. On June 1,

1860, he issued the first official warning of an approaching storm. Bis-Balot sought to find rules for weather forecasting based on local indicators. He observed pressure differences between nearby places. Based on that, in 1857, he came up with a well-known law (rule), which says: if the wind blows at our back, then the lower pressure is on our left side. The stronger the wind, the greater the pressure difference on the left side compared to the right side. Based on this, he constructed the "aerocline-scope." This instrument was a kind of semaphore placed on the seashore. Information about the direction and intensity of the pressure difference on the left and right sides (pressure gradient) would be visible from passing ships. From this, the strength of the wind normal to the pressure gradient could be determined.

One rule based on which Biss-Balot forecasted was: "A westerly storm can be expected whenever the barometer at Maastricht or Vlissingen shows a value 4 mm higher than at Groningen or Den Helder." This forecasting concept is based on the dependence of wind speed and pressure gradient (the pressure difference between two places divided by the distance between those places).

France

In France, unlike the tragic stagnation in England, the weather forecast is gaining importance. The meeting Leverje had with Napoleon III in 1863 in Biarritz contributed to this. From the middle of that year, with Napoleon's approval, the observatory began issuing regular weather forecasts. On September 11, 1863, the bulletin of the first synoptic maps in the world was printed. One such map is shown in Fig. 7.4. Lines of equal pressure expressed in mm (height of the mercury column measured in mm) are drawn on the map.

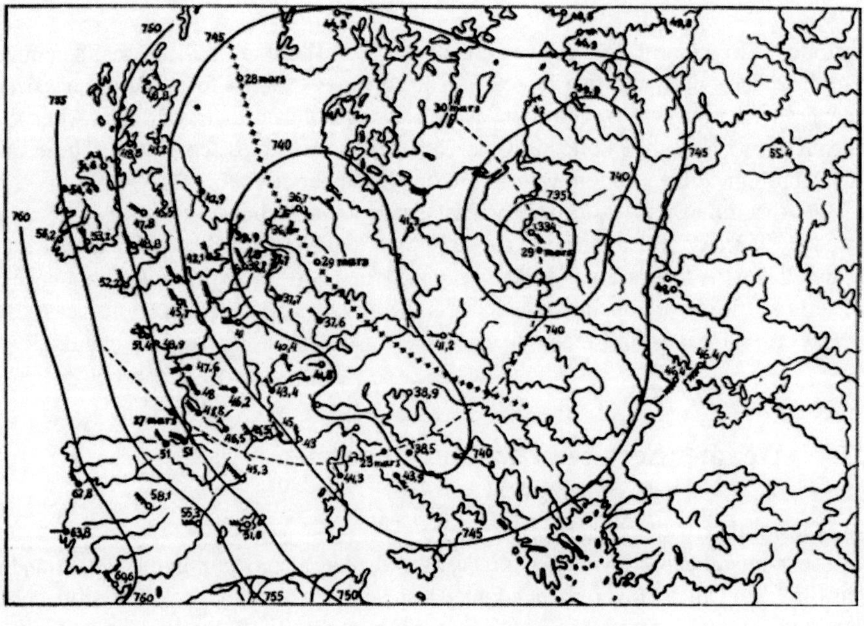

Fig. 7.4 Synoptic map of the Paris Observatory from September 11, 1863

In the nineteenth century, almost all countries accepted the weather forecast based on the method of the Paris Observatory. The method consisted in drawing maps of isobars (lines of equal pressure), and based on that, the movement of the closed isobar with the lowest pressure (cyclone) was forecasted. It was observed that sudden changes in the weather were associated with cyclones, which were then called "storms." Because of the sudden changes in the weather brought by storms, they were a subject of special interest. An effort was made to find connections between weather characteristics and the shape of isobars, as well as the path of storms (cyclones).

Russia

The weather forecast service in Russia began with the publication of daily weather bulletins in 1872 (Khromov 1932, 1934). Bulletins were issued under the supervision of Rikachev. The first bulletins were made for the European part of Russia based on data from 26 stations. Of course, this is a small number of stations for such a large territory. Wild emphasized: "... The importance and benefit of these newsletters, in my opinion, are not sufficiently appreciated even by the experts themselves." The number of stations on the basis of which weather bulletins were made grew, both from Russia and from abroad. One synoptic weather map that was made at the Main Physical Observatory is shown in Fig. 7.5.

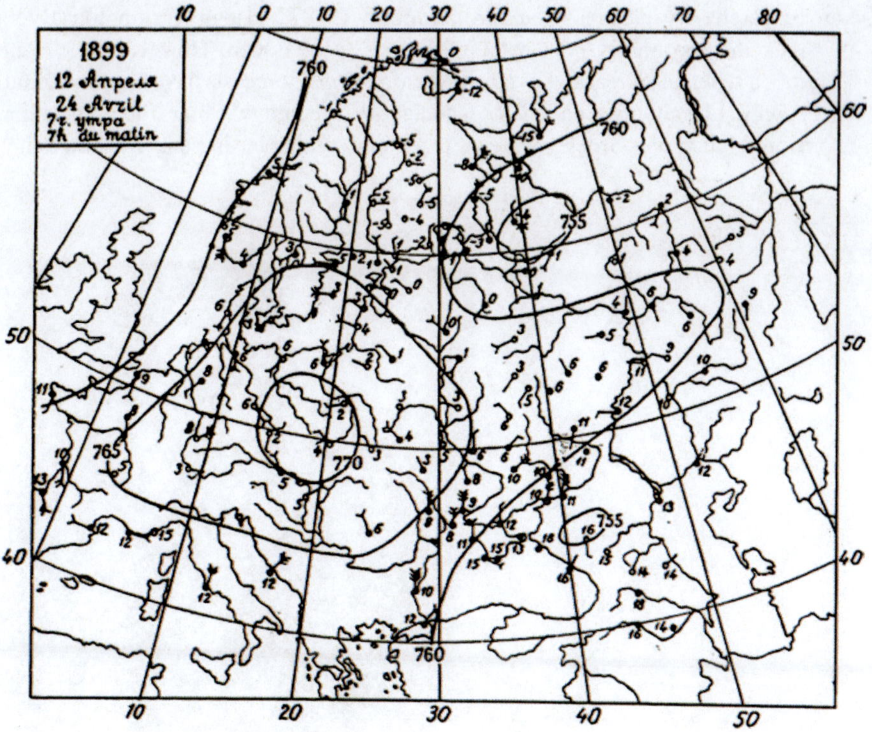

Fig. 7.5 Weather map for 4/23/1899 years, from the Main Physical Observatory in Petrograd

The USA

The weather service in the USA was established a little later than in Europe. After the Civil War (1861–1865), several people came up with that idea (Whitnah 1961; Kutzbach 1979; Kimball 1981). Director of the Cincinnati Astronomical Observatory, Cleveland Abe (1839–1916) began collecting telegraphic weather reports in 1869 (Cohen 1950). He was also interested in theoretical meteorology, which Ferrell dealt with. In 1870, Abe actively supported the establishment of the Time Bureau. He published a large number of works in various fields of meteorology, from climatology, and long-term weather forecasting, to the history of American meteorology. On the initiative of Abe, in 1902, the publication of the meteorological magazine "Monthly Weather Review" was started. This magazine is published now.

The man who is responsible for the organization of the storm forecasting service on the Great Lakes is Lafram (1811–1875). He was an amateur geologist and meteorologist. In December 1869, he also collected information on damage to ships sailing the Great Lakes. In 1869 alone, 1914 vessels were damaged, with a total damage of more than four million dollars. Lafram, through Congressman Pena, submitted to Congress an estimate of the cost of organizing a storm warning service throughout America. Congress passed a resolution on February 4, 1870, establishing a weather service within the US Army (the Reporting Service). The head of the service was General Mier (Fig. 7.6).

From November 1, 1870, three synoptic maps were produced daily. At first, they were prepared only in Chicago, and from January 1, 1871, also in Washington (Byers 1976). In the beginning, no weather forecasts were issued. However, since late November, when the Great Lakes storm season begins, forecasts have been essential. That's why Lafram was called to Chicago to prepare weather forecasts. Abe, Thomson, Maury, and Greeley were engaged in the same business in Washington.

Fig. 7.6 General Albert Mier

Already at the time of the establishment of the weather service, in 1870, the US Congress set the task before it (i.e., before the Secretary of War): "... To provide such stations, reports, and reports as may be required for agronomy and other commercial needs." From January 1, 1872, the Reporting Service began to receive reports on the water level in the rivers, and from the spring, weather forecasts were added on the expected fall or rise of the water level. From 1873, Mier began producing the "Farmers' Bulletin," which consisted of a short report on the state of the weather and some kind of forecast. The newsletter was distributed by mail. Since 1887, the "Weather and Crop Bulletin" has been produced.

Already in 1890, the representation of the meteorological service in the American economy was quite large. Then the newly formed "US Weather Bureau"—the American weather service—was separated from the military reporting service. In addition to other tasks, this service was assigned: "... To responsibly predict the weather; issues announcements of approaching storms and issues appropriate signals to indicate storms and floods of interest to agronomy, commercial affairs, and navigation; to inform cotton growers about temperature and precipitation, as well as to warn of frosts and penetration of cold waves with appropriate signals." As early as 1892, there were 9300 subscribers who received forecasts and warnings. The American weather service did not gain popularity for its extremely accurate forecasts (accuracy was the same as others, from 77 to 78%). This was, first of all, thanks to the quick delivery of information to various users who were scattered all over America. That popularity enabled the development of a wide range of service activities. So, already at the end of the nineteenth century, there was a very close connection between synoptic meteorology and its application in America.

The first weather maps were drawn by hand, and from May 1871, they were reproduced by a lithographic process. In October 1871, only isobars were drawn (plotted) on them, and later isotherms were also included. Their daily bulletin consisted of "weather reports, probabilities, and facts." Here the word "probability" referred to the weather forecast. As early as 1871, meteorological telegrams were received from Canada, and in January 1874, Mier established a connection with stations from all over the world. Weather maps are then issued for the entire northern hemisphere. These maps were known as "International Weather Maps."

Due to being busy with practical work, it was not noticed that there is not a single scientific institute or observatory for meteorology in America. This shortcoming was somewhat overcome when it was necessary to solve some practical problems. At that time, a small group known as the "Study Room" was formed within the Military Reporting Service. The group was led by Abe and consisted of Waldo, Hazen, Russell, and later Ferrell. They solved various practical problems related to spring frosts, hurricane paths, floods, etc. Hazen dealt with the reduction of cellular pressure to sea level. This made it possible to compare the pressure from stations located at different altitudes. Russell chose statistical procedures for forecasting the penetration of cold air. Abe emphasized that a higher level of scientific education, high scientific standards, and modern laboratories must be introduced. Little was done of that until the end of the First World War. That's when military and civil aviation developed. For these purposes, it was necessary to have good weather forecasts, including cloud and fog forecasts. From that new demand came scientists with works that significantly improved meteorology.

England

We saw that weather forecasting activity ceased after Fitz Roy's suicide. But everyday life made me forget about that event. The Meteorological Committee has received numerous requests from various quarters to start issuing weather forecasts again. In response to that, in December 1867, forecasts began to be issued again. Fearing the failure experienced by FitzRoy, the Committee tried to issue forecasts without synoptic charts. They used only empirical Bis-Balot rules. It was realized that this was not enough, so they started drawing maps again in 1869. From March 23, 1872, maps even began to be printed in the form of a "Daily Weather Bulletin."

At first, forecasts were issued irregularly. Later, morning forecasts were made for Government offices and daily newspapers, and forecasts were made during the day based on daily tickets for other users. At the request of the "Times" editor, evening maps were made (6 p.m. Greenwich Mean Time), based on which forecasts were issued and printed in the morning edition of the paper. For this special forecast, the paper paid 500 pounds sterling annually. Forecasts were also issued for English ports with warnings of approaching storms. The accuracy of warnings increased from the initial 68% (in 1870) to 83% (in 1876).

Unfortunately, the observation quality and network density were very poor. The 8 h were often unsynchronized, and the 24-h measurement was insufficient. All this made it difficult to get a real picture of the model of atmospheric processes. Those old charts in particular did a poor job of showing cyclones and anticyclones over the Atlantic. However, it should be pointed out that Admiral FitzRoy, as the first head of the Meteorological Service in London, very nicely depicted the weather systems of extratropical latitudes. His cyclone model, published in 1863, contains a cold current from the north in the rear part and a warm current from the south in the front part. That model did not experience a more complete affirmation, although it is a very correct model, due to his premature death (in 1865). Somewhat at the same time, another significant figure of English science, Francis Galton (a relative of Charles Darwin and later a pioneer of statistical genetics) tried to present meteorological data from all 300 meteorological stations of Europe in a concise manner. He presented his method in the paper Meteorographica (1863). Unfortunately, after 1870, Galton left active work in meteorology, and his beautiful ideas were not followed by anyone else. Practically no one else in Europe presented meteorological data in such a rational way even after 50 years.

Norway

Mon, director of the Meteorological Institute of Norway, began sending weather telegrams to the ports in 1866, based on the reports he received from meteorological stations. Later, the Institute received numerous telegraphic reports from abroad. During 1914, the number of such reports was 60. Weather reports were received in 1907 from as far away as the Faroe Islands and Spitsbergen. In 1869, they began to use the announcement of the arrival of storms that they received from England. But such announcements were unsuccessful. Until 1882, the Institute was content to produce daily weather charts, and only occasionally (no more than five times a year) did they send warnings to the coast services.

In 1882, they started sending weather forecasts by telegraph to several places in Norway. In 1890, Stortini formed a fund for the preparation of a study on the possibility of establishing a weather forecast service, not only in Christiani (Oslo) but also in Bergen, Trondheim, and the Lafontaine Islands. That project was not implemented immediately. It was not until 1903 that a meteorological observatory was opened in Bergen. The main job of the director of the Feuna observatory was to issue storm warnings for the southern part of Norway. About 40 warnings were issued annually. Since 1912, the Institute has regularly published the newsletter "Daglig Veir – Buddstikke." It published data tables from the stations, weather maps, and forecasts. In 1918, the Weather Forecast Institute for the Western part of the country was founded in Bergen. However, the contributions made there to world meteorological science far exceed the original local purpose of the Institute in Bergen. Due to the great importance of the obtained results, it will be discussed in a separate chapter.

Germany

Political division and a relatively weak fleet slowed the creation of a weather forecasting service. In Berlin, in 1865, a special storm monitoring service "Zentralstelle für Sturmwarnun-gen" was established under the direction of Dove. In a 10-year period, this service issued only nine warnings.

The establishment of a real weather forecast service in Germany is linked to G. von Neumayer (1828–1909). He completed his studies in Munich. He later worked in Australia as a gold prospector. In 1856, he founded the meteorological observatory "Flagstaff Observatory" in Melbourne (partly with the German capital). There, at the "Verein" meeting of the "Union of Germans" in 1861, Neumeier gave a speech in which he emphasized: "... We must join the ranks of maritime nations." The idea of superiority at sea suited the German Empire well. Thus, by the decree of January 9, 1875, the "Deutsche Seefarte"—German Shipping was founded in Hamburg (Fig. 7.7). The director was Neumeier, who showed an extraordinary ability to acquire money.

The meteorological telegraph services were first managed by Keppen, and in 1878, by Van Weber. The first complete bulletin with weather maps and weather forecasts was published on February 16, 1876. Everything was subordinated to the needs of ship commanders in the North Sea and the Baltic.

In Germany, there was more interest in weather forecasting in agriculture than in the Navy. At the request of farmers, small-time bureaus were established in Leipzig (1878), Munich (1881), Baden (1884), etc. A large service for weather forecasting was founded in 1906, modeled after the American service, in 16 cities. This service had the task of delivering weather conditions and forecasts to all who are interested but also to exchange observed data with each other via telegram. This created a large number of amateur meteorologists, especially among farmers. Such expansion was not successful, since the preparation of forecasts required a higher level of knowledge and experience.

Fig. 7.7 The building of the German Navy in Hamburg

Austria

In Austria-Hungary, a telegraphic collection of weather reports began in 1865. Storm warning services were established in Trieste and along the Adriatic coast, according to the forecasts of the Paris Observatory. Since 1877, daily forecasts have been issued with a weather map of Europe. That year, a special summer telegraph service for farmers was established at the proposal of the Agricultural School in Neutyschen. Later, special weather forecasting centers were established in Budapest, for the merchant navy in Trieste and for the military navy in Pula.

Italy

The time service was organized in 1865 at the Ministry of the Navy in Florence. The director was the famous scientist and senator Mateuci. He and Secchi well noted that storms in Italy come from the northwest, from the area of the British Isles. Therefore a plan was made to make special observations at Italy's expense in Ireland, in order to give good warnings of the coming of storms. In 1880, the weather forecast service was moved to Rome, within the Central Meteorological Service of Italy.

Denmark

The Meteorological Institute in Denmark was founded in 1872. The director was Hoffmeyer (1836–1886). He was one of the very capable meteorologists—forecasters of that time. He created and published synoptic maps of the entire Northern

Hemisphere, which were known as Hofmeyer maps. He organized the printing of the meteorological "Northern Bulletin," which covered the Scandinavian countries. For a long time, regular forecasts were issued with only three designations: "good weather," "unstable weather," and "ugly weather."

Spain

The Meteorological and Astronomical Observatory in Madrid began collecting telegraphic data from the stations in 1864. They also published a weather bulletin modeled after the Paris Observatory. Telegrams about the expected weather were sent to the main Spanish ports.

Portugal

In Portugal, as in Spain, the service began in 1865. "Traffic lights" were introduced in 1871, on which signals for the announcement of storms, which were prepared at the Lisbon Observatory, were announced. The observatory began to receive telegraphic weather reports from the Azores in 1893.

Sweden

In Sweden, the collection of telegraphic reports began in 1873. A newsletter was printed that included a "Weather Outlook" section. The Swedish Meteorological Institute has long avoided preparing storm reports. They started with such information only in 1905.

Serbia

The observatory in Belgrade started at the end of 1902 to issue a "general weather forecast"—a kind of today's meteorological report (Fig. 7.8). Forecasts were issued from 1902 and regularly printed in "Serbian Newspapers." They were compiled on the basis of telegraph dispatches from 43 meteorological stations from the surrounding countries and 7 stations from Serbia.

Japan

The first attempt to exchange weather telegrams between the cities of Nagasaki, Shanghai, Hong Kong, and Amoy was made in 1873 via Formosa (Taiwan). Nevertheless, the weather forecast service in Tokyo publishes the first synoptic maps on February 16, 1883. Cards were made only once a day at the beginning and later three times a day (6, 14, and 21 h Kyoto time). They issued weather forecast bulletins emphasizing the intensity and location of low- and high-pressure centers. Based on these global reports, one meteorological station in each of the 15 areas of Japan produced more detailed forecasts.

The first signal tower for announcing storms was built in 1883, in Shinagawa. Two types of weather were announced: "bad weather" and "before the storm." Later they used a whole system with various flags to announce wind, precipitation, temperature, etc.

ИЗВЕШТАЈ ОПСЕРВАТОРИЈЕ

11—12. јан. 1903.

Прогноза времена. — Хладно време; местимично снегом.

ИЗВЕШТАЈ ОПСЕРВАТОРИЈЕ

4. септембра 1903.

Прогноза времена. — Топло време, местимично слаба киша и грмљавина.

ИЗВЕШТАЈ ОПСЕРВАТОРИЈЕ

2. новембар 1903.

Прогноза времена. — Облачно, хладно време, са местимичним падежима.

Fig. 7.8 Weather forecast from the Observatory Report for 1903. The reports were published in the "Serbian Newspaper" in 1902

7.5 International Synoptic Codes

The weather services of different countries used different ways to encrypt the meteorological data that was transmitted by telegraph. Here we will only highlight the efforts that have been made to agree on international keys (codes). The first codes used Beaufort designations of two letters each (see figure with Beaufort wind speed scale). Data encryption became especially important when data began to be sent by telegraph. Since 1858, they have been using the codes of the Paris Observatory or FitzRoy, which he described in the book *Practical Meteorology* (Russian translation of the English edition of *The Weather Book*).

The first international meteorological key, or Utrecht key, comes from Wilde and Scott. It was approved for use by the Standing Committee, in 1874, in Utrecht. All European countries, except England, used the following key:

BBBDD FwTTT BBBDD FwTTT T'T'T'RR MMmmS.

England used a slightly modified key:

BBBDD FFwTT BBBDD FFwTT T'T'RRR MMmmS.

Groups of 5 digits were used, and the meaning of the labels is as follows:

BBB pressure at the time of observation; DD wind direction (from 00 to 32); F Beaufort wind speed; w time at the moment of observation; TT temperature; T'T' wet bulb temperature; RR amount of precipitation; MM maximum temperature; mm minimum temperature; Water surface condition.

Later, the keys were supplemented, especially with a group of labels for cloudiness, pressure change between two observations, and weather conditions between two observations.

During World War I, aviation developed rapidly, so the Commission for Meteorological Telegrams from 1920 to 1921 worked on the development of a new key for them. The key was adopted by the International Committee in 1921. Small changes followed in 1923. In it, special attention was paid to the detailed description of phenomena: A distinction is made between showers, uniform, and pouring precipitation. For clouds, in addition to the type, the height of the cloud base is also determined, as well as the visibility. Such a synoptic key was adopted at the meeting of the directors of meteorological services in 1929 in Copenhagen. Symbols for entering data on synoptic maps were also adopted then.

References

Barlow, N. 1958. *The autobiography of Charles Darwin 1809–1882*. London: Collins.

Byers, H.R. 1976. The founding of the Institute of Meteorology at the University of Chicago. *Bulletin of the American Meteorological Society* 57: 1343–1345.

Cohen, I.B. 1950. *Some early tools of American science*. Vol. 201. Harvard University Press.

Crawford, M., and D. Marsh. 1989. *The driving force, food, evolution, and the future*. Harper & Row.

FitzRoy, R. 1863. *The weather book: A manual of practical meteorology*. London: Longman.

Frisinger, H.H. 1967. Isaac Greenwood: Pioneer American meteorologist. *Bulletin of the American Meteorological Society* 48 (4): 265–267.

Gaigerov, S. S. 1949. *O trudakh Lomonosova v oblasti meteorologii i aerologii*. Meteorologiya i Gidrologiya, No. 4.

Khromov, S. P. 1932. *Ot Dove k sinoptike* (From Dove to Synoptic Meteorology). Zhurnal Geofiziki.

———. 1934 *Vvedenie v sinopticheskii analiz* (Introduction to Synoptic Analysis).

Kimball, H.H. 1933. Recent advances in the science of meteorology and in its practical applications. *Bulletin of the American Meteorological Society* 14: 3–7.
———. 1981. Pioneer: The first American meteorological doctorate. *Bulletin of the American Meteorological Society* 62: 362–367.
Kutzbach, G. 1979. One hundred and twenty-five years of meteorology at the University of Wisconsin. *Bulletin of the American Meteorological Society* 60: 1166–1171.
Tikhomirov, E. I. 1932. *Fits-Roi i sovremenennaya meteorologiya* (Fitzroy and Modern Meteorology). Meteor. Vestn.
Whitnah, D.R. 1961. *A history of the United States weather bureau*, 267. University of Illionis Press.
Williams, J. 1992. *The weather book*. Vintage Books.

Chapter 8
Exploring the Free Atmosphere

8.1 Introduction

Travelers have long known that clouds, stormy weather, fog, and cold winds occur in mountainous areas (Lewis 1995; Laskin 1996). The shepherds who grazed their flocks in the high mountains in the summer also knew very well the difference in climate between the mountainous and lower regions. These differences began to be interpreted using scientific methods only in the eighteenth century (Ludlum 1984).

The famous Russian scientist Lomonosov wrote in 1753 that "there is a difference in density and heating between the air at ground level and at height." It was clear to him that the higher parts of the atmosphere are less heated by the Sun than the lower, near-surface air. The detailed characteristics of mountain areas were particularly well studied by the famous Swiss naturalist Saussure (1740–1799). In addition to the Alps, he also visited other mountains. In 1772, he climbed Mount Vesuvius and in 1783, Mount Etna. On his mountain tours, he always made numerous meteorological measurements. Thus, on August 3, 1787, Saussure climbed Mont Blanc, the highest peak in Europe (4810 m above sea level). Only one man before him had climbed that inaccessible snowy peak. Along the way, he made numerous measurements. Among other things, he found that there was a lot of carbon dioxide in the air. He measured that at the top the boiling point of water was 68.993 °R. At the same moment, his son was measuring at the foot of Mont Blanc, in Chamonix, and in Geneva, Senebier. The following year, Saussure spent 17 days (July 3–19) on the Ghent pass (3370 m) near Mont Blanc. Then he measured daily changes in pressure and temperature and observed the difference in the blueness of the sky. He found that the pressure change at the height is of the opposite sign to the change at ground level. He also found that the temperature decreases with height by 1 °R per 100 toises (0.65 °C per 100 m).

Others have carried out similar measurements. Thus, by measuring in the Pyrenees in 1802–1804, Raymond calculated that the decrease in temperature with

M. Ćurić, V. Spiridonov, *History of Meteorology*,
https://doi.org/10.1007/978-3-031-45032-7_8

height was 0.49 °C/100 m. The first permanent mountain station was set up on the St. Gotthard Pass (2100 m) within the Mannheim Society. The measurements were taken by priests in the hostel belonging to the Capuchin Order. When Saussure visited the station in 1783, he found that the barometer was not well placed in relation to the scale, so it was showing up to two lines higher. The great German naturalist A. Humboldt paid great attention to the climatic characteristics of the mountains during his travels in South America from 1799 to 1804. According to the types of vegetation, he determined the climatic zones of the Andes. Through measurements, he established that the temperature drops faster with height in the area of high peaks (Chimborazo, 6310 m) than in the mountain massifs around Quito and Bogotá. Through these measurements along the mountainsides, the first knowledge about the change of meteorological parameters with height was obtained.

8.2 Measurements Using Kites

Kites played an important role in the development of aerological measurements, as they were later called. Legends say that the dragon was found by Architas of Tarantum (around 400–365 BC). Since many centuries ago, dragons have been used as toys in China, Japan, and Thailand but also in New Zealand. The use of kites for scientific purposes dates back over 250 years. Thus, in 1749, Franklin used a kite to measure the electricity associated with cumulonimbus clouds. At the same time, from 1748 to 1749, Willis (1714–1786), a professor from Glasgow, was the first to tie a thermometer to a kite. The measurement results were not satisfactory. On July 1, 1754, in front of the Russian Academy of Sciences, Lomonosov presented a machine he had found himself, the wings of which were driven by a clockwork spring. He called it the "airport machine." Meteorological instruments were placed in it and were supposed to be raised by a spring drive. Unfortunately, the spring turned out to be too weak to carry such a load.

In 1756, Euler wrote a theoretical paper on kites, but he failed to find a general rule for how high a kite can be raised at a given wind speed. In 1823, Captain Perry tied a thermometer to a kite on the Arctic island of Iglolik. Admiral Beck, commander of the ship "Terror," worked in a similar way during the Arctic expedition of 1836–1837.

Meteorologists in America (Marvin, McCaddy, Roth, and Clayton) often made measurements using kites (Reichelderfer 1928; Loomis 1841). Great attention was paid to the construction of kites. Some of them are shown in Fig. 8.1.

The first systematic measurements using a kite were made at the "Blue Hill" observatory in America. A light thermograph was attached to a balloon for the first time on August 4, 1894. A little later, they did the same with the thermoanemograph and in 1896 with the barothermograph. Many technical problems were solved to get the best kite that can carry thin printers to a height. By February 1897, 112 successful kite flights had been made. The highest height reached was 2665 m, and on July 19, 1900, the kite climbed to a height of 4820 m above sea level.

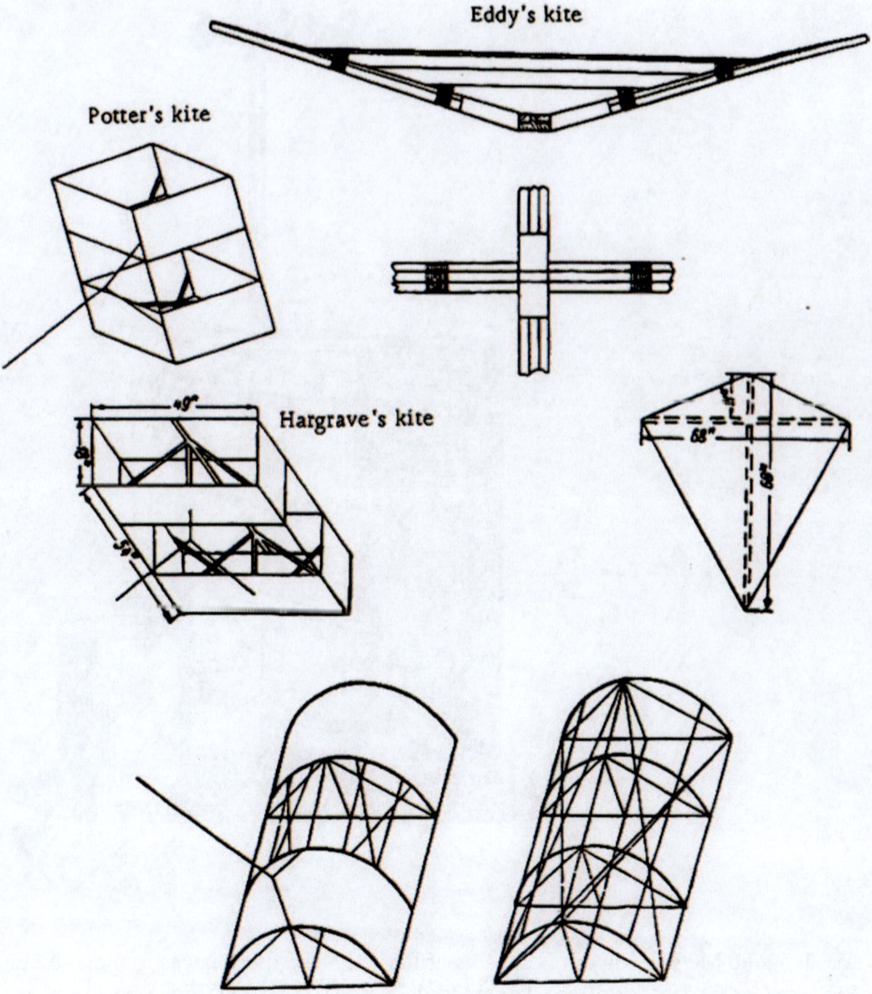

Fig. 8.1 Different types of "weather" kites

And at the Pavlovsk Observatory, they worked on aerological measurements using the power of a kite. The famous constructor Kuznetsov worked there. By 1897, they had made more measurements with different types of kites. The heights reached were 2–3 m. With later modifications, they reached heights of up to 4500 m (in 1903). At that time, kite measurements were also carried out in other Russian cities (Kazan, Uljanin). The Russians also successfully launched kites from the hills in the Gulf of Finland (in 1904). Kuznetsov constructed a meteorograph, a recorder (writer) of several meteorological elements that was attached to a kite (Fig. 8.2).

There have been many successful measurements using kites in various parts of the world. There are no specifics why they should be mentioned. However, it is necessary to mention the activities that were carried out in this area at the Observatory

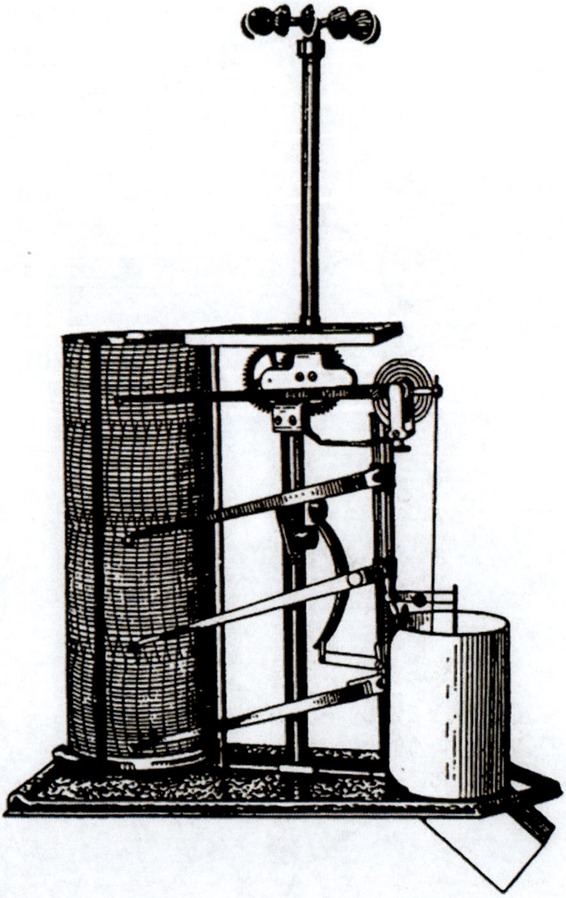

Fig. 8.2 Kuznetsov's kite meteorograph

for Dynamic Meteorology in Trapps, near Paris. The institution was founded by the French scientist Leon Philippe Tisserand de Bor (1855–1913) (Fig. 8.3).

He invested almost all the wealth he had there. Tisserand de Bohr has been measuring with kites since 1897. When Roth achieved the height record with a kite at the "Blue Hill" observatory (4820 m), de Bohr successfully achieved the record of 5200 m above Trappes.

The first kites in Germany were used for military purposes. That is why a large sum of money was invested in this activity of the Prussian Meteorological Institute. A total of 5691 kites flew from Lydenberg. Important scientific discoveries were made with dragons at the beginning of the twentieth century. From 1904 to 1905, Prince Albert of Monaco organized kite measurements in the Mediterranean and Atlantic. Hergsel, president of the International Aerological Commission, managed the measurements from his yacht "Princess Alice." In 1905–1906, Roth and Tisserand de Bor performed various aerological measurements from the ship "Otaria" in the tropical part of the Atlantic. They assumed that these measurements

Fig. 8.3 Roth and Tisserand de Bor (right)

could show some specifics. And, indeed, for the first time, meteorological measurements established the so-called inversion above the trade winds.

For centuries, meteorologists have observed the movement of smoke from the Cotopaxi volcano in Ecuador and from West Indian volcanoes. From that and ground measurements, they created a diagram of the flow of the atmosphere in tropical areas. In the northern hemisphere, relatively constant northeast winds (trade winds) blow in that area on the ground floor, below a height of about 3 km. Above that height, there is a zone where southwest winds (antitrade winds) blow. Measurements on "Otaria" and "Prince Alice" from 1904 to 1906 showed the difference in trade winds over the eastern and western parts of the Atlantic. A trade wind inversion has been detected in the eastern Atlantic. The layer of air with the trade wind humid northeast (NE) wind is shallow, from 300 to 1500 m. Above that height, the temperature rises to 30°, and the humidity drops to almost 0%. Tisserand de Bohr established that such an inversion does not exist in the western part of the Atlantic.

At the end of the nineteenth and the beginning of the twentieth century, measurements using kites made it possible to reach some important conclusions. Thus, in 1897, meteorologists in America established six profiles of temperature change with height. For example, the third type is characterized by a temperature inversion in foggy weather from 100 to 400 m and then decreases much more slowly than the adiabatic change. This situation is related to a warm front. The fifth type corresponded to a cold front, or the conditions in which cumulonimbus clouds develop. It is characterized by a rapid decrease in temperature with altitude. Clayton stated before the scientific society in Boston, according to measurements by kites, that when a cold wave comes from the NE or E, it begins to come above the ground, first as a thin layer, and then thicker. This is a correct conclusion that meteorologists, unfortunately, neglected for a long time, until after the First World War, when the model of cyclone fronts was formulated.

8.3 Measuring with Manned Balloons

The first measurements of meteorological elements with height in the free atmosphere (away from high mountains) were made using balloons (Hughes 1981; Hughes 1970). The first manned balloon flight was carried out by the Montgollier brothers on September 19, 1783 (Teisserenc de Bort et al. 1904). They ascended in a balloon made of cardboard filled with warm air, from Paris' Place de Versailles. The balloon was 17.4 m high and 12.5 m wide. It was until then the biggest attraction in the world, which attracted almost all Parisians (Fig. 8.4).

Fig. 8.4 The first balloon flight in Paris

In the same year (1783), Pilatre de Rosier and the Marquis de Arland took off from the Boulogne Forest Park in the Montgollier brothers' balloon. A large number of observers also gathered. King Louis XVI and Marie Antoinette were among those present. The king demanded that two convicts make the flight (because he considered it a life-threatening flight), but Rosier refused because he did not want to leave the glory to them. The balloon landed successfully, and the king presented Rosier and Arland with decorations. After that, they were celebrated as national heroes.

Also in 1783, on December 1, Jacques Charles (1746–1823), a well-known professor from Paris, brought instruments with him. He flew in a rubber balloon called "Globe," whose diameter was 9 m. The balloon is filled with hydrogen. The filling lasted for 4 days, during which 540 kg of iron filings and 270 kg of diluted sulfuric acid were consumed. His friend Nicola Roberts flew with Charles. The balloon took off from Tuileries Park. After a flight that lasted 25 min, they landed in a meadow, about 40 km northwest of Paris. At the highest altitude, the temperature was −8.8 °C. According to the barometer, it was at a height of 3467 m. Later, Charles alone rose to an altitude of over 1 km, where the temperature was −5 °C (13 °C lower than ground temperature).

Balloon soundings attracted the attention of both scientists and the general public. Thus, Euler, just 1 day before his death (August 18, 1783), dealt with the problem of the ascent speed of an "aerostatic machine." Lavoisier prepared the experiment in detail for the Montgollier brothers. Scientists were willing to pay huge sums of money for each flight to balloon manufacturers. So for the flight over London, on November 30, 1784, the American doctor John Jeffrey paid 100 guineas to the French balloonist Blanchard.

Jeffrey described the balloon flight and the preparation process in detail. A special barometer with a scale of 18 inches (609 mb) was made. Geoffrey, on Cavendish's advice, had brought several bottles filled with distilled water, in order to bring samples of the air from the heights. At the appropriate height, he would spill the water (then the air would enter) and cap the bottle. Before the main balloon took off, he released a smaller balloon (without a crew) to determine in which direction the balloon would rise (to properly position the theodolites). Geoffrey and Blanchard climbed to a height of 2740 m, where the temperature was −1.9 °C (at ground level it was 10.6 °C). On the second flight, January 7, 1785, Geoffrey and Blanchard took off from Dover, crossing the English Channel.

From 1803 to 1804, a certain Robertson made several flights from Paris, Hamburg, Riga, and Petrograd. For him, it was more of a circus than a research job. Despite this, he was hired by the Russian Academy of Sciences in Petrograd to make several flights for scientific purposes. Zakharov (1765–1836), who later became a member of the Academy, flew with him on July 14, 1804, in Petrograd. The purpose of the flight was to examine the state of the atmosphere with altitude. It was supposed to be different from what Saussure and Humboldt got. Zakharov took air samples and examined the propagation of sound, the charging of objects, and the behavior of a magnetic needle. He found no differences in height compared to ground measurements. In addition, the standard meteorological characteristics of the air were measured. He measured that the northeast wind at ground level gradually changed to easterly at some height. This is in agreement with the much later derived law on the change of wind with height in the boundary layer of the atmosphere.

In Paris in 1804, Laplace and Bertolet proposed to the Academy of Sciences to conduct some scientific research using balloons. It was necessary, among other things, to verify Robertson's claim that the magnetic force decreases with increasing height. Thus, on August 24, 1804, two young scientists, Gay Lissac and Biot, made a balloon ascent to a height of 3980 m. They showed that Robertson's data was incorrect. They found that the air is drier with altitude. During the second climb, they noticed that the tank of the thermometer, which was attached to the cabin, was heating up significantly from the sun's rays. This meant that mistakes were made in meteorological measurements, which had not been taken into account until then. Much later, in 1850, two French scientists, Baran and Bixio, measured a temperature of −39 °C at an altitude of 7 km, which surprised them. When descending through the cloud, they made measurements, when there was no influence of direct solar radiation.

During his second flight, on September 16, 1804, Gay Lysack reached a height of about 7 km. In addition to the measurements needed to investigate the nature of the atmosphere, they carefully observed the behavior of the animals they brought with them (Fig. 8.5). They took air samples.

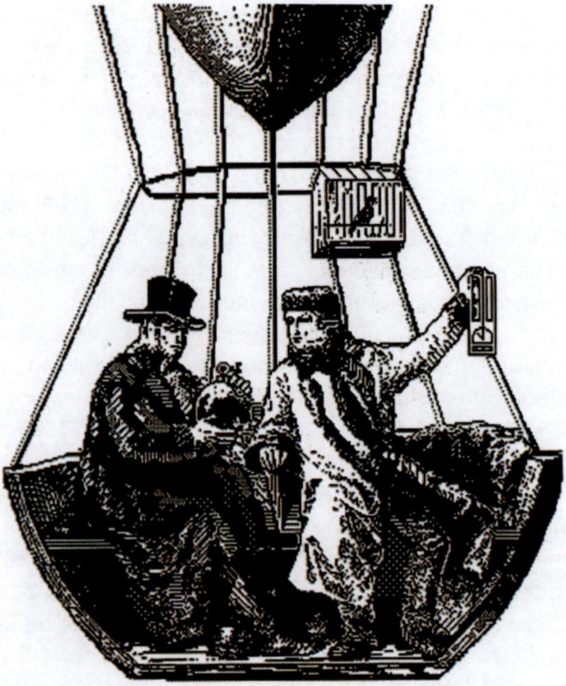

Fig. 8.5 Gay Lisak and Biot during a balloon flight

Later analysis showed that the composition of the air at ground level is the same as at height. Oxygen and nitrogen are mixed in the same ratio. The temperature decreased from the ground floor to 7 km by 37 °C. It is interesting that it roughly coincided with Laplace's calculation. Based on measurements of the astronomical-refraction, he calculated that the temperature in that layer should drop by 40 °C.

Balloon flights have continued to this day, and they are always attractive for something: flying around the Earth in a balloon without descending, achieving height records, etc. All in all, today, balloon flight does not have the main purpose of measuring the state of the atmosphere. However, such a flight always has some scientific interest. We will therefore cite the example of US Army captain Joseph Kittinger, who in 1960 reached a height of 34.3 km over the Mexican desert with a balloon. He was sitting in a gondola attached to a balloon. He was dressed in a pressure-controlled suit (like in passenger planes). From the highest height, Kittinger jumped out of the gondola. When it was 6 km above the ground, it was falling at a speed of 380 km/h. Then the parachute opened automatically. After 9 min, he landed safely on the ground.

Twenty years later, a young American tried to fly to a height greater than that at which passenger planes fly (about 10 km). With the help of several balloons filled with hydrogen, he reached a height of about 10 km, where the temperature was below −30 °C. Almost frozen, the young man barely managed to puncture some of the balloons with a shot to begin his descent toward the ground. Along with this barely surviving the adventure, he was arrested upon his return to earth for violently violating US airspace regulations.

After Gay Lissac's flights, further use of balloons for scientific purposes was no longer interesting. From 1838 to 1843, Rush spent three summers in England. By ascending on September 10, 1839, he was the first to experience what is now called "airplane icing." The objects were covered with a layer of ice. For the history of meteorology, it is interesting to mention that in 1862 Welsh, the director of the famous Kew Observatory in England made four flights in a balloon. It reached a height of about 7 km. He was the first to use an aspiration psychrometer (dry and wet thermometer with ventilation) in flight. Between 1862 and 1866, Glacier (a famous meteorologist) made 28 flights over England. He collected a huge amount of data. He took 18 thermometer readings per minute. He constructed a specially sensitive thermometer for these needs, which is known as a "grid thermometer."

On September 5, 1862, Glacier reached the highest altitude with co-pilot Kouhvel, but it almost led to their death. Namely, at a height of 8840 m, Glacier lost consciousness due to lack of oxygen. During this time, Couchwell did not notice that the balloon had risen to a height of 11,300 m. And Couchwell completely lost his strength. With his hands, he could not tighten the rope used to close the valve that regulates the descent. Yet somehow he managed to do it by pulling on the rope with his teeth. Then the balloon began to descend. At the last moment, they managed to save themselves from a horrible death. And after that Glacier made several flights. It is not known exactly to what height the balloon rose on September 5, 1862. All measurements were recorded up to a height of 8,840 m, because after a height of 11,300 m, the balloon still had to rise, while Kouhvel struggled to close the valve.

And at the Main Physical Observatory in Petrograd, Rikachev performed aerological measurements with a balloon (Molchanov 1933). He used a thermometer with a spiral-shaped tank to show the temperature of the environment more quickly. Gleshir proposed this thermometer to Rikachev. He had his first flight on May 21, 1868. The balloon was piloted by the Frenchman Banel. The flight from June 1, 1873, is particularly interesting. The balloon took off from a meadow near the Pavlov Military School in Petrograd. It reached a height of 4046 m above sea level, where the temperature was −8 °C. When rising, the balloon was first carried to the west and then returned to the southwest. Because of this return, the balloon did not get very far from the city. The balloon's position was monitored in detail using theodolites from Pulkovo, Kronstad, and Petrograd. Accurate measurement of the balloon's position using a theodolite made it possible to calculate the height of the balloon, and the wind speed could be calculated from the horizontal displacement. Thus, the accuracy of the visually determined speed recorded by Rikachev could be assessed. At that time, the relationship between altitude and pressure was known. Thus, the height could be determined from the barometric condition, but this estimate was not as accurate as that obtained with theodolite.

Within the framework of the Russian Technical Society, in 1880, a section for aerostatics (the so-called VII section) was founded. The famous Russian scientist Mendeleev (author of the famous Periodic Table) also participated in numerous flights. He took measurements from a balloon in 1887 during a solar eclipse. The balloon took off from Klin and reached a height of 3350 m.

It is very instructive at this point to say something more about how this great and generally known scientist viewed the development of meteorology. This can be shown through his two attitudes. The first is illustrated by the following example: Mendeleev, based on Glacier's balloon measurements of temperature and pressure with height, wrote a formula about the relationship between temperature at height h, th, and pressure ph (and all of this expressed using those values on the ground floor, marked with the index zero).

$$t_h = c + \frac{t_0 - c}{p_0} p_h$$

About this formula, there was a big discussion between Rikachev (deputy director of the Observatory) and Mendeleev. Despite the visible shortcomings of this formula, it shows the main thing that Mendeleev wants to change in meteorology. He wants to find a law from the mass of data, a law from which a simple number can be obtained. That is why Mendeleev says: "Numbers, and only numbers, will lead meteorology to true science!". He wanted to derive universal laws of time from numbers. Second, Mendeleev sought to point out the importance of the universal laws of nature. He says in one discussion:

"I want to direct your attention to the main problem of meteorology – the battle between science and superstition. Many still believe that the weather is the result of something random, some unknowable and unpredictable situations, and are surprised when it is said that there are some external causes that determine the state of the weather. This is happening because scientific ideas about the weather are not yet well developed... The age is coming when not only accurate weather forecasting will be possible, but the weather will be completely controlled. In the future, man will be in a situation to fight for meteorological elements that suit him."

These words of the wise Mendeleev, spoken more than 120 years ago, oblige meteorologists to fight against superstition. Therefore, he noticed that the ultimate goal of meteorologists is not just to forecast the weather well but to be able to change (modify) the weather at will.

Balloon measurements were especially popular in France. Flamarion, de Fonviers, and Tissandire did this for a particularly long time. They were called the "triumvirate." During the ascent, they observed many interesting processes in the clouds, the wind, and optical phenomena. Thus, for example, on August 16, 1868, Tissandir circled the coast above Calais. He established that the wind blew from the sea toward the land up to a height of 400 to 500 m, and from 600 to 1600 m the wind blew from the opposite direction. It was a well-known land-sea circulation cell. Tissandir found a strong temperature inversion above stratus clouds. In 1873, he measured a temperature of −2 °C in the cloud and +17.5 °C above it. Later, many other aeronauts (people who fly in balloons) received similar data. It was an unexpected and unusual occurrence. Then Tissandires and Fonviers established that the current of air under a thick layered cloud is completely opposite to the movement inside the cloud. The discontinuity (rapid change) of the flow direction was so pronounced that one of the ropes holding the nacelle to the balloon broke.

The "Triumvirate" described its flight results in 1870 in the voluminous book *Journey in the Air*, and a little later, in 1878, in the book *History of My Ascensions*. Along with scientific results, they contributed to the popularization of height measurements. De Fonvier became secretary of the International Aerological Organization.

Lifting by balloons was an expensive and often dangerous business. That is why scientists tried to find other methods to study the free atmosphere. That did not come true immediately, so climbing up the mountainsides to the highest mountain peaks became relevant again.

The word "aerology" is very old. It was used in 1642 by a certain Domenico Panirolo in a book published in Rome entitled *Aerology, or the Discourse of the Air, a Useful Treatise on Health*. However, Keppen gave this word today's meaning. At the conference of the International Commission of Scientific Aeronautics (later called the Aerological Commission) held in Milan in 1906, Keppen proposed "that the field of meteorology that uses aeronautical measurements to study the free atmosphere be called Aerology." Since then, this word has spread rapidly, and scientists have used it.

8.4 Atmospheric Sounding Using Aircraft

Immediately after 1910, sounding measurements using airplanes began. In 1915, the Danish Meteorological Service introduced routine soundings of the atmosphere up to a height of 4.5–5 km. Chief aviator J. Bakens was trained to fly, even though nimbostratus was without special instruments for blind flying. A meteorograph was hung from the wing of the plane, which recorded atmospheric pressure and temperature. After reaching a height of 4–5 km, the pilot would quickly land in order to provide fresh data. This method of probing was maintained between the two world wars. Before that, the meteorographer was attached to unmanned balloons, which reached a height of up to 18 km. By the way, the balloon was found by Montgollier, way back in 1873.

Even after the First World War, airplanes were used as a means of sounding. The question arises why they used such a complicated and expensive tool. There are several reasons for this. First, the use of aircraft provides more reliable and detailed data than other means. The thickness and visibility of the clouds could be determined by flying the plane directly through them. The measurement data were usable within 1 h of the measurement, unlike the balloons, which took several days to be used. Kites and tethered balloons posed a real danger to communication and power lines. Kite cables were also a danger to aircraft.

Therefore, in the Dutch town of Sustenberg, in May 1919, regular aerial soundings began. Fokker-Wulf VII aircraft was used. Meteorologists were tied to the planes. By 1929, more than 3000 flights were made here, 1800 of which reached a height of over 5000 m (but they did not fly at a height higher than 5480 m).

Since 1916, airplanes have been used for these purposes in Russia as well. This was done at the initiative of Friedman, who had in mind the study of winds and turbulence. At the suggestion of Assmann and Wegener, aerial soundings have been carried out in Germany since 1912. This was especially interesting during the First World War. Systematic measurements were organized by Wegener in 1922 from the Lindesberg observatory. At about the same time, similar measurements were made in France, Sweden, Norway, and other countries. Aircraft measurements are less and less used after 1930 when the use of radiosondes begins.

8.5 Unmanned Balloons

It is an old idea that meteorological measurements are made by unmanned balloons. One such plan was presented by Lomonosov in 1754. In 1783, in front of the Paris Academy of Sciences, Lemonier demonstrated how to attach a thermograph to a balloon. Both of these ideas were not realized.

During the Franco-Prussian War (1870), the French tried to send a written message using unmanned balloons from the besieged cities of Paris and Metz. Much later, meteorological instruments were sent into the free atmosphere by balloons. In March 1892, Hermit (nephew of a famous mathematician) began experimentally releasing small balloons (volume about 1 m^3) from Paris, which he called "ballons perdus"—lost balloons. Attached to each balloon was a label asking the finder to inform the Hermit in Paris of the place where the balloon was found and the method of landing. In this way, he was able to reconstruct the path of the balloon.

In connection with the trajectory of small balloons, there is an interesting story from our region. It tells about the incredibly long journey that one balloon traveled, and about the incredible, but completely true combination of various events. The story begins and ends in the mountain village of Grčić (1117 m above sea level), 25 km from Ljubovija. Tiosava Lejić was born there. Radiša lost her father early (he died in 1949) and her mother Ljubica in 1956. She finished primary and secondary economic school, thanks to the Milutinović couple, teachers with whom she helped in the house while studying. A newspaper ad took her to London to look after the children, while their parents worked. And there she studied while working. She soon married the Englishman Ken Mantel, a geologist. They gave birth to three children, sons Sasha and Misha and daughter Tanja. They served in various countries of the world, but their husband Ken liked the village of Grcic the most. There they built a nice house (cottage) on Tiosava's land near the spring of Vinca water.

In the summer of 2000, while they were staying there, Comrade Stanko Pavlović came to their house. He showed Ken a piece of paper that said London and some other content in English. The neighbor said he found that pa-pir near Ken's cottage. When the Mentell family returned to London, they discovered, using the phone and the address on paper, that it was a company that made balloons. According to the number with which the balloon was marked, it was established that it was a balloon that was released from the yard of the school in London where their son Misha attends. What a surprise it was for the balloon maker, for the balloon to travel so far before bursting. For the producers, and even more so for the Mentels, it was a surprise that a piece of balloon sent from Misha's school with a sticker fell next to their house in the village of Grcic.

That knowledge was a good advertisement for the company for the quality of the balloons. They raised the Mentel family, and with that money, they bought material for filling the teeth of elementary school students in the village of Rogačici near Bajina Bašta. What a collection of true, and statistically speaking, incredible events!

It should be noted that this was not the first gift to the Mentel family to health institutions in Ljubovia and Bajina Bašta. The gifts followed when Ken's democratic government from London, as unquestioning followers of the largest democracy in the West, in 1999 "mercifully" destroyed the nation to which his wife Tiosawa belongs with aerial bombs and tomahawk missiles.

Hermit went on to construct larger paper balloons. After several unsuccessful launches, finally, on September 17, 1892, he launched a paper balloon with a volume of 26 m^3. The balloon carried a mercury barometer with a recorder. After a short flight, the balloon landed because of the rain that started to fall. Later he made a balloon of 15 m^3 and a smaller one of 0.37 m^3. They were made of lacquered paper or of animal membranes (bladder). The balloons are filled with gas. One balloon was successfully launched on October 11, 1892. It reached a height of 1200 m. Then on November 2, 1892, he launched a balloon with a barometer. It reached a height of 8700 m and descended 150 km east of Paris. They found a simple way to show minimum and maximum temperature and minimum pressure. Hermit, together with Besançon, launched a balloon with a volume of 113 m^3 on March 21, 1893. They called it "Aerofil I." Unfortunately, at an altitude of 12,500 m, at a temperature of −51 °C, the ink in the thermograph froze, so that in the further ascent up to about 16 km, there was no temperature registration. This showed the advantage of these balloons compared to those with a human crew. They were also very cheap compared to the manned ones. Later Aerofil II reached a height of 15.5 km, where the temperature was −70 °C.

Scientists have noticed that it is possible to measure high altitudes with these balloons. That is why this method of measurement quickly spread to other countries (Russia in 1893, Germany in 1894, etc.). The first simultaneous soundings were made in the night between June 14 and 15, 1893, from Stockholm and Berlin. Later, more cities were included (Petrograd, Gothenburg, and Warsaw). At the International Meteorological Congress in Paris, held in 1896, it was presented to the International Aerological Commission to standardize and regulate all such research on a scientific basis. The members of the commission were Rikachev, Pomoretsev, Roth, Hermit, Asman, Ketelet, Besanson, Jobert, Andre, and Erk. Hergsel was president and Fonvier's secretary. Glacier and Tisander were honorary members.

The Commission's first job was to organize international soundings by balloons and aerostats. It is the first step toward synoptic aerology, the simultaneous analysis of the state of the atmosphere over large areas. The first such measurement was carried out on the night between November 13 and 14, 1896. The measurements were performed at night, due to the reduction of the effect of direct solar radiation on the instruments. Taking air samples of about 6 liters at an altitude of 15 km showed that the composition of the air is the same as at ground level.

Measurements in the upper atmosphere allowed meteorologists to understand the shape of clouds. Stratus and stratocumulus clouds were observed to form below the inversion, but this could not be explained for a long time. Only later was it shown that the inversion acts as a turbulence cap. In 1908, Herven confirmed the idea that altocumulus castellanus formation requires a large vertical temperature gradient in the air layer.

In the summer of 1908, an Italian aerological expedition was sent to the equatorial part of Africa. They were part of the crew of the ship "Caprera," which was sailing near Zanzibar. The measurements were unsuccessful due to bad weather but also due to poor technical means. On an international day, July 30, they released a tethered balloon (which was held on a cable) from the Palazzo. It reached a height of 4910 m. During the night between July 31 and August 1, they released another tethered balloon. However, the cabal broke, and he became "free." It climbed to a height of 6630 m and then fell to the ground, together with the meteorographer. It was found on the island of Pemba. The pilot balloon measurements were not successful either.

In the same year, a German aerological expedition was sent to Uganda. The team led by Berson was well-equipped. They came to Lake Victoria via the railway line, which had just opened at that time. That was their base. From July 25 to November 3, they made 24 measurements with a tethered balloon and about 70 pilot balloon measurements. They also made 70 measurements with kites on Lake Mombasa and in Dar el Salem (Fig. 8.6).

The expedition established a strong tropical monsoon (summer monsoon of the Northern Hemisphere) through measurements. They also observed westerly winds at heights of 12–20 km above equatorial Africa, which was previously unknown.

Fig. 8.6 Launching of a tethered balloon on Lake Victoria, August 1908

Measurements in Batavia from March to September showed the presence of easterly winds at a height of 18–23 km.

Berson's expedition studied in detail the daily shift of winds from and over Lake Victoria. They found that at an altitude of 16–17 km, the vertical temperature gradient decreases sharply. It was apparently the tropopause—the transition layer between the troposphere and the stratosphere.

8.6 Pilot Balloons

It has been observed that smaller unmanned balloons can be used to measure wind speed. Such balloons are called "pilot balloons." They are filled with light gas, and their position is monitored using a theodolite. Before that, air speed at altitude was measured by tracking the movement of clouds. It is the so-called nephoscopic wind measurement method. Various methods were developed in the nineteenth century for nephoscopic measurements. The movement of clouds can be monitored on horizontal mirrors, or with special theodolites. In 1894, Pomorcev constructed a special theodolite for determining the direction and speed of cloud movement.

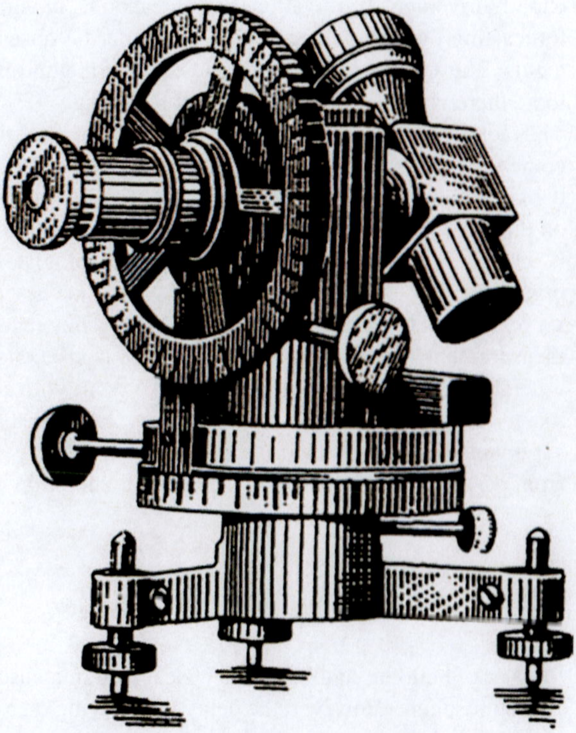

Fig. 8.7 Kuznetsov's theodolite

Pilot balloon measurements were organized by Tieseran de Bor in Trapps between 1899 and 1902. He was able to follow them even at a distance of 80 km. This monitoring was made easier when the ordinary astronomical theodolite was changed to a specially constructed one for this purpose. One of these was constructed by Kuznetsov (Fig. 8.7).

The measurements allowed Tiseran to conclude that meteorological variables, especially temperature, change with height. He called the lowest layer of the atmosphere, where the temperature decreases with height, "troposphere," after the Greek words "tropos," which means—change, and "sphere"—ball. He also called the layer of the atmosphere above the troposphere, where he expected the "stratum"—a layer composed of lighter gases, the stratosphere.

Measuring with pilot balloons, due to its simplicity and inexpensiveness, was popular since 1906. These measurements have been continued until today, practically without any changes.

From 1911 to 1917, pilot balloon measurements were made mainly for the needs of aviation. That is when aviation began to develop suddenly, for which the wind was the most important element, in addition to cloudiness, of course, which was easily determined visually. In particular, military aviation was developed. However, we should not forget that this measurement also served to improve weather forecasting. For example, in Germany in 1911, Bethel published a paper entitled "Results of

Cloud Movement Observations at Hildesheim," in which he gave practical rules for forecasting the arrival of cyclones based on the observation of cirrus cloud movements. The wind carries such clouds, so if the wind is measured using a pilot balloon, there is good data for weather forecasting.

Tracking cirrus clouds allowed sailors to predict the arrival of so many unwelcome typhoons. So Kartsev, the first officer of a Russian merchant ship, observed the movement of cirrus clouds in the vicinity of Taiwan on long voyages, and based on that, he forecasted the vertical speeds of the typhoon.

During the First World War, many meteorologists served in air-naval units and organized pilot balloon measurements. So Rikachev and Friedman volunteered to serve in a unit based in Jablan, near Warsaw. They regularly performed pilot balloon measurements there. They found that there is a secondary maximum wind speed at a height of 300–500 m. Molchanov also dealt with methods for correcting errors made during pilot balloon measurements. He also, based on the analysis of measurements, found that the wind changes with height in a clockwise direction in the front part of the cyclone and in the opposite direction in the rear part of the cyclone.

8.7 Upper Air Sounding

The use of balloons and kites was widespread in measuring the characteristics of the free atmosphere. However, the delay in receiving data and often their complete loss required the search for new measurement methods. The beginning of this new approach was made by the Russian scientist Popov, the inventor of the radio. On May 7, 1902, he gave a lecture in Petrograd, entitled "On the atmospheric electric field intensity recorder for sounding balloons and kites." This work was published in the journal "Meteorological Journal" in 1906. He explained the principle of functioning of the new device: "The sheets of the electroscope are connected to the radioactive collector; if there is an electric field around the electroscope, the sheets will receive the collector potential and will spread. By touching the electroscope, they descend. The number of discharges per unit of time will be proportional to the intensity of the electric field around the collector. Recording the number of discharges is very important for instruments that are on balloons and kites."

With this in mind, in 1923, Pavel Molchanov began to construct a radiosonde for measuring temperature and air pressure with height. Molčan's work ended in 1930 when he successfully launched the first radio station from the Pavlovsk Observatory. A radio transmitter and decoder connected to the balloon carrying the meteorographer allowed aerologists to obtain data without delay. With radiosondes begins a new era of using aerological data in synoptic meteorology, aviation, etc.

Today, radiosonde balloons reach a height of about 30 km. The height depends on the quality of the balloon (its elasticity, because it expands with height). Data on pressure, temperature, and humidity are transmitted from a probe attached to the balloon via radio to a receiving station on the ground.

The radiosondes are released twice a day (at noon and at midnight), at about 950 places around the world. Today, it is an integral part of regular meteorological measurements within the World Meteorological Organization.

References

Hughes, P. 1981. Francis W. Reichelderfer, part I: Aerologists and air devils. *Weatherwise* 34: 52–59.
———. 1970. *Mythical meteorology.* ESSA Magazine.
Laskin, D. 1996. *Braving the elements, the stormy history of American weather.* Doubleday.
Lewis, J.M. 1995. LeRoy Meisinger, part I: Biographical tribute with an assessment of his contributions to meteorology. *Bulletin of the American Meteorological Society* 76: 33–45.
Loomis, E. 1841. Meteorological observations made at Hudson. *American Journal of Science and Arts, Ohio* 40.
Ludlum, D. 1984. *The weather factor*. Houghton Mifflin.
Molchanov, P. A. 1933. *Institut aerologii GGO za pyatnadtsat'let* (The Aerological Institute of GGO over Fifteen Years) – Meteor. Vestn., No. 1.
Reichelderfer, F.W. 1928. Postgraduate course in aerology and meteorology for naval officers. *Bulletin of the American Meteorological Society* 9: 149–151.
Teisserenc De Bort, L., and I.H. Hildebrandsson. 1898 1907. *Les bases de la météorologie dynamique*, I–VII.

Chapter 9
Early Theories About Cyclones and Anticyclones

9.1 Introduction

The first barometer measurements showed that air pressure is constantly changing. Experience has been gained on the connection between strong winds and barometric conditions. Thanks to this, Otto von Gerik, the mayor of Magdeburg, became famous by announcing the coming of a storm on December 9, 1660, based on his water barometer showing. The storm did indeed appear 2 h after his announcement. For almost two centuries, the barometric condition in one place was used for weather forecasting. However, it was noticed that the pressure field must be available in a wider area.

But before cyclones were viewed through measured data, there were many beautiful descriptions of storm and cyclone weather. As early as the ninth century, Arab travelers to China recorded the devastating effects of typhoons in the Far East. Later, in the fifteenth century, European navigators (primarily Columbus and Vasco da Gama) described an encounter with a tropical storm—a hurricane. The word hurricane itself means strong wind. In the Mayan language, it is said "huran—vukan" and in Spanish "huracan." Columbus was able to predict the arrival of a hurricane over the island of Haiti in 1502, using his long experience as a navigator. A nice description of the same atmospheric formation that appeared on July 4, 1687, in the southern part of the China Sea comes from Dampier, an adventurer, and pirate. His records, although not from scientists, were very detailed, and meteorologists were happy to read them. He described the "eye" of the storm, strong winds from opposite directions, etc. From the same area, there are descriptions of storms given by Hirt from Formosa (Taiwan) in 1694. He described them in "Annals." The word "tai" in that area is used to describe a destructive wind that "lasts for several days blowing from different points of the compass." Hirt describes the tai in the Annals: "If it blows from the north, it changes suddenly, then it blows from the east if it blows from the east, it changes, then it blows from the south, and so on until it

M. Ćurić, V. Spiridonov, *History of Meteorology*,
https://doi.org/10.1007/978-3-031-45032-7_9

makes a complete circle." It does not stop until the wind makes a full circle. "It is interesting that the word Tai feng (or Tai Fung) in the modern Chinese language means wind 'with a source from all points of the compass.'"

In 1698, Captain Langford described hurricanes around the Antilles. He presented a report on them before the Royal Society in London. And he, like Columbus before him, was able to predict the hurricane that occurred on August 16, 1667, near Nevis Island (an island from the Lesser Antilles). He then suggested to the British detachment how to avoid destruction. In addition to predictions, Langford went a step further and tried to describe the cause of hurricanes. He claimed that hurricanes are formed as a result of the confrontation between the northeast wind and the trade winds.

9.2 Theories About the Origin of Storms

Here we will describe the first significant theories of cyclones, which appeared in the first half of the nineteenth century (Shaw 1926; Kutzbach 1979). Before that, it should be noted that the word "cyclone" was not yet used but "storms." Incidentally, for the first time, the word "cyclone" was described by Captain Henry Piddington (1797–1858), the manager of the museum in Calcutta, in his book *Book for Sailors*, which was published in 1842. He derived the word from the Greek word "kyklos," which means "circle," to depict the winding current of air. Later, the famous meteorologist Abercrombie, in his book *Weather*, published in 1888, introduced the cyclonic pressure field in addition to other characteristics of the pressure field.

At that time, three Americans particularly stood out in formulating the cyclone theory: William Redfield (1798–1857), J.P. Espy (1785–1860), and Robert Hare (1781–1858). They not only had conflicting theories but also strongly attacked each other's personalities. As for the storms, they had different experiences. In any case, they wanted to explain many observations. In September 1821, as an engineer, William Redfield (Fig. 9.1) made an official trip from Massachusetts to the area that had previously been affected by a strong storm. He noticed that in western Massachusetts, trees were uprooted and blown in the opposite direction from where they were near his house. This led him to conclude that the storm was rotating counterclockwise, moving northeast. The amount of data he collected about the storm of 1821 was published only in 1831. He plotted the winds on a map of the area and placed transparent paper over it with concentric circles drawn on it. He noticed that the winds are tangential to the circles. This led him to theorize that the storm is a circular vortex with cyclonic rotation. Assumed that in the southern hemisphere, the rotation of the storm is in the opposite direction. Redfield's other important claim was related to the displacement of tropical hurricanes. He argued that they originate between the Equator and the Northern Return and move in a WNW (west-northwest) direction up to 30°N. From there, they turn toward the north and northeast. Redfield argued that since the diversion, "hurricanes expand in diameter and weaken in intensity."

Fig. 9.1 William Redfield

Now it seems that his claims are easily visible. But it was not visible to his opponents, such as Espy, who studied law and was employed by the Franklin Institute. He was interested in Daniel's and Dalton's works on hygrometry. In the field of meteorology, Espy introduced the laws of nature into research. He applied the concept of adiabatic processes (which had just been discovered at the time) to explain convective processes and to explain phoenixes, daily wind changes, and storms. From 1840 to 1841, Espy visited Europe. There he was warmly received, although Europeans looked down on American science. On February 15, 1841, he was invited to give a lecture on storms at the French Academy of Sciences. Then he was presented to those present as an outstanding scientist. A special Committee of the Academy (consisting of Arago, Puglia, and Babine) was established, which judged Espy's work on storms as brilliant.

In 1845, Espy was appointed chief of the Meteorological Bureau at the US War Department. Later, he prepared about 1100 different synoptic maps that he used to analyze the characteristics of storms. Then he discovered the so-called center line of the storm, which in current terminology is called a front. All his hair dryer analysis, daily wind changes, etc. were practically forgotten.

As for Espy's theory of storms, he gradually left out of consideration (intentionally, or he did not understand) Hadley's principle of the deviation of the movement of the body from its original direction due to the rotation of the Earth. It was formulated later (in 1836) by Coriolis and in 1838 by Poisson. Forces will be discussed in the next chapter. Espy did not accept any real reason for rotating storms, namely, forces. He formulated his theory of motion under the influence of "centripetal" force. He argued that centripetal force (a force acting toward the center of rotation of a body) moves air toward the center of low pressure. Espy sought to explain that strong upward motions in the center of a cyclone cause radial motions toward the center from all sides. He did not recognize the rotational component of motion. Espy was right in one part (convergence and response in the form of upward movements). He correctly followed the processes in ascending movements "When air rises, water vapor condenses and clouds are formed." "Air is raised until its temperature is higher than the temperature of the surrounding air." We see that both theories, Redfield's and Espy's when combined, give a complete picture of storm movements.

Their third opponent, Robert Hare, argued that the buildup of electricity in the atmosphere works against the force of the Earth's gravity, causing it to move toward the center of the storm and then lift the air. This theory proved to be completely wrong, and others were left to unify Espy's and Redfield's views, having failed to do so themselves.

When the swirling nature of storms became apparent in the mid-nineteenth century, Espy's authority began to decline further. Thus, in 1855, Henry Piddington wrote: "A certain Professor Espy, who formulated the radial theory of storms, had never been at sea, nor had he looked at ship's logs or charts of different areas, where the wind was never radial." "I haven't heard anything about that theory for a long time."

The circular and centripetal theory captured the attention of meteorologists of that time. One of them is Loomis (1811–1889), a young professor from Hadron, Ohio. He visited Europe from 1836 to 1837. In Paris, he listened to the lectures of famous scientists (Araga, Poulier, Biot, Poisson). He was familiar with the Coriolis theory of deflection in relative motions (which Hadley had previously established empirically). In a paper from 1846, Loomis pointed out that storms are neither circular nor centripetal and that one can only speak of "movements directed towards the center with the tendency of the circular movement of the sun" and that "the air flows from all directions towards the center of the barometric digression, along bent paths" (Fig. 9.2).

Redfield had many supporters for his geographical characterization of storms. In 1838, he sent a proposal to all civilian and military ships sailing in the Indian Ocean to collect information about storms and send them to Captain Piddington in Calcutta (Donn 1946; Hughes 1970). From 1839 to 1855, Piddington wrote about 40 studies of hurricanes in the Bay of Bengal. He wrote, as already mentioned, in 1842, the first book with instructions on how to sail when a storm is approaching. He called the book *The Sailor's Hornbook for the Low of Storms*—a beginner's book for sailors about the laws of storms. Two transparent cornice tiles were stuck to the book.

To these, conflicting theories about the cyclone should be added one rather successful theory originating from Dove.

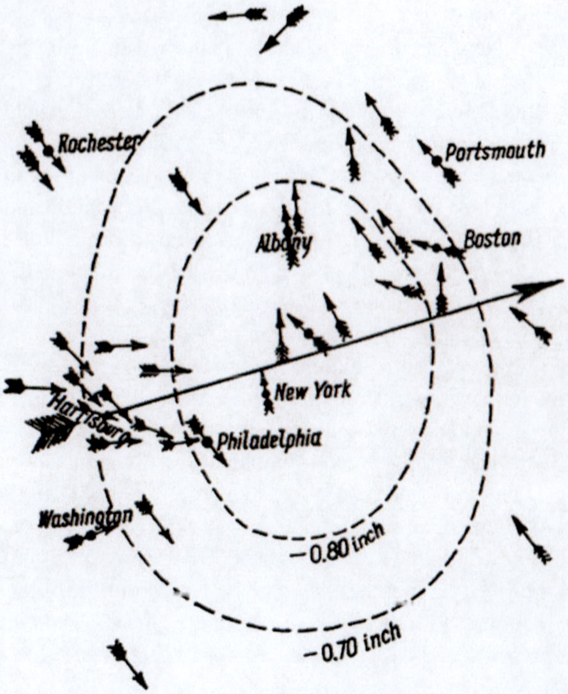

Fig. 9.2 Chart of the cyclone of February 16, 1842, near New York

Heinrich Wilhelm Dove (1803–1879)

(continued)

(continued)

He was born in Lignica and studied in Berlin. At the age of 23, in 1826, he began teaching physics in Königsberg, and in 1829, he became a professor at the University of Berlin. From 1849, after the death of Mallmann, he became the director of the Prussian Meteorological Institute, where he remained until his death. He was a person of great personal qualities. He was very eloquent and able to win listeners over to his ideas and used many generalizations and figurative comparisons. He used intuition, which he occasionally developed into fantasy. In the 50s and 40s of the nineteenth century, he influenced the opinion of many, so he is called the "father of meteorology." His ideas were accepted by most European meteorologists. It is not true, as few thought, that he was a genius that no one understood.

Dove considered the storm study method using synoptic charts, which began around 1860, to be inadequate. In 1868, he wrote: "... It must be emphasized that the cartographic representation of storms using isobars leads us to the wrong idea that almost all storms are cyclonic in shape." In 1865, he also wrote a wrong statement about the fen as a wet wind, despite already many records that it is a dry wind. Dove also did not accept the metric system of measurements. Because of this, he lost the influence he had, and toward the end of his life, he avoided contact with meteorologists. He died in 1879, and his work outlived him.

Dove was engaged in the analysis of meteorological phenomena already after 1820. Dealing with climatological analyses, he drew maps of isotherms for each month and for the whole world. He tried to recognize the fundamental problems of meteorology. Regarding cyclones, he concluded that the atmosphere is a large arena in which two currents collide. In 1828, he wrote a paper in which he pointed out that storms have a circular motion. He concluded this based on tracking one, which passed over Europe on the 24th of December 1821. During 1837, Dove considered the confrontation of polar and tropical air currents. Tropical is warm and humid, and polar is cold and dry. Midlatitude cyclones are created when currents occasionally "conflict." Dove also distinguished three types of precipitation: updraft precipitation, precipitation within the rising air above the cooling Equator, and "transitional precipitation," when warm and cold air mix.

9.3 Cyclone Path

Meteorologists who dealt with synoptic maps followed the movement of storms and cyclones. Thus Redfield found in the 1840s that the trajectories of simple cyclones are parabolic. Similarly, Algue (1856–1930), director of the observatory in Manila, found the paths of typhoons.

The study of the paths of cyclones of moderate latitudes began with Vladimir Kepen (1846–1940) (Fig. 9.3). In 1875, due to a disagreement with Wild, he left Russia and moved from Petrograd to Hamburg (Tikhomirov 1929; Kashin 1957).

Fig. 9.3 Vladimir P. Keppen

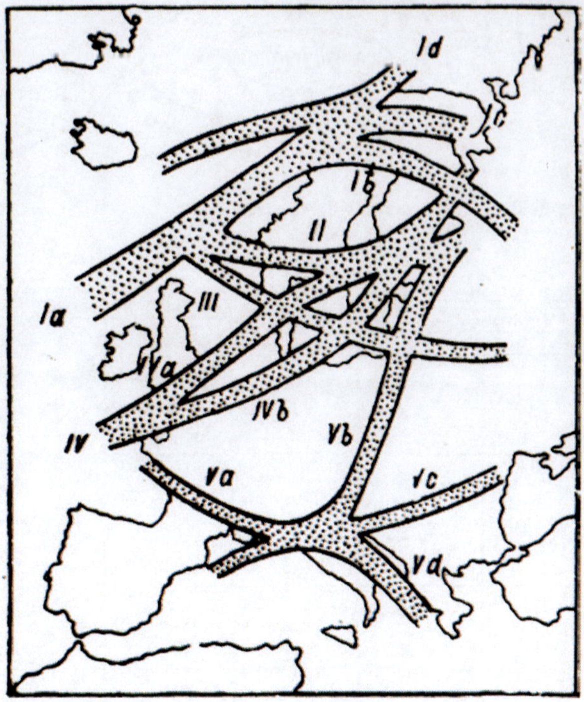

Fig. 9.4 Van Beber main cyclone tracks over Europe

There, he continues to deal with cyclone trajectories. Based on a scant amount of data, Keppen found some mean cyclone paths. His successor, Van Beber, started to deal with that business. Based on synoptic maps (Beber was head of the synoptic department in Hamburg from 1878) from 1876 to 1880, Beber drew the main paths of cyclones in Europe (Fig. 9.4). Later, from a longer period (1876–1890), he received practically the same trajectories. He tried to formulate the typical weather that accompanies cyclones on certain paths. However, a small number of insufficiently detailed data did not allow him to find anything other than the relocation of low-pressure centers.

In the popular book *Weather*, in 1888, Abercrombie presented a pressure field classification of seven basic forms: cyclone, anticyclone, ridge, valley, saddle, partial cyclone, and rectilinear isobars. He illustrated these fields on a map dated February 27, 1865, covering the Atlantic, North America, and Europe (Fig. 9.5).

The path of tropical cyclones over the Indian Ocean was analyzed by Pydington. It was a significant contribution to the study of tropical cyclones in this area (Fig. 9.6).

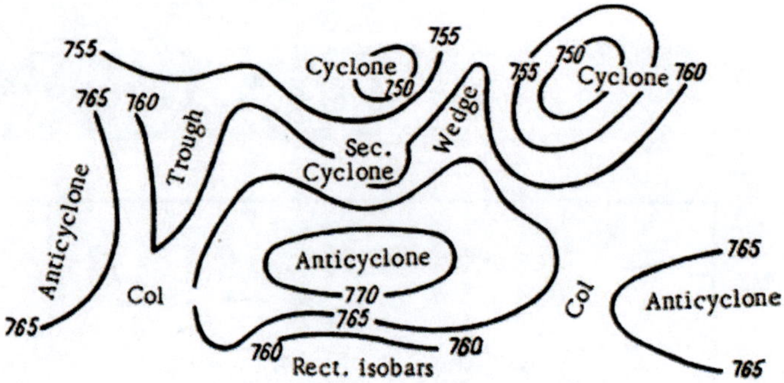

Fig. 9.5 Pressure field patterns according to Abercrombie

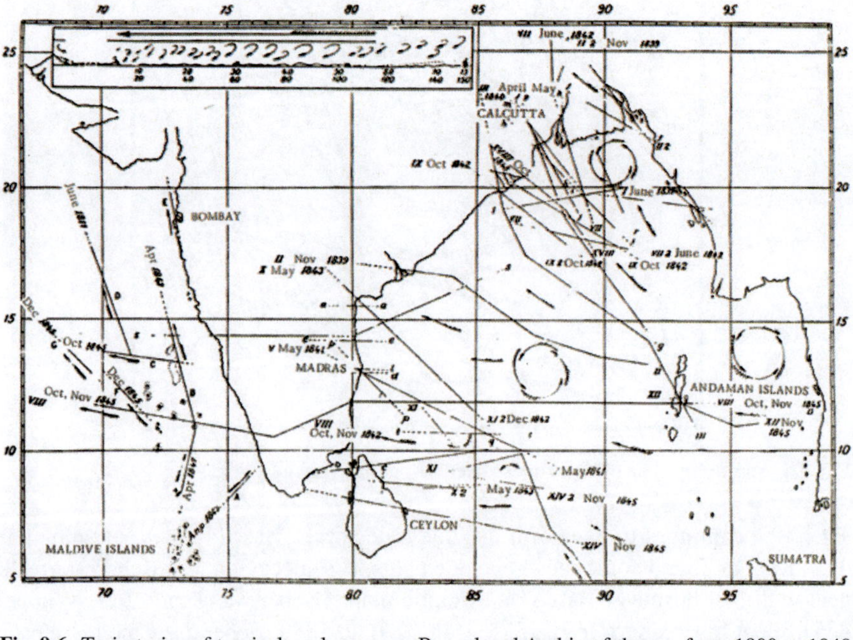

Fig. 9.6 Trajectories of tropical cyclones over Bengal and Arabia of the sea from 1800 to 1848, according to Piddington

9.4 Anticyclone

For many years, meteorologists did not pay attention to anticyclones, because the weather that accompanied the anticyclones did not change suddenly and did not pose any danger. That is why their scientific analysis was delayed at least 20 years after the cyclone. By the way, the word "anticyclone" was introduced by **Galton** in 1863. An anticyclone is characterized by weak winds, weak downdrafts, clear weather, and clockwise circulation (in the Northern Hemisphere).

The notion that an anticyclone is a creation of secondary importance in synoptic meteorology was refuted by Tisserand de Bor (1855–1913) from France. In 1880, he was the first to say that it is "the center of atmospheric activity." Namely, in 1879, the winter in Western Europe was extremely cold. For example, in Paris in December 1879, the temperature was lower than −26 °C, in Krakow lower than −30 °C, etc. De Boer tried to find a reason for such weather. That is why he looked at mean pressure charts for months. He concluded that the weather in Europe during the entire season is determined by the intensity and displacement of the "center of atmospheric activity." He also concluded that for Europe, the two most important centers of activity are those over the Azores and Siberia and the cyclone over Iceland. According to de Bor, the time in Europe depends on their dominance. He also called the anticyclone the "weather regulator."

After de Bohr's analysis, Hildebrandsson and other colleagues began to regularly monitor changes in anticyclone activity. This represented a new turning point in the creation of modern synoptic meteorology. It was the first step toward long-term weather forecasting based on scientific foundations.

References

Donn, W.L. 1946. *Meteorology with marine applications*. New York: McGraw-Hill Book Co., Inc.

Hughes, P. 1970. *Mythical meteorology*. ESSA Magazine.

Kashin, K.I. 1957. *Razvitie sinopticheskoi meteorologii za 40 let* (The development of synoptic meteorology over 40 years) – Meteorologiya i Gidrologiya, No. 11.

Kutzbach, G. 1979. *The thermal history of cyclones, A history of meteorological thought in the 19th century*. American Meteorological Society.

Shaw, N. 1926. *Manual of meteorology*. Vol. I. Meteorology in History.

Tikhomirov, E.I. 1929. *Meteorologicheskie nablyudeniya v Rossii v XVIII v.* (Meteorological observations in Russia in the 18th century) – Meteor. Vestn.

Chapter 10
Recognition of Forces in the Atmosphere

10.1 Introduction

In the second half of the nineteenth century, forces were taken into account in theories about the formation of storms. Therefore, it is necessary to give a brief overview of the contributions of scientists to this issue (Zinszer, 1944).

Even in ancient times, thinkers tried to figure out what causes the wind. Then the Greeks believed that the wind blows from the cave where the god Aeolus (*Αιολοζ*) resides. An altar was erected in Zion to pray to the winds. A special fortress was built in Delphi in honor of the wind. The stormy north wind Bura, represented in the form of a dark bird, faithfully guarded Athena. His friendly appearance destroyed part of the Persian fleet of the ruler of Xerox at Cape Sepia. In honor of Bura, an altar was built on the Ilisos embankment. According to Aristotle, any movement that is not natural occurs under the influence of a force as a driving factor. These opinions have not changed for over a thousand years. In the Middle Ages, there were still those who believed that the wind was caused by the flapping of angels' wings.

Galileo understood that the speed of a body does not change as long as external factors of acceleration or deceleration are excluded. So he knew that acceleration is proportional to the force. Wilhelm Leibniz (1646–1716) studied moving bodies. In doing so, he introduced the term "vis viva" (living force) for the quantity we now call kinetic energy. He defined it as the product of mass and the height to which a body reaches under the influence of some force. He, like Galileo, also showed that height is proportional to the square of velocity. Almost a century later, Gaspar Coriolis (1792–1843) divided Leibniz's expression by two and arrived at the current definition of kinetic energy, $\boldsymbol{E}_{\mathrm{k}} = \boldsymbol{m}\boldsymbol{v}^2/2$, where $\boldsymbol{m}$ is the mass of a body moving at speed $\boldsymbol{v}$. Christian Huygens (1629–1695) analyzed the behavior of bodies during collisions. He concluded that the sum of the "living forces" of the body before the collision is equal to their sum after the collision. In this way, he found the fundamental law of conservation of energy.

M. Ćurić, V. Spiridonov, *History of Meteorology*,
https://doi.org/10.1007/978-3-031-45032-7_10

A significant contribution in that area was made by Lazare Leonard Cornot (1753–1823), who was the Minister of War during the French Revolution. He introduced into science the term "latent vis viva," that is, known today as potential energy ($E_p = \boldsymbol{mgh}$, where m is the mass of the body located at a height h above a certain reference height and g is the acceleration due to the Earth's gravity). Cournot pointed out that any object that rises to a certain height (from its initial height) gains "latent living force" which it converts into "living force" (kinetic energy) when it falls. The English scientist Thomas Young (1773–1829), a professor at the Royal Institute in London, 1807, proposed to replace the term "living force" with the word "energy." Then, finally, William Rankine (1820–1872) introduced the names "kinetic" and "potential" energy.

10.2 The Beginnings of the Development of Hydrodynamics

The beginnings of hydrodynamics are associated with Evangelista Torricelli (1608–1647). In his work "De motu Gravium" he considered the speed of water escaping through a small opening made in the bottom of the vessel (Darrigol 2005). If the opening is made so that the jet moves upward, Torricelli noticed that the jet does not reach the height that the water has in the vessel. He concluded that it comes from the medium's resistance to water movement. If the resistance were zero, then the maximum height of the jet would be equal to the height of the water in the vessel. He also found that the velocity of the jet, v, at the outlet, is equal to the velocity that the drop would acquire if it were falling freely from the height of the top of the liquid to the bottom, h, and is $v = 2gh$.

One of Torricelli's great followers in the development of fluid science was Daniel Bernoulli (1700–1782). He published his well-known book "Hydrodynamics" in 1738. In it, he formulated the principle known as "Bernoulli's law" or "Bernoulli's equation." That principle was later further explained by his father Johann I Bernoulli in the book "Hydraulics."

Daniel Bernoulli (1700–1782)

Daniel Bernoulli was born on February 8 in Groningen, the Netherlands. He is the son of Johann Bernoulli (1667–1748). Daniel is the most famous member of a large, well-known family in the field of science. He taught mathematics like his father and older brother, Nicholas II. In 1724, Daniel wrote the book "Excercitationes Mathematicae," which attracted a lot of attention. In 1725, Daniel and his brother Nicholas II accepted the invitation of the newly founded Academy of Sciences in Petrograd to move there to work. Daniel worked there from 1725 to 1733, and then returned to Switzerland, to Basel. He finished writing "Hydrodynamics" in 1724 but published it only in 1738. His father later published "Hydraulika" but shifted the printing date to 1732, a rare instance where a father refrained from manipulating the timing to falsely claim precedence over his son. In the 10th chapter of "Hydrodynamics," Daniel formulated the kinetic theory of gases. He died in Basel on March 17, 1782.

It should be emphasized that Daniel Bernoulli was a member of a famous family of mathematicians, physicists, and other scientists of the natural sciences. His father Johann I and his uncle Jacob are the founders of differential calculus. Jakob also dealt with probability. Daniel's two brothers and numerous nephews were also mathematicians or scientists in the field of natural sciences.

The modern version of Bernoulli's law states that in the movement of fluid without friction, which is also invariant with time, the sum of potential, kinetic, and pressure field energy remains constant. This practically means that if the air is moving slower in the horizontal direction, it will cause more pressure than the faster air. By means of this principle, Bernoulli was able to explain many phenomena related to the movement of fluids. Thus, if one end of a sheet of paper is placed in a book, the other will relax along it. However, if we blow into the book while holding it in front of us, the page will rise, because the higher air velocity above the page corresponds to less pressure than below the page, which raises the page. Airplanes in the air are maintained on the same principle. Airplane wings are made so that air flows faster above their upper surface than below the lower one, and this pushes the plane up.

In 1687, Isaac Newton wrote a paper known as the "Principles" for short. In it, he laid out the laws of motion, known as the first, second, and third laws of motion. According to Newton's second law, acceleration occurs only under the influence of a force and in the direction in which the force acts. Forces are of the most diverse nature. Here, the chronology of the forces occurring in the atmosphere will be described. But before that, something must be said about the mathematical concept that made it possible to study forces (Köppen 1934).

To understand the movement of air, imagine any part of it and apply Newton's law of motion to it. Movement is caused by forces of which there is more than one in the atmosphere. Forces, air movement speed, and acceleration are quantities different in essence from pressure, density, and temperature (Kármán 1967). These other characteristics are the so-called scalar quantities (to describe them you need to know only one number, e.g., 20 °C), while the first characteristic ones are called vectors. In meteorology, vectors are expressed by means of three scalar quantities, components representing parts of the force directed in three different directions (two in the horizontal direction and one in the vertical direction). The term "vector" was introduced by Simon Stevin, aka Stevinus (1548–1620), engineer, and high officer in the Dutch army. In the book on statics and hydrodynamics, published in 1586, he introduced the concept of vectors and stacking (addition) of vectors using the parallelogram method.

10.3 Gravitational Force

Gravitational force is the result of the attraction between air particles and the Earth. The problem of gravity was considered by Robert Hooke in 1682 in his lecture "The Nature of Comets" before the Royal Society. This was 5 years before Newton published his "Principles." Hooke presented the following thought: "By gravitation, I mean that force which causes bodies to move towards each other until they unite." In 1687, Newton formulated the law of gravitation in his "Principles," as follows: "There is a gravitational attraction between any two bodies in space, and this force of attraction is proportional to their masses, and inversely proportional to the square of their distance." After publishing this in the "Principles," Hooke claimed that

Newton had stolen the idea of said law from him. Let us recall that he accused Newton and Leibniz of stealing the work on infinitesimal calculus (which Leibniz began to publish in 1684, and Newton only in 1704). Nevertheless, the opinion remained that the book "Principles" is one of the greatest scientific works ever created by the human mind. Few works, except perhaps Darwin's *On the Origin of Species*, have had such a profound influence on modern thought. The ratio between the product of the mass of the body and the gravitational force represents the gravitational constant. It was accurately determined by Henry Cavendish in 1798, measuring the attraction of two pairs of lead balls.

Henri Kevendish (1731–1810)

He was born on October 10, in Nice. His father, Lord Charles Cavendish, was the son of the second Duke of Devonshire, and his mother was the daughter of the Duke of Kent. Charles was a renowned experimenter and a distinguished fellow of the Royal Society. He often took Henry to Society meetings and dinners at the social club. Henry was elected a Fellow of the Royal Society in 1760. He almost never missed the meetings of the Society. He was interested in different fields of science. He was a very withdrawn and reserved person. He would visit everyone he met. Of the women, he only allowed his maid to talk to him and leave food for him at the end of the table. His house was full of equipment for performing various scientific experiments: telescopes, chemicals, and vessels for chemical experiments. Henry isolated hydrogen and determined the gravitational constant. Despite his great work, he published only a few works during his lifetime. Many of his inventions remained unknown until James Maxwell, a century after Henry's death, published his notes.

10.4 Pressure Gradient Force

The pressure gradient force acts from higher to lower pressure and is proportional to the pressure difference divided by the distance between the places where such pressure prevails. Because of this force, the liquid from the container with the deeper liquid will move to the container with the shallower, if they are connected at the bottom until the liquid level is equalized. Due to the force of the pressure gradient, the passengers (who, of course, died) were pushed out of the plane flying over Africa in 1996 through the open doors of the plane. Because, in the plane, the pressure that prevails on the ground floor is maintained, which is significantly higher than that at the height of the flight. When Blaise Pascal claimed in 1648 that air pressure decreases with height, many wondered how the air would not be dragged into space under the influence of the pressure gradient force. Pierre Simon Laplace (1749–1827) was the first to explain why the air does not leave the Earth. In the paper "Celestial Mechanics," published in 1823, he explained that the vertical component of the pressure gradient force is mostly in balance with the gravitational force. From this balance arises the well-known "hydrostatic law."

The difference between the vertical component of the pressure gradient force and the force of gravity (more precisely, the force of the Earth's gravity) is called the "thrust force." It occasionally, in certain places, raises or lowers the air. Thrust, as we mentioned earlier, was first explained by Archimedes of Syracuse (287–212 BC).

The effect of thrust force can be easily demonstrated in various experiments. One, originating from Descartes, is called the "Cartesian grebe." For the experiment, it is necessary to have one plastic bottle half-filled with water, and one plastic straw 2 cm long, closed on one side. A little water is poured into the straw so that it can swim when submerged (if it does not swim, add or remove water from the straw). Close the bottle and squeeze it. The water level rises, and the air above it is compressed. The increased air pressure will act through the opening of the straw, which will begin to sink. When we release the bottle, the air pressure will decrease and the straw will emerge.

10.5 Frictional Force

If the molecules or particles of a gas or liquid move at different speeds, they create a frictional force on the surface that separates them. In the "Principles," in 1687, Newton pointed out the following: "Resistance arises due to sliding in a part of the fluid and is proportional to the difference of velocities between the two parts of the fluid." That force is now called frictional force, or shear stress. The meaning of the action of this force is illustrated in Fig. 10.1.

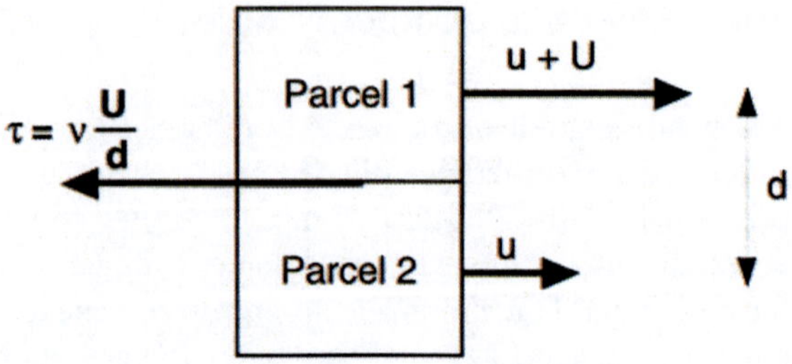

Fig. 10.1 Frictional force τ

10.6 Centrifugal Force

In 1659, Christian Huygens described the centrifugal force that occurs during rotation and acts in the field, in the direction normal to the axis of rotation. The word itself comes from the Latin word "fugio" (flee from) and "centrum" (center)—forces acting from the center. Huygens found that the acceleration created under the action of this force is proportional to the square of the velocity and inversely proportional to the distance of the body from the axis of rotation. This force appears in the so-called relative motion systems. When the movement of air is viewed in relation to a system that is attached to the Earth, then a centrifugal force occurs due to the rotation of the Earth around its axis.

10.7 Coriolis Force

Another force that occurs in rotating systems is the Coriolis force. That such a force exists was intuitively explained by Halley when he analyzed the winds over large areas. He concluded that the path of air in the Northern Hemisphere bends to the right, due to the rotation of the Earth. In 1835, Gaspar Gustav Coriolis (1792–1843) developed a theory that explains the behavior of moving bodies relative to a rotating surface (Burstyn 1966). He found that this force acts to the right in relation to the direction of movement in the northern hemisphere and that it is proportional to the speed of movement. In the southern hemisphere, it acts to the left.

Gaspar Gustav Coriolis (1792–1843)

He was born on May 21 in Paris. His father was a supporter of Louis XVI. Later he served in engineering. Coriolis attended Napoleon's polytechnic school in 1808. Later he served in engineering. He became an assistant at the Polytechnic School in 1816. He never married, due to poor health. He prepared a paper on the existence of relative acceleration for the Academy of Sciences on June 6, 1831. From 1832, Coriolis was an assistant to Claude Louis Navia (who derived the equation of fluid motion) in the course of Applied Mechanics at the School of Pont Chaussee. He succeeded Navia in 1836 as a member of the mechanical section of the Academy of Sciences. His most significant contribution to science was the description of the force in a rotating system. He also proposed to take half of the "life force" as an expression for kinetic energy and defined work as the product of the force and the distance traveled.

Coriolis explained that this force is the result of the interaction of two factors. The first factor, which has the strongest effect, is the one that originates from the component of body movement in the south-north direction, and the second from the component of movement in the west-east direction. The first factor occurs because the speed of any point of the Earth, due to its rotation, is the highest at the Equator and changes with latitude (to zero toward the pole). So, a body, or a small element of air, which moves toward the north, has a greater and greater speed toward the east (due to the maintenance of its momentum) in relation to the ground below it. Therefore, the relative motion of the body will be curved to the east (to the right in the northern hemisphere). Similarly, a body moving from the south toward the Equator maintains its speed of movement toward the east, which is decreasing in relation to the ground below it. Its path curves toward the west (to the left of the direction of movement). The second part of the Coriolis force originates from the fact that an increased centrifugal force acts on the particle, moving toward the east, due to rotation. One component of that force acts in the vertical direction, and the other in the north-south direction. When a body in the northern hemisphere moves toward the east, this causes the body to turn south (to the right) and due to the vertical component of the force, the body is apparently lighter than if it were at rest.

The Coriolis force is relatively small compared to other forces acting in the atmosphere. However, when the movements are fast, or when they last for a long time, this effect is very noticeable.

Simeon Denis Poisson (1781–1840), a French scientist, proved in 1836 that a grain fired from a cannon, due to the Coriolis force, will deviate from the direction where the target is located, by some amount to the right, in the northern hemisphere. At the time, it was not considered a significant shift. However, 80 years later, during the First World War, a German marksman on a cannon with which he was shooting at Paris, proved in practice that this force is very significant. Namely, in March 1918, shells were fired from a specially made 210 mm cannon (howitzer) toward Paris, from a distance of 110 km. The shell's trajectory took a surprising turn to the right, landing on the ground 1 km to the right of the target. A similar thing was observed during the British-German naval battle near the Falkland Islands (50°N) during the First World War. British marksmen used target charts that were only valid for the northern hemisphere. Because of this, the shells fell into the sea 100 m to the left of the German ships. In both cases, this happened due to ignorance of the nature of the Coriolis force. Of course, if every soldier who is training to be a gunner, regardless of the level of education, be taught to correctly use the tables in which the contribution of the Coriolis force to the trajectory of the shell is calculated.

10.8 The First Equation of Fluid Motion

In 1755, Leonard Euler, using Newton's second law and the concept of partial differential equations, wrote the equation of motion for a fluid (gas or liquid) for the first time. He assumed that the movement is carried out without the effect of the force of friction, that is, that the fluid element (particle) accelerates only under the influence of pressure gradient force and gravitational force (Euler 1755; Brewster 1872).

Euler's concept of the equations of motion, in 1827, was supported by the French engineer Claude Louis Navi (1785–1836) as well as his countrymen Simeon Denis Poisson (1831), Auguste Cauchy (1841), and Saint-Vevant (1843). They added the frictional force to Euler's equation of motion. In 1845, the Englishman George Stokes (1819–1903), a professor from Cambridge, worked on a similar problem, replacing Isaac Newton in that position. Stokes later claimed that his research was entirely independent, as he had no knowledge of French literature. He also introduced some assumptions by which he differed from the French. This is why the system of equations used in meteorology is often called Navier-Stokes equations.

Leonard Euler (1707–1783)

He was born in Basel, Switzerland, on April 15, 1707. He studied at the University of Basel, where there were 100 students and 19 professors. One of his professors was Johann I Bernoulli, who got him interested in mathematics. Euler received his master's degree in 1723 when he started working at the theological department of the university. When the Academy of Sciences was founded in Petrograd in 1725, Johann I Bernoulli and his sons Daniel and Nicholas II were invited to work there. At Daniel's insistence, in 1726, Euler was also invited as an associate. Euler left Switzerland for Petrograd on April 5, 1727, to which he never returned. He arrived in Petrograd on May 24, 1727, where, as a professor of mathematics, he replaced Bernoulli, who had returned to Basel. He married and had five children. In 1740, the Prussian King Frederick the Great decided to reorganize the Berlin Scientific Society (founded by Leibniz). Euler was offered a place at the Academy, which he accepted and left for Berlin on July 25, 1741, where he lived happily for 25 years.

Euler was a very prolific author. He prepared 380 works, of which 275 were printed. In 1752, he wrote the work "Principia Matus Fluidorum," published in 1761. In it, he derived the equation of continuity and the general equation of motion. At the invitation of Catherine the Great, on June 9, 1766, he left Berlin and came back to Petrograd. He accepted the invitation because he did not get along well with Frederick the Great in Berlin. On his way to Petrograd, he stayed in Warsaw for 10 days, at the invitation of the Polish king Stanisław Poniatowski. He arrived in Petrograd on July 26, 1766, with all members of his family. His eyesight suddenly weakened, due to numerous observations of the Sun, until he became completely blind in 1767. Despite this, he was still very active in science. His wife died in 1773. Three years later, he married again, to her half-sister Salma. At 5 pm on September 18, 1783, he suddenly suffered a stroke. Nevertheless, he said: "I'm dying" and at 11 pm he breathed his last.

He left a visible mark in almost all fields of science, especially in mathematics. He found the number e ($e = 2.7182\ldots$), the imaginary unit, the sign f for the function, etc. He published his theories in the field of meteorology in 1755 in "Letters to a German Princess."

Euler was religious. The famous French philosopher Denis Diderot once challenged him to a duel about the existence of God in Petrograd. Euler expressed his arguments in support of the existence of God in the form of insignificant algebraic equations. Diderot, who did not understand mathematics, remained without comment, and, feeling defeated, quickly returned from Russia.

A somewhat different type of equation was presented in 1760 by Joseph Louis Lagrange (1736–1774). They are named after him—Lagrange's equations. Later, in 1774, Pierre Simon Laplace (1749–1827) formulated the general equations of fluid motion in Euler's and Lagrange's forms. He derived them in Cartesian rectangular (Cartesian) and polar coordinate systems. Then, other equations (for fluid continuity and the thermodynamic equation) were added to the Navier-Stokes equations of motion. This system of five equations describes the five basic characteristics of air: the three components of velocity, pressure, and temperature.

10.9 Laminar and Turbulent Motion

The British scientist Osborn Reynolds (1842–1912) studied the characteristics of fluids in motion in more detail. Through a long tube, he pointed out the colored liquid from the tank. He noticed that at low speeds the colored liquid trails were parallel to each other. He called such a movement "laminar." By increasing the movement speed to a certain amount, the highlight would be messy. He called it disordered or "turbulent" motion. Turbulent motion is very complex, so even today it is the subject of theoretical analyses.

The British scientist Sir Horace Lamb (1849–1934) 2 years before his death (1932) pointed out that when he goes to "the other world," he would like God to teach him about two things: quantum electrodynamics and turbulent fluid movement. He also mentioned that he is optimistic only about the first topic, while for the second he thinks that even divine advice will not solve all the problems related to turbulence.

Reynolds described the characteristics of turbulence through a number, named after him the Reynolds number. That number represents the ratio of inertial and viscous forces in the fluid. When the frictional force prevails at very weak motion, the Reynolds number is small, and then laminar motions occur. At very fast movements, or at low viscosity, the Reynolds number is high, and then turbulent movements occur. He determined that this transition occurs at a value of Reynolds number of 1400.

Reynolds' original apparatus with which he experimented is now in the hydraulic laboratory of the Technical Department of the University of Manchester. Recently, using it, they tried to repeat Reynolds' experiments. In the meantime, the traffic has become so intense that the large oscillations it creates have made it impossible to reach the earlier results.

10.10 The First Real Textbook on Dynamic Meteorology

Until the end of the First World War, a lot of theoretical approaches were developed to solve various problems of meteorology. There were textbooks that summarized some parts of that knowledge. However, it can be said that only **Felix Exner** (1876–1930), Fig. 10.2, wrote the first real textbook on dynamic meteorology, in 1917.

When we say dynamic meteorology, we mean the interpretation of the eight-second characteristics of the atmosphere using the basic laws according to which the air in the atmosphere behaves. The basis of dynamic meteorology is the laws of hydrodynamics and thermodynamics.

Fig. 10.2 Felix Maria Exner

Felix Exner is the son of the prominent Viennese physiologist and academic S. Exner. He completed his studies in Vienna, where he received his doctorate, in 1900. He started working at the Austrian Institute for Meteorology and Earth Magnetism in 1901. He became the secretary of the Institute in 1907 and was appointed director in 1917. Together with Pernter, in 1907 he wrote the famous paper "Meteorological optics." Exner began writing dynamic meteorology in 1910 and finished the work after several years of work. The textbook was printed in 1917.

When writing the textbook, Exner set the goal "to summarize the previous theoretical knowledge in the field of meteorology." In the textbook, he laid the foundation for modern synoptic meteorology. For the first time in this textbook, he introduced the problem of turbulence and separation surfaces (which were later called frontal surfaces). Exner was close friends with Margules, who was among the first to deeply understand the importance of theoretical considerations in meteorology. That is why Margules' results were also found in Exner's textbook.

In the period from 1890 to 1925, great progress was made in the field of dynamic and synoptic meteorology, as well as in climatology. Numerous scientists contributed a lot to the progress, of the so-called Vienna Schools, Julius Hahn, Joseph Pernter, Wilhelm Trabert, Felix Exner, Wilhelm Schmidt, Heinrich Ficker, Albert Defant, and, of course, the greatest theorist, Max Margules.

References

Burstyn, H.L. 1966. The deflecting force and Coriolis. *Bulletin of the American Meteorological Society* 47 (11): 887–891.

Euler, L. 1755. General principles of the motion of fluids. *The Histoire de l'Académie des Sciences et des Belles-Lettres de Berlin* 11: 274–315.

———. 1872. *Letters of Euler on different subjects in natural philosophy addressed to a German Princess*, 2 vols. Trans. D. Brewster. New York: Harper & Brothers.

Kármán, T. 1967. *The wind and beyond*, 326. Little, Brown and Co.

Köppen, W. 1934. *Einstehung der meteorologischen wissenschaft*. Wetter.

Zinszer, H.A. 1944. Meteorological mileposts. *Scientific Monthly* 58: 261.

Chapter 11
Later Theories of Cyclones and Anticyclones

11.1 Introduction

Dove's theory about the origin of the cyclone and the character of the weather in it was widely accepted around the middle of the nineteenth century. Despite this, immediately afterward, it was spoken of as a primitive and unsatisfactory theory.

In the second half of the nineteenth century, the convective theory of cyclones was popular (Kutzbach 1979; Gill 1982; Avila 1991; Neumann 1999; Jones 2000; Evans and Hart 2003; Rasmussen and Turner 2003; Houze et al. 2009). It was introduced by Espy and later popularized by Ferrell. Thus, in 1856, Ferrell wrote: "…Hurricanes are said to be created by the collision of two opposing currents, during which a rotational moment of air occurs around the point of collision… It is very likely that they receive their initial impulses in this way… Obviously, hurricanes and normal storms can form and intensify under the influence of some operating forces (Charney and Eliassen 1964; Holford 1977; Shapiro 1982a, b; Möller and Montgomery 1999). Then when these forces disappear, the force of friction returns the atmosphere to a state of rest."

Ferrell further argued that the forces can be produced by the condensation of water vapor, which lifts the air to the center region of the hurricane. This updraft causes the air in it to become thinner, which causes air from the environment to flow toward it. Under the influence of the deviating (Coriolis) force of the Earth's rotation, the incoming air forms a characteristic cyclonic vortex. Condensation of water vapor in the updraft creates precipitation, and the cyclone gains energy from the latent heat of condensation. This is the essence of the convective theory.

Ferrell saw the initial impulse of the development of the updraft in the collision of two currents, while later Mon, Buchan, Loomis, and others linked the beginning of the cyclone development to the local heating of the near-surface layer (land or sea). This is the theory of "local pressure reduction." They believed that pressure depression can also occur due to local wind strengthening, the so-called dynamic pressure reduction.

M. Ćurić, V. Spiridonov, *History of Meteorology*,
https://doi.org/10.1007/978-3-031-45032-7_11

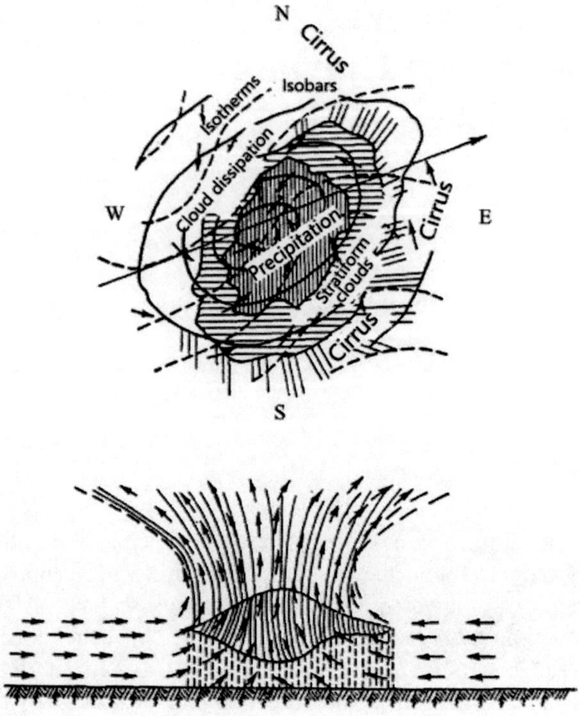

Fig. 11.1 Cyclone diagram according to Mon

Despite serious shortcomings, the convective theory was widely accepted in the 1870s. It was most applicable in tropical areas and in symmetrical cyclones. For example, in 1872, in his work "On Wind and Weather," Mon showed a vertical section of a symmetrical cyclone (Fig. 11.1). In that paper, Mon pointed out that: "We must accept that there are forces that pull the air towards or away from the center, around the center or in a vertical direction."

In 1889, Pomortsev, in his paper "Review study on weather forecasting," gave a nice description of cyclones. He says: "If the air in a certain place is strongly heated, it will move upwards, where such heating does not exist. At that place, the pressure is lowered…If the air moving towards the center of the cyclone is moist, it will cool down during the ascent and release part of the water vapor and its latent heat, which gives a further impulse to move upwards."

Several scientists (Ferrell, Buchan, and Marie-Devi) believed that the low pressure in the cyclone is due to the condensation of water vapor leaving a kind of vacuum. Pomoretsev did not agree with that. In this period, there were also several mechanical theories. That is how the French meteorologist Marie-Devy (1820–1893), a good writer but not always a precise scientist, wrote a paper in 1866 (and a second edition in 1877) "Meteorology of the Movement of the Atmospheric and the Sea." In it, Devy hypothesizes that a cyclone is a large rotating disc-like mass of air at a certain latitude. According to him, the centrifugal force pulls the air up and toward the axis of the disk, while at the ground, the air moves in a spiral from the center. Later, he presented a model of a cyclone that is composed of two disks that are located one above the other. Apparently, he felt that the ground layer of air has a different movement than the one at the height, which is the forerunner of the ground boundary layer theory.

A more complex analysis of the origin of cyclones and the laws of their movement was given by the Russian scientist Braunov (1852–1927) (Fig. 11.2).

At the Main Physical Observatory in 1882, he analyzed the movement of 48 cyclone centers based on synoptic maps from the period 1875–1878.

He observed the temperature and humidity field near each pressure minimum. He found that there is an area of warm air that extends in the form of a tongue to the center of the cyclone. This is what we call the "warm air sector" today. It is noticeable that the horizontal temperature gradients are mostly large and that "... During the day, the center of the cyclone moves more or less in the direction of the isotherms so that the warmer air is on the right side." He even precisely found that in the middle of the path of the center of the cyclone, they overlap an angle of 28° in relation to the isotherms (the paths are shifted by this angle in the opposite direction of movement of the hands of the clock in relation to the isotherms). This rule about the movement of cyclones has become very well known. Braunov also gave a completely new explanation of the movement of cyclones. He showed that the center of the cyclone must move in the direction of the flow of warmer and more humid air and that the pressure decreases the most in the direction normal to the isotherms and

Fig. 11.2 Petar Ivanovich Braunov

is directed from the warmer to the colder area. According to this, cyclones usually move from west to east. He also found that the higher the temperature gradient, the faster the cyclones moved. In this way, he explained why winter cyclones move faster and why they have lower pressure in the center. According to his theory, the anomalous movement of cyclones, those that move from NE to SW, and where the warmer area is on the NW side, is also explained. Similar conclusions were made by Braunov for anticyclones.

It is obvious that Brownov made a significant advance in terms of knowledge about the formation and movement of cyclones in relation to the simple convective theory. **Sprung** wrote about this in the famous book "Lehrbuch der Meteorologie," which he published in 1885 (Hann 1883, 1926).

11.2 Contributions by J. Hahn

As for Hahn's contribution to establishing the nature of cyclones and anticyclones, he may be said to have further developed Dove's theory (Graedel and Crutzen 1995; Saucier 1989). As early as 1875, he pointed out: "… Most pressure minima in storms can be explained mechanically. It represents a return to the old theories (Zurückgreifen), and for that reason, I can be classified as a reactionary." Reviewing the work of two Indian scientists (Elliott and Blanford), Han concluded the following: "… Small differences in pressure can cause such destructive storms because the kinetic energy of large air masses is concentrated and the air begins to rotate around the axis of the vortex." Hahn argued that the initial impulse in the formation of hurricanes is a small difference in pressure between the two areas.

Julius Ferdinand von Hahn (1839–1921)

He was born on March 23, 1839, near Linz, Upper Austria. He studied high school in the Kremsminster Monastery. He liked physics and mathematics the most. He studied at the University of Vienna. Immediately after that, he became a lecturer at the universities of Innsbruck and Vienna, and in 1873, he became a professor at the University of Vienna. He became interested in meteorology around the middle of the nineteenth century. He participated in the founding of the Austrian Meteorological Society in 1865 and worked in the magazine *Zeitschrift d. Österreich Gesellschaft fur Meteorologie*, which was later renamed *Meteorologische Zeitschrift*. It is a very important magazine to this day. All the most important works in the field of meteorology from the second part of the nineteenth century and the first part of the twentieth century were published there. After Jelinek's death in 1876, Hahn became the director of the Austrian Meteorological Institute. He performed that job until 1900. He was elected a member of the Vienna Academy of Sciences in 1892. Immediately after that, he was the secretary of the Academy. Hahn is certainly one of the greatest meteorologists of the nineteenth century. He was an opponent of the convective cyclone theory.

Hahn conducted a series of experiments to determine the nature of the "fen" wind, the change in temperature and humidity in the updraft, and the nature of precipitation. There is almost no field of meteorology where Hahn did not make a decisive contribution. In 1872, he published an important scientific work on physical-geographic concepts "Die Erde als Ganzes..." In synoptic meteorology, he published the dynamic theory of cyclones. In climatology, he collected and correlated a large number of climatological data from various areas of the world. His extremely voluminous book *Handbuch der Klimatologie* saw four editions (the first in 1883 and the last in 1934). It served as an elusive guide through climatology. Hahn also did a large number of climatological studies for the orographically complicated areas of Austria. In doing so, he provided a well-known method for reducing climate names to a longer period. He prepared the meteorological section for the "Physical Atlas," published by Berghaus in 1887.

In his work, Hahn always first collected data, then correlated it, and finally drew some kind of pattern. He strongly emphasized the importance of comparing knowledge from different fields of science. He almost never made mistakes. Hahn left the University of Vienna in 1908 but continued to edit the journal "Meteorologische Zeitschrift." He published his capital work *Lehrbuch der Meteorologie* (first edition in 1901, second in 1906, and third in 1914). After that, he did not publish any significant work. He died on October 1, 1921.

Han's views on the structure of cyclones and anticyclones led to the formation of a new program of measurements at high-altitude meteorological stations in Europe. From January 23 to February 3, 1876, there was high air pressure over Austria and Switzerland. It was very cold with fog and heavy frost. Analyzing this situation, Hahn noticed that the high-pressure aloft was accompanied by clear, calm, and warm weather. At mountain stations between 1500 and 2000 m high, the temperature did not drop below 0 °C, while in the valleys, it dropped to −20 °C. A similar situation existed in December 1879. Hahn showed that such high pressure in the anticyclone cannot be explained only by thermal factors. He was looking for a dynamic factor that contributes to this. In his analysis, he later especially used data from Zonblik, the old highest observatory in Europe (3106 m), which started operating in 1886. In honor of the centenary of Zonblik's work in the foothills, in Rauris, the International Conference on Alpine Meteorology was held in September 1986.

Hahn analyzed the minimum pressure values during the winter of 1886–1887 over Austria. He compared data from Zonblick and a nearby station at the foot of Bad Ischl. The drop in pressure with height was an average of 20 mm and temperature by 8.7 °C. He concluded that such a pressure drop cannot be explained only by what happens in the air column between Bad Ischl and Zonblick but that the entire air mass, which is significantly higher than Zonblick, has a temperature that is below-average values. Such cooled air cannot be explained only by the lifting and cooling of the air but also by some nonthermal formation. From the analysis of the measurements, Han drew a very important conclusion: "... That the pressure in the anticyclone and cyclone is a consequence not only of the temperature conditions but also of the type of movement of the air masses inside them. Temperature conditions depend on the type of motion, and are their consequence, just as dry air follows the absence of clouds."

On the characteristics of an anticyclone, Hahn concluded: "... Since the air temperature rises faster in the descending air than it decreases in the rising air, there must be a higher temperature in the area of descent than in the nearby area where the air rises." In a report to the Academy of Sciences dated April 17, 1890, Hahn concluded: "... Prevailing views on the causes of the appearance of anticyclones have proven to be wrong, although some still support them, such as, for example, Ferrell in the latest work. On the other hand, the observations confirm the ideas of those, such as this author, that cyclones and anticyclones are only part of the process of the general circulation of the atmosphere, i.e. of kinetic energy that is created from the temperature difference between the pole and the Equator." This conclusion about the importance of the horizontal temperature gradient is similar to the one made earlier by Brown. In the following years, prominent meteorologists clashed over this idea. Deschevren, Tisserand de Bor, and Berson supported Han, while Americans Harrington, Roche, and Clayton, together with Englishmen Shaw and Daines, were against his thesis.

In the meantime, Hildebrandson showed in 1883 that in the front-day part of the cyclone at all levels, a southerly relatively warm flow prevails, while in the back part, a northerly cold flow prevails. These results, together with Hahn's conclusion, were used by Ecoli (1848–1923) who was director of the Swedish Meteorological Service from 1913 to 1918. He claimed that: "a cyclone is an eddy consisting of two approximately opposite currents, cold from the west and warm from the east." This is in complete agreement with Hahn's work which he published in the *Meteorologische Zeitschrift* in 1905. The dynamical theory of cyclones and anticyclones represented an important step forward toward scientific truth, no matter how many adherents the convective theory had.

11.3 Margules' Cyclone Theory

The energetic approach to the development of cyclones was represented by Max Margules, a talented meteorologist and theorist, who was well ahead of his time (Margules 1903–1905, 1906). His character and the difficult circumstances in which he lived forced him to stop his scientific activities at an age when he could have done much more and to fully develop his ideas about the energy of storms. His fate is only slightly less tragic than that of Fitz Roy.

Margules' work was not widely accepted until after his death. Every weather prognosticator is familiar with the Margules formula, which calculates the slope of the frontal surface. However, his consideration of certain time processes is of incomparably greater value. Margules started by solving theoretical problems that could be applied in practice under certain conditions. It can be said that, although incomplete, he made a significant contribution to the development of theoretical meteorology.

In addition to Margules and Hahn, other prominent meteorologists worked in Vienna between 1890 and 1925. Joseph Pernter, Wilhelm Trabert, Felix Exner, Wilhelm Schmidt, Heinrich Ficker, and Albert Defant stand out among them. They form the famous "Vienna School of Meteorology." All of them significantly contributed to the development of dynamic and synoptic meteorology, as well as climatology.

Max Margules (1856–1920)

(continued)

(continued)

He was born on April 23 in Brad, western Ukraine. He studied at the University of Vienna. From 1877 to 1879, he worked at the Meteorological Institute (service) in Vienna. He spent 1879 at the University of Berlin. From 1880 to 1882, Margules worked at the University of Vienna as a lecturer. Due to some mistakes, he had to leave the University and end his university career, which he loved very much. He returned again to the Meteorological Institute (service) of Austria. He worked there from 1882 until his retirement in 1906. At the Meteorological Institute (service) of Austria, Margules worked on some climatological projects. However, he was not satisfied with that, so from 1890 to 1893, he wrote a series of articles on inter-day pressure changes. Toward the end of 1890, he organized a series of measurements around Vienna with barographs and thermographs. He compared those measurements with measurements at altitude stations.

Margules disagreed with the concept that the pressure gradient is responsible for the movement of the atmosphere. He believed that the energy of the storms was so great that the energy of the pressure field alone could not be responsible for it. He argued that the potential energy of the air and the asymmetric distribution of temperature in the storm serve as a source for the energy of motion. In this sense, in 1906, he wrote the work *Theory of Storms*.

He presented the ideas in the paper in a confused manner and in rather bad German. That is why no one from scientific circles accepted him. This offended him very much, so he retired at the age of 50 and stopped any scientific work.

He barely survived on a small pension, especially during the years of the First World War. His pride did not allow him to ask for help from his friends from the Austrian Meteorological Society. In 1919, the Society awarded him a silver Han medal, along with a small monetary portion of the award. Margules accepted the medal but refused to accept the prize money. He died of exhaustion (of hunger) on October 4, 1920, in Perstdorf, near Vienna.

11.4 Method of Analogies

All theories of the development of cyclones and anticyclones had the ultimate goal of arriving at reliable methods for good weather forecasting. In this sense, in this chapter, a methodology for forecasting the development of cyclones and anticyclones, which is known as the "method of analogies," will be presented. The method consists in recognizing in historical data the weather situation that is most similar to the "current" one, from which future development is to be forecasted.

The basis of this method is very old and comes from Gabriel Gilbert, a so-called meteorologist from Normandy. He spent his boyhood days on the banks of the English Channel. From his grandfather and father, he inherited a love for observing the weather. Gilbert began his scientific activities when he was elected secretary of the meteorological commission in the department of Cannes, where he spent most of his life. Already in 1886, he presented a paper with the sonorous title: "A New Method of weather forecasting" (Walker 1993). His many years of experience enabled him to win first prize in a unique weather forecasting competition organized in Liege, Belgium, in 1905. The competition was organized by the Meteorological Society of Belgium. In 1909, Gilbert published a comprehensive book *Nouvelle methode de prevision du temps*—a new method of forecasting the weather. With the numerous rules he formulated, he tried to create the impression that the problem of weather forecasting had been solved. Of his 25 rules, the most important are 2: The first is "if the wind in the center of the cyclone is weaker than normal then the minimum pressure will decrease even more." The second rule is: "The center of the cyclone (minimum pressure) moves in the direction where the wind is weaker than normal."

In the book, Gilbert also presented the essence of the method of analogies. He writes: "By analyzing the similarity between some type of pressure field with isobaric maps from a period of about half a century, a number of similar or even identical situations can be found. In this way, a prognostic rule can be found for the further development of such a situation."

The idea of synoptic analogies was developed even earlier, in the 1870s, in the Department of Reporting in America. In Sweden, before the end of the nineteenth century, Ecoli, guided by the same idea, prepared a "Dictionary of weather maps." Ecoli worked for 10 years on the preparation of the catalog of weather maps and did not complete them. It was difficult to find synoptic situations that have, for example, even distributions of precipitation in Sweden. Therefore, he abandoned this method and tried to forecast the weather using the "isallobar method" (lines of equal pressure change at some time interval, usually 3 h).

The catalog of maps from the period of 15 years (1890–1904) was completed in 1905 in Pula (then Austria) by Captain Kesić. He recognized similar maps according to the position of anticyclones over Europe (their nature) and the position of cyclone centers. The catalog was relatively easy to use.

11.5 The Weather Prophets

At the end of this chapter, it is useful to look at the behavior of the so-called *weather prophet*. These are people who do not know nor are they interested in the accepted scientific methodology in this area (Freier 1992; Lee 1976; Griffiths 1977; Nicholson 1981). But they believe they have found a new method to accurately forecast the weather for a week, a month, a year, or even longer. There were such prophets ("clairvoyant Zorkas") in all areas and in all historical periods. Their work is based on a wide range of methods, from those that represent pure fantasy and gross quackery to those that have some grain of truth.

Prophets do not have basic knowledge of meteorology, and they are not interested in measured meteorological data. They are made up of people of different professions and ages. Some that are more widely known will be mentioned. Demczynski was a railway engineer, Falb a minister, Mattie a peasant, Lamarck a famous naturalist, Morrison an officer and astrologer, etc. In the nineteenth century alone, there were more than a hundred such prophets.

The question arises why it is necessary to talk about them here. I think it should be spoken for several reasons: First, because there is a more permanent lesson to be learned from this. Second, prophets are primarily affirmed by the media, which speaks volumes about the state of awareness in this area, because the attitude of an ignorant and aggressive person is more interesting to them than that of established scientists from a narrower scientific field. Affirming the prophets speaks of the level of development of the scientific views of the people who believe in the prophets. The appearance of prophets is particularly pronounced in some periods of the development of meteorology. These are usually periods when there is a general interest in long-term forecasts, occasional weather phenomena, etc. The activities of prophets are generally very negative for the profession. But, in healthy environments, they can in a certain sense also be the initiators of the development of science. Because they exert such a significant influence, often disrupting decision-making processes, special scientific projects are organized to address and resolve specific scientific problems.

What is it that makes prophets engage in weather prophecy? If it is about healthy people (which is not often the case), then it is mainly their opinion that they are right and that they are doing great good to the people. Often there is also a material factor involved. So the Englishman Murphy was very lucky to guess that on the 20th of 1838, it would be extremely cold in London (−20 °C). Because of this, he became a famous prophet overnight. He managed to sell 45 editions of his "Almanacs" (in which he predicted the weather) and earn 3000 pounds sterling. Demchinsky and Falb also became known beyond the borders of their countries (Russia and Germany). Often the prophets were under the protection of some kind of scientists (not scientists from meteorology, but from other fields).

The weather forecasts of the prophets were mainly based on monitoring the behavior of animals and plants, the movement of the moon, the position of the heavenly bodies, climatological averages, and different types of periodicity. A popular method of forecasting was based on monitoring the behavior of spiders. Thus, Disjonval, formerly a member of the French Academy, and later adjutant of General Pichereau, successfully guessed that there would be severe frosts in December 1794. Because of this, General Pichereau successfully crossed the frozen Massa with his army, and so on December 28, 1794, occupied the entire Netherlands. The fame of Disjonval was great. Then, in the revolutionary enthusiasm (of the French Revolution), a new "science" was created called "araneology"—arachnology (from the Greek word **αραχνη**, or Latin aranea). Between 1798 and 1801, his work was translated into several languages.

Many scientists and inventors have predicted the weather by tracking the moon. One of these was the famous naturalist Lamarck (1744–1829). He printed weather forecasts in his Meteorological Yearbooks for the years 1800–1811. And the German Falb (1838–1903) was a famous "lunar prophet." He observed critical days of solar activity, snowfall in the mountains during the summer, heavy rains, frosts, etc. Falb was very popular in the press. The controversies among scientists with his views did not diminish his popularity.

Demczynski (1851–1915), a railway engineer, wrote in 1901: "Since the influence of the Moon on the weather has been demonstrated, I decided to start publishing a special magazine (Demczynski's climate journal). Various papers as well as monthly or month-and-a-half forecasts are printed in it. He gave the forecast in the form of graphs for many places in Russia and the rest of Europe. Later, he included the USA, Japan, and India in his forecast. His forecasts were even published in the British newspaper "The Daily Mail." It is difficult to say that he had any clear principle on which to base his forecasts. He noted that after every 652–656 days, that is, every other lunar year, the synoptic charts are repeated. He considered that the sums of temperatures for the 8th, 9th, 10th, and 11th lunar months are constant. He also claimed that the atmosphere and changes in atmospheric pressure are strongly influenced by electromagnetic effects.

His activities met with fierce resistance from the official meteorologists—forecasters. They performed an analysis of his forecasts and determined that they were completely arbitrary and based on nothing. Its significance lies in the fact that it encouraged official meteorologists to engage in long-term weather forecasting.

Weather prophets, regardless of the method they use, where they are from, what time they work, and what social position they belong to, have one common feature. They think that official weather forecasts contain gross errors and that they are called upon to solve "an underdeveloped branch of science like meteorology." The prophets, carried away by the glory attached to them by the primitive public, do not shy away from challenging and insulting scientific gems. Thus the French prophet, the farmer Mati, mocked the famous scientist Leverier, director of the Paris Observatory. In 1863, Mati wrote the pamphlet "Lessons in Cosmology addressed to M. Leverier." Such "lessons" are written to this day in all environments, including ours. Prophets survive, thanks to the unscrupulous media, which drove one great FitzRoy to commit suicide in 1865. Those "lessons" are fed by an uneducated public that always embraces sweet-talking frauds, charlatans, and ignoramuses rather than wise, patient, and measured scientists. Prophets can always give such a public what they expect—weather forecasts for a much longer period of time than those based on scientific methods. Because nothing prevents them: neither morality, nor ignorance, nor scientific methodology. There are no obstacles before the prophets, because they are charlatans. They are not ashamed of anything.

References

Avila, L.A. 1991. Atlantic tropical systems of 1990. *Monthly Weather Review (MWR)* 119: 2027–2033.

Charney, J.G., and A. Eliassen. 1964. On the growth of the hurricane depression. *Journal of the Atmospheric Sciences* 21: 68–75.

Evans, J.L., and R.E. Hart. 2003. Objective indicators of the life cycle evolution of extratropical transition for Atlantic tropical cyclones. *Monthly Weather Review (MWR)* 131: 909–925.

Freier, G.D. 1992. *Weather proverbs*. Fisher Books.

Gill, A.E. 1982. *Atmosphere-ocean dynamics*, 662. Academic.

Graedel, T.E., and P.J. Crutzen. 1995. *Atmosphere, climate, and change*. Scientific American Library.

Griffiths, J.F. 1977. A chronology of items of meteorological interest. *The Bulletin of the American Meteorological Society* 58: 1058–1067.

Holford, I. 1977. *The guinness book of weather facts & feats*. Guinness.

Houze, R.A., W.-C. Lee, and M.M. Bell. 2009. Convective contribution to the genesis of hurricane Ophelia (2005). *Monthly Weather Review* 137: 2778–2800.

Jones, S.C. 2000. The evolution of vortices in vertical shear. III: Baroclinic vortices. *Quarterly Journal of the Royal Meteorological Society* 126: 3161–3186.

Kutzbach, G. 1979. *The thermal theory of cyclones. A history of meteorological thought in the nineteenth century*. Lancaster: Lancaster.

Lee, A. 1976. *Weather wisdom, facts and folklore of weather forecasting*. Congdon & Weed.

Margules, Max. 1903–1905. Uber die Energie der Stürmer (On the energy of the storms). *Jahrbücher der K K. Zentralanstalt für Meteorologie und Erdmagnetismus, Jahrgang 1903, Wien 1905* (*Yearbook 1903 of the K K. Central Institute for Meteorology and Earth Magnetism, Vienna, 1905*) 40: 1–26. See also the Translation by C. Abbe in the *Smithsonian Miscellaneous Collections* 51 (4): 533–595, 1910.

———. 1906. Zur Sturmtheorie. *Meteorologische Zeitschrift (Meteorological Journal; Journal of the K.K. Austrian Society of Meteorology)* 23 (11): 481–497. https://archive.org/details/meteorologische01unkngoog.

Möller, J.D., and M.T. Montgomery. 1999. Vortex Rossby waves and hurricane intensification in a barotropic model. *Journal of the Atmospheric Sciences* 56: 1674–1687.

Neumann, C.J. 1999. *Tropical cyclones of the North Atlantic Ocean, 1871–1998*, 193. National Oceanic and Atmospheric Administration.

Nicholson, S.E. 1981. Climate and history, Wigley, Ingram and Farmer, National Museum of natural history. In *Ice Age mammals and the emergence of man*. Elephant Press.

Rasmussen, E.A., and J. Turner. 2003. *Polar lows: Mesoscale weather systems in the polar regions*, 612. Cambridge University Press. ISBN 978-0-521-62430-5.

Saucier, W.J. 1989. *Principles of meteorological analysis*. Dover Publications.

Shapiro, L.J. 1982a. Hurricane climatic fluctuations. Part I: Patterns and cycles. *Monthly Weather Review (MWR)* 110: 1007–1013.

———. 1982b. Hurricane climatic fluctuations. Part II: Relation to large-scale circulation. *Monthly Weather Review (MWR)* 110: 1014–1023.

Von Hann, J. 1883. *Handbuch der klimatologie*. Stuttgart: Verlag Von J. Engelhorm.

———. 1926. *Lehrbuch der meteorologie*. 4th ed. Leipzig: Herm. Tauchnitz.

Walker, J.M. 1993. The meteorological societies of London. *Weather* 48: 364–372.

Chapter 12
Atmospheric Motion

12.1 General About Atmospheric Motion

Wind, the local movement of air or globally, over large areas, arises as a tendency to equalize the difference in temperature and pressure caused by the different amounts of heat from the Sun received by certain parts of the Earth, as well as by the different disposition of that heat. For these reasons, the forces that move the air appear. The nature of air movement was observed by people in various historical periods, but the British scientist George Hadley (1686–1768) was the first to clearly perceive and describe it in the middle of the eighteenth century (Kington 1988; Sorbijan 1996; Thaxter 1999). However, even he failed to fully understand the connection of global movement with the rotation of the Earth. When air moves from the center of high to low pressure to neutralize the established pressure difference, it is forced, due to the Earth's rotation, to change its direction of motion by a certain amount. French scientist Gustave Gaspar Coriolis explained how much it turns, and what that turning depends on, only in 1835. In fact, he provided a general theory of relative motions that was later applied (primarily by Poisson in 1836) to motions observed relative to the Earth.

Local air movements, for example, between the land and the sea, also occur as a consequence of efforts to equalize the differences arising from the properties of water, which heats up less than the land for the same received amount of heat. Then, during the day, the air above the heated ground heats up more than the air above the water and begins to rise. The cooler air above the water reacts to this and moves toward the "abandoned" area above the ground. At altitude, movement occurs in the opposite direction to this. This is how the regulated air circulation appears, which we feel every summer when we find ourselves on the seashore. Similar movements occur in areas where the orography (relief) shields certain parts of the soil from the direct influence of solar radiation. This happens in areas with wider valleys. There is the so-called valley or mountain circulation.

M. Ćurić, V. Spiridonov, *History of Meteorology*,
https://doi.org/10.1007/978-3-031-45032-7_12

We know all this today. However, about 250 years ago, it was not fully understood (Walker 1845; Gregory 1930; Namias et al. 1935; Scot 1976). Thus, the French scientist Jean le Rond Dalamber (1717–1783) pointed out that solar radiation affects only the heating of the soil, and it is not the primary cause of the creation of wind. In 1746, Dalamber received the prize of the Berlin Academy of Sciences for the best paper dedicated to the mathematical description of the atmosphere. The jury was chaired by Leonard Euler (1707–1753). The paper was entitled "Reflection sur la Cause Generale des Vents"—Description of the cause of the appearance of wind. The work was dedicated to Frederick the Great, ruler of Prussia. In the paper, Dalamber introduced the notation for "partial derivatives" (Fig. 12.1).

Jean le Rond Dalamber (1717–1783)

He was born in Paris on November 17. He is the illegitimate son of Madame de Tensin, a famous salon hostess and cavalry officer Chevalier Detouche-Canon. At the time of his birth, his mother left him on the steps of the Church of Saint Jean le Rond in Paris. When his father found out where the baby was, he took him and gave him to be raised in the house of the craftsman family Rousseau. His father provided for his education. Although he never married, Dalamber lived for years with his lover, Juliet de Lespinas. Dalamber became a leading figure in the French intelligentsia. He was coeditor of the newly created Encyclopedia. He published several works in mathematics, mechanics, music, philosophy, and astronomy. He was small in build, but that is why he was recognizable by his deep penetrating voice. He was considered a man who had a sense of humor. He was a good speaker. Dalamber died on October 29, 1783.

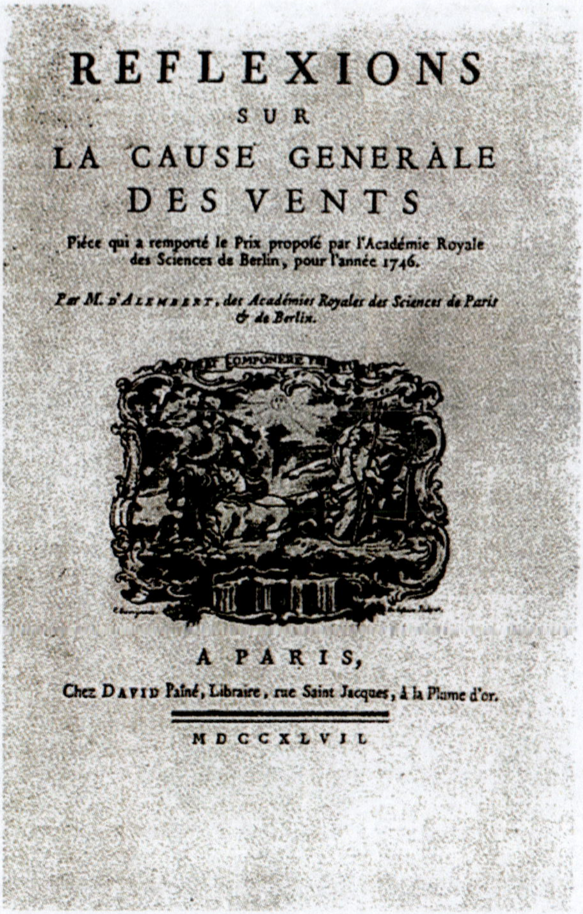
REFLEXIONS
SUR
LA CAUSE GENERALE
DES VENTS
Piéce qui a remporté le Prix proposé par l'Académie Royale des Sciences de Berlin, pour l'année 1746.
Par M. D'ALEMBERT, des Académies Royales des Sciences de Paris & de Berlin.
A PARIS,
Chez DAVID l'aîné, Libraire, rue Saint Jacques, à la Plume d'or.
MDCCXLVII.

Fig. 12.1 Title page of Dalamber's work

He also tried to formulate general equations for atmospheric motions. However, his conclusion was wrong. For example, he argued that movement in the atmosphere is due to the tidal effect due to the attractive forces between the Sun and the Moon. He also partially analyzed the winds above the ground (Fig. 12.2).

In 1926, Lewis Richardson pointed out that in the atmosphere there are movements spread over areas of several thousand kilometers to those that are on a millimeter scale. Movements on the largest scales are called the "general circulation of the atmosphere". It is conditioned by the difference in temperature and pressure above the entire Earth, as well as the rotation of the Earth. Land and oceans additionally change the primary flow and influence the development of the secondary movement. Local orography determines (local) tertiary movements. A cascading process occurs in which eddies of the largest (global) scale create eddies of smaller scale, and thus the movement of molecules, through which, due to friction, the

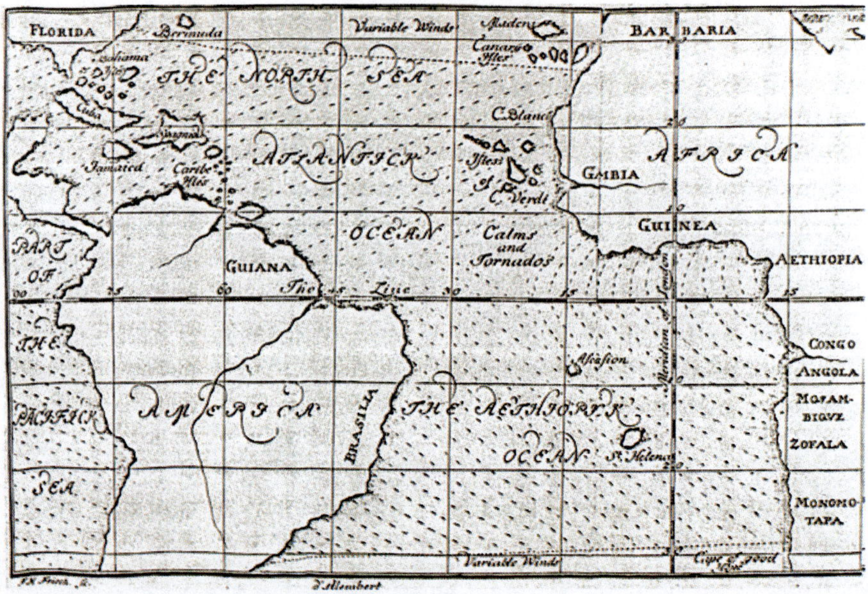

Fig. 12.2 Dalamber wind map

movement is lost and turns into heat. Richardson expressed this poetically, but quite accurately: "Great eddies have little eddies, which feed on their speed (of the larger eddies); and small eddies have smaller eddies and so on down to viscosity."

12.2 General Circulation of the Atmosphere

Movement in the atmosphere on the largest scale was first explained by Edmond Halley (1656–1742). He formulated the theory of tropical winds in 1686, after spending 2 years on the island of St. Helena. According to his theory, near the Equator, air heated by the Sun's energy rises, and air flows in its place to compensate for that rise (Hadley 1735). Halley said that in the tropics, the wind blows from east to west, following the movement of the Sun. Halley's wind map is shown in Fig. 12.3.

Edmond Halley (1656–1742)

Edmond Halley was the British Astronomer Royal from 1720 to 1742. In 1705, he predicted the return of a comet (called Halley's Comet after him). She did recover in 1958, 16 years after his death. He was a friend of Isaac Newton. He was the main man who found the means to publish Newton's book *Principles*. Halley made a great contribution to the science of weather. He particularly well described the causes of trade winds and monsoons. He was the first to show those winds on a meteorological map. Halley recorded his weather observations while traveling to compile a star catalog of the southern hemisphere. Based on this and on the basis of the information gathered from the sailors, he was able to get an idea of the global movements in the atmosphere.

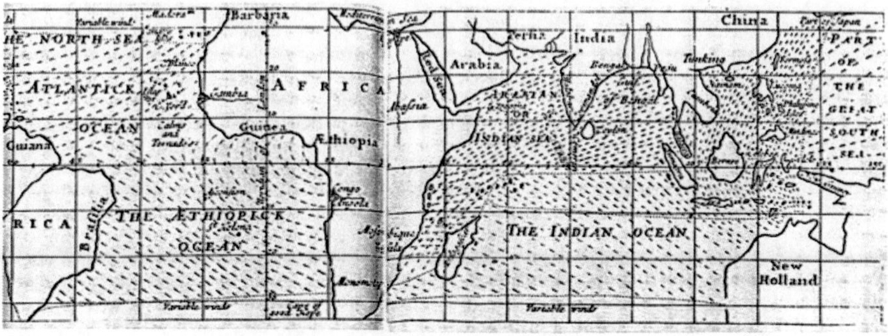

Fig. 12.3 Halejeva karta opšte cirkulacije. Halley's General Circulation Chart

Fig. 12.4 Hadley's model of the general circulation

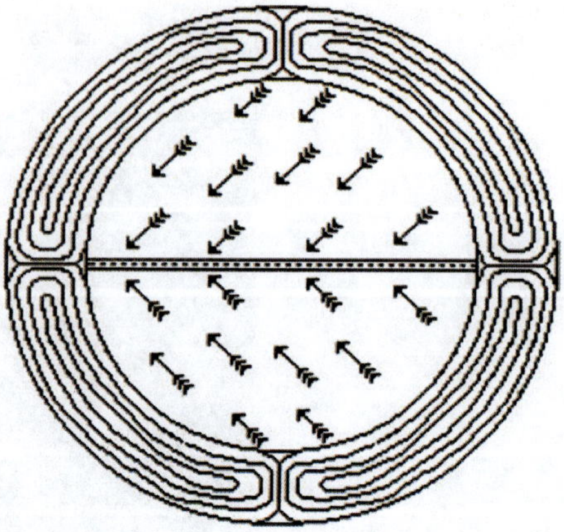

Halley's model of the general circulation of the atmosphere was simplified due to insufficient data. That painting was repaired 50 years later by George Hadley (1686–1768). He was a lawyer by profession but devoted himself to natural sciences. In 1736, Hadley presented a paper on tropical winds before the Royal Society in London. He pointed out that warm air rises above the Equator and moves toward the poles. There, it cools down and returns toward the Equator (Fig. 12.4). Its circulation cell is symmetrical over the northern and southern hemispheres. He correctly concluded that the direction of movement must be shifted from the north-south direction due to the rotation of the Earth.

Jaime Coffin (1806–1873) published the monograph "Winds of the Northern Hemisphere" in 1853. In it, he presented a series of wind maps obtained from 579 stations. He showed that there are three areas of higher and lower pressure that alternate from the Equator toward the North Pole. In each of these areas, there is one circulation cell in the vertical plane that is in the direction of the meridian. A few years later, the Russian climatologist Alexander Voeykov (1842–1916) presented wind data based on measurements in 3223 places around the world, including ship measurements collected by Maury (1806–1873).

These empirical data served the American William Ferrell (1817–1891) for theoretical work in this area. Based on the analysis of a large number of data on wind, ocean currents, and pressure fields, Ferrell observed in 1855 that there are well-defined and stable pressure zones around the Equator, 30° and 60° of both latitudes. Based on this, he modified Hadley's theory. According to Ferrell, there are three circulatory cells above each hemisphere (Fig. 12.5).

The modern understanding of the general circulation of the atmosphere is completely consistent with Ferrell's (Halley 1686a, b). A circulation cell from the Equator to the subtropics (30° both latitudes) is called a Hadley circulation cell. From the subtropical high-pressure belt, air moves poleward, turning more and more to the right due to the Coriolis force. That cell is called Ferrell's and the third (from 60° to the pole) is called a polar cell. In Ferrell's cell (in moderate latitudes),

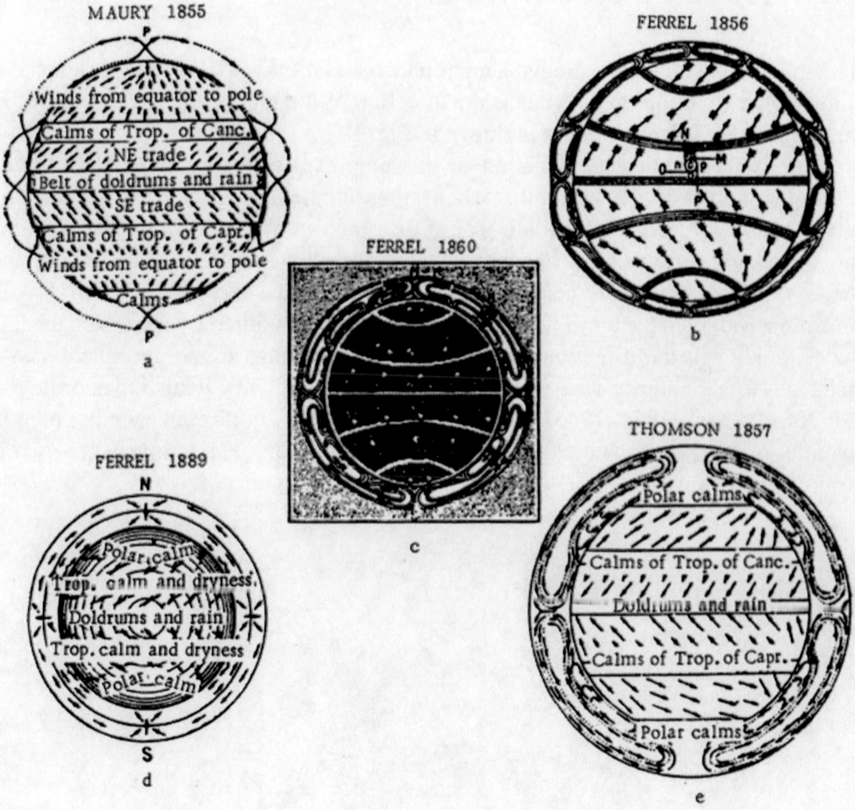

Fig. 12.5 Scheme of the general circulation of the atmosphere by various authors

the air on the ground floor moves mostly from the same direction, and at the height, westerly winds prevail. Between the Ferrell and the polar cell is the polar front. In the tropopause, the meandering westerly current prevails above the polar front. In some parts, called "jet streams," the air has a speed of about 150 km/h. Such an area is several hundred kilometers wide and several kilometers deep. This current was discovered by American pilots during the Second World War flying in B-29 planes during the bombing of Japan. This was explained in 1951 by the Finnish meteorologist Palmen.

The meandering westerly current was studied by Carl Gustav Rosby (1898–1957), a Swedish meteorologist. In 1939, he was the first to mathematically describe the importance of the meandering westerly current for heat transfer in the meridional direction. It is called Rosby and waves after him. It is one dominant dynamic feature of the atmosphere. Based on the behavior of Rossby waves, long-term weather forecasts can be made.

12.3 Pressure Field and Winds

In 1857, the Dutch meteorologist Christoph Buys Ballots (1817–1890) introduced a simple empirical rule that connects the direction of the wind at the ground with the pressure field. The connection is shown in Fig. 12.6.

The first theoretical explanation of this empirical rule was given by William Ferrell around 1859. He found that when a pressure field is established, the force of the pressure gradient will initially cause the air to move toward the low-pressure area. Then the Coriolis force appears, which gradually bends the path of air particles to the right in the northern hemisphere. When the balance between these two forces is established, then the air moves parallel to the isobars, with low pressure on the left side (looking in the direction of movement). Ferrell also found an equation that relates velocity and pressure distribution (pressure gradient). British meteorologist Sir Naper Shaw (1854–1945), Fig. 12.7, in 1916, that equilibrium wind, which is parallel to the isobars, he called the "geostrophic" wind. He composed the word from the Greek words "geo"—Earth and "strefein"—to turn.

William Ferrell (1817–1891)

He was born in Fulton, Pennsylvania, to a farming family. From 1846 to 1857, he was a teacher in Montana, Kentucky, and Tennessee. He started doing scientific work quite late. He was encouraged to do so by Laplace's work on "Celestial Mechanics." Ferrell made some significant additions to Laplace's work. He became acquainted with Maury's scheme of the general circulation of the atmosphere in 1856. Between 1859 and 1860, he too sought to show a general circulation based on complicated hydrodynamic equations of motion for the atmosphere, taking into account the rotation of the Earth. In 1859 (well after Coriolis), he independently came up with the concept of the "deviation" force due to rotation. Ferrell was involved in the analysis of the heat balance of the Earth's surface, tidal waves, currents, and storms. He derived a formula for the change of geostrophic wind with height (thermal wind) and formulated the convective theory of cyclones. From 1867 to 1882, Ferrell was a member of the US Geodetic and Coastal Survey, and from 1882 to 1886, he worked in the "Reporting" service. He died on September 18 in Maywood.

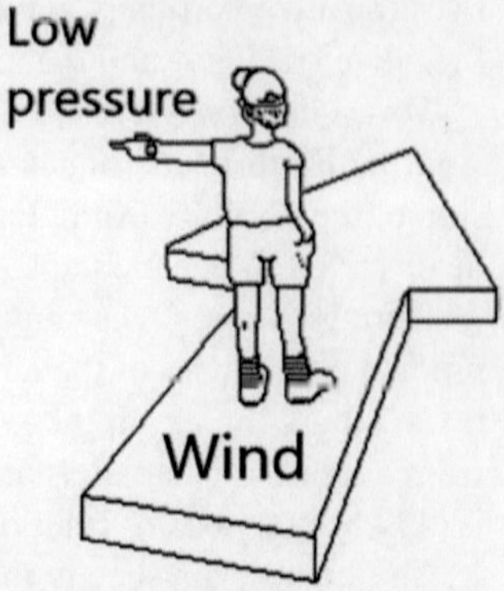

Fig. 12.6 Buys Ballots' Law for the northern hemisphere

Fig. 12.7 Naper Shaw (left) with W. Bjerknes

Due to friction, the wind in the near-surface layer up to a height of about 1–1.5 km is not geostrophic (because it is partially directed toward the area of low pressure). From that height, the actual wind almost exactly matches the geostrophic wind.

12.4 Atmospheric Vorticity

Leonard Euler (1707–1783) and Jean le Rond Dalamber (1717–1783) mathematically derived a term from the equations of motion that were later called the "vorticity term." Almost all movements in the atmosphere are more or less eddy (Fig. 12.8). Thus, in a cyclone, the air rotates in a counterclockwise direction and gradually rises (positive vorticity), while in an anticyclone, it rotates in the opposite direction (negative vorticity) and the air descends.

Vortex occurs as a result of individual parts of the air moving at different speeds. And the Earth itself rotates around its axis. That vorticity is always positive. The sum of the vorticity of the Earth and the vorticity of the air relative to the Earth can be observed. It is called absolute vorticity.

During rotation, the angular momentum of the movement is maintained (the product of the mass of the air, its speed, and the distance from the center of the vortex axis). As the air on the ground floor converges during cyclonic vortices (passes from a larger to a smaller circle of rotation), the speed of movement of the air that is closer to the axis of rotation increases (Fig. 12.9).

The highest air speeds in eddies occur in tropical cyclones. There, speeds are higher than 300 km/h. In tornadoes (small vortices) below the base of thunderclouds, velocities can be even higher than these. In our country, fortunately, such vortices occur very rarely, but in the USA, it represents a real sea for residents over large areas, especially in the state of Oklahoma.

Determining the speed of tornadoes is very difficult, because they destroy everything they come across, including installed instruments. However, in the nineteenth century, the meteorologist Elias Loomis (1811–1889) estimated those speeds in a

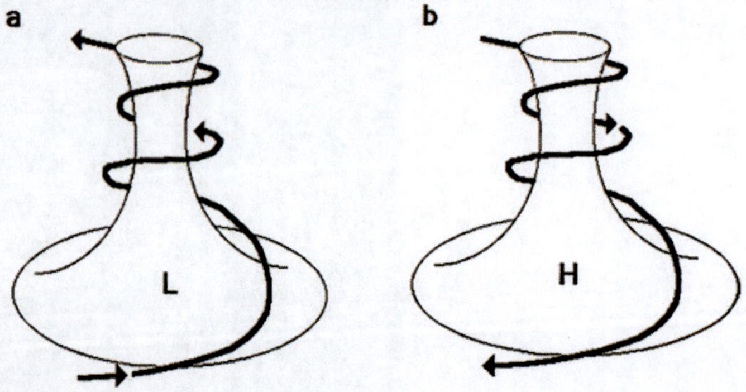

Fig. 12.8 Vortex in (**a**) cyclone; (**b**) anticyclone in the northern hemisphere

Fig. 12.9 Convergence and speed increase toward the axis of rotation in positive vorticity

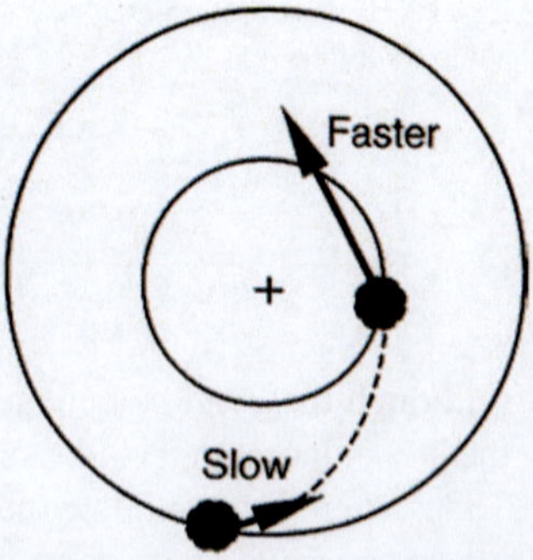

very humorous way. He noticed that the strong winds that occur in a tornado remove feathers (completely pluck) from geese, chickens, turkeys, etc., which are caught in their vortex. Guided by this knowledge, he performed the following experiment. Instead of cannonballs, he fired slaughtered poultry (geese, chickens, etc.) from a cannon aimed vertically. Loomis measured the ejected bodies to reach speeds of about 340 mph (~540 km/h). During this, the bodies were torn into smaller pieces that could barely be found when they fell to the ground. Before tearing the body, all the feathers are plucked from the body. From this record, Loomis concluded that the wind speed in the tornado was less than 340 mph, because the tornado only removed the feathers, without tearing the poultry's body to pieces.

The first pilots to fly through the center ("eye") of a tropical cyclone—hurricanes—were Captain Joseph Dankworth (pilot) and Lieutenant Ralph Ohar (navigator). On July 27, 1943, they flew an AT-6 light single-engine plane over a hurricane that was over the Gulf of Mexico and Texas. The flight was carried out as part of the regular monitoring of hurricanes in the season of their occurrence, with the aim of improving their forecasting.

12.5 Convective Movements in the Atmosphere

Vertical movements of air are called convection. Convection occurs when a layer of air (or any fluid in general) heats up from the bottom or cools down from the top. As a result of such movement, wonderful images of convective clouds appear in the form of a honeycomb (or a grouping of coffee powder on the top of a cup, Fig. 12.10).

This phenomenon was first dealt with around 1900 by the famous French experimenter Henri Benar (1874–1939). The first theoretical explanation of convective movements was given by Lord Rayleigh in 1916.

Benares cells are easy to spot in a cup of coffee. When we pour hot coffee into a cup, and immediately add a teaspoon of cream, a cellular structure in the form of a honeycomb will appear on the surface. In nature, convection almost always occurs when the air in the morning begins to warm up from the substrate. However, if there are no clouds at the point of rising air, such cells are not visible to the naked eye

Fig. 12.10 Convection in a cup of coffee

(although they can be detected by measurement). Similar pictures were noticed by the Nobel laureate Irving Langmuir (1881–1957) in 1938 on the surface of the lake. Today, there is a complete theoretical solution to this phenomenon. Mathematically, it is expressed by very complicated partial differential equations of the sixth degree.

12.6 Atmospheric Boundary Layer

Ludwig Prandtl (1875–1953) (Fig. 12.11) a German scientist, concluded in 1904 that fluid flow near a solid substrate can be divided into two sublayers. The first is thin, next to the base, called the "ground boundary layer," in which friction plays a decisive role, and the second layer is above it, where friction is negligible. In the atmosphere, there is a ground boundary layer with a thickness of 1–1.5 km.

In 1940, Prandtl was the first to derive the equation of fluid movement in the ground boundary layer. Together with Theodore von Kármán (1881–1963), as shown in Fig. 12.12, they developed the first semi-empirical theory of turbulence. Later, that theory was generalized and applied to turbulence in the atmosphere, especially in the ground boundary layer. Karman, a famous American scientist (of Hungarian origin) in the field of mechanics and fluid dynamics, also dealt with the statistical theory of turbulence.

Among those who dealt with the fundamental statistical characteristics of turbulence, we should mention A. Friedman (1888–1925) (Fig. 12.13). He is a famous Russian scientist who worked on various problems of nature, from cosmological (his model of the expansion of the universe) to the derivation of dynamical equations with turbulent terms. In 1924, he derived various correlation functions. Despite his short life, he made a great contribution to science. Dynamic equations of motion with turbulent terms are called Friedman-Köhler equations.

The wind speed increases from the ground toward the upper boundary of the boundary layer, and in terms of direction, the wind turns to the right in the northern hemisphere. This law was discovered by Fridtjof Nansen (1861–1930) and Walfrid Wagner Ekman (1974–1954) (Fig. 12.14).

Nansen was a famous Norwegian Arctic explorer. He received the Nobel Prize in 1922 for this work. Analyzing the data of the Norwegian Polar Expedition from 1893 to 1896, Nansen concluded that icebergs on water do not move in the direction of the wind, but at an angle of about 40°. Seeking an answer to why this happens, Nansen discussed it with Ekman. They concluded that each layer of water moves under the influence of the layer above it. In motion, there is a balance between three

Fig. 12.11 Ludwig Prandtl

Fig. 12.12 T. von Kármán

forces, the pressure gradient force, the friction force, and the Coriolis force. As the frictional force in the air decreases with height, the new equilibrium slows down so that the wind must veer to the right relative to the layer of air below it. Thus, if the peaks of the velocity vectors at various heights in the ground layer are joined, a spiral line is obtained, which is called the "Ekman spiral." Ekman published these results in two papers. The first was published in 1902, and the second, entitled: "On the Influence of the Earth's Rotation on ocean currents," was published in 1905.

The speeds of ascending and descending convective movements in a cloudless environment in the ground boundary layer are about 1 m/s. However, in developed cumulonimbus clouds, they are 10–50 times stronger. Because of this, there have been numerous plane crashes with a large number of victims. These currents are

Fig. 12.13 A. Friedman

Fig. 12.14 Walfrid Wagner Ekman

called "downbursts." When an airplane flies through such an area, it encounters strong updrafts and then strong downdrafts for a short distance. Since the plane is close to the ground, the pilot does not have time to adjust the plane's performance and therefore the crash is inevitable.

References

Gregory, R. 1930. Weather recurrences and weather cycles. *The Quarterly Journal of the Royal Meteorological Society* 41: 103–120.

Hadley, G. 1735. Concerning the cause of the general trade winds. *Philosophical Transactions of the Royal Society of London* 34: 58–62.

Halley, E. 1686a. An historical account of the trade winds and monsoons, observable in die seas between and near the tropics; with an attempt to assign their physical cause. *Philosophical Transactions of the Royal Society of London* 16: 153–168.

———. 1686b. On the height of the mercury in the barometer at different elevations above the surface of the Earth; and on the rising and falling of the mercury on the change of weather. *Philosophical Transactions of the Royal Society of London* 16: 104–116.

Kaemtz, L.F. 1845. *A complete course in meteorology*. Trans. C.V. Walker. London: Hippolyte Bailliere.

Kington, J. 1988. *The weather of the 1870s over Europe*. Cambridge: Cambridge University Press.

Namias, J., et al. 1935. *An introduction to the study of air mass analysis*, 129. The American Meteor Society.

Scot, D.F.S. 1976. Memoir of Luke Howard. In *Luke Howard (1772–1864) his correspondence with Goethe and his continental journey of 1816*. York: William Sessions.

Sorbijan, Z. 1996. *Hands on meteorology*. The American Meteor Society.

Thaxter, W.J. 1999. *The history of weather*. New York: Nova Scientific Publishing.

Chapter 13
Bergen Synoptic School

13.1 Early Bergen Weather

As we have seen, at the end of the nineteenth century, significant progress was made in the measurement system, especially by direct aerological sounding of the atmosphere. Since 1900, these measurements have been carried out continuously within the framework of the International Aerological Commission. Another significant advance occurred in the field of theoretical hydrodynamics. That new era in the development of theoretical meteorology was started by Wilhelm Freeman Bjerknes (1862–1951) with his famous circulation theorem. He talked about it to his students in theoretical physics classes at the University of Stockholm in April 1897. Then he published it in 1898, and in its complete form in 1902. It should be noted that in Krakow in 1896, Silberstein had already published one of the two circulation theorems. However, he viewed the baroclinic state only as a transitional state, without speaking of implementation. Bjerknes, in contrast to him, spoke about the

M. Ćurić, V. Spiridonov, *History of Meteorology*,
https://doi.org/10.1007/978-3-031-45032-7_13

application of the circulation theorem, and for that reason, it can be said that he is the real inventor of that theorem.

Wilhelm Friedman Bjerknes (1862–1951)

He was born on March 14, 1862, in Christiani (Oslo). His father, Karl (1825–1903), was a lecturer of applied mathematics at the University of Christiani. After completing his studies in Norway, Bjerknes continued his studies in classical mechanics and electromagnetism in Paris. He attended classes with the famous French mathematician Poincar (1845–1912). From Paris, he moved to Bonn, Germany, where he became an assistant to the famous scientist Heinrich Hertz (1857–1894). Since 1902, he has been dealing with geophysics, practically meteorology. From 1893 to 1907, he was a professor at the "Hochschool" of Engineers and a professor at the University of Stockholm. Between 1912 and 1917, he worked at the University of Leipzig. He returned to Norway in 1917. With his son, Jakob, and assistants Halvor Solberg (1895–1974) and Thor Bergeron (1891–1977), he founded the Bergen School of Meteorology, which made a significant contribution to modern synoptical meteorology. He was not only a great scientist but also exceptional in his human qualities. He was an extremely gifted speaker and an outstanding organizer. A lot of the best qualities are combined in one person. A talented theoretician and an outstanding practitioner. Despite his colossal contribution to science, he did not receive the Nobel Prize, thanks to a minor scientific figure, the physicist Osen, who was a member of the Nobel committee. Bjerknes died on April 9, 1951, in Oslo.

During his stay in Stockholm, from 1893 to 1907, Bjerknes often discussed his pioneering plans with his friend Nils Ekholm, who had already drawn some synoptic maps in 1891. This inspired Bjerknes to take one step forward and make the classical hydrodynamics of ideal fluids the hydrodynamics of real fluids, oceans, and atmospheres. Bjerknes created a research plan to put complex meteorological problems, especially weather forecasting, on a scientific basis. The plan ended in 1904. That plan was so ambitious and promising that it received the support of the Carnegie Institute in the USA for its implementation. The support was long-term and lasted from 1906 to 1941. The help consisted of the fact that he could finance talented young researchers who would work on the Project under the direction of Bjerknes. The first step in this direction was the establishment of the Geophysical Institute at the University of Leipzig.

13.2 Leipzig School

Based on his program from 1904, Bjerknes started with the realization of the program already in Stockholm. Then, together with J.V. Sandstrom, he published the textbook *Dynamic Meteorology and Hydrography*. He continued this work after 1907 in Christiani (Oslo). There, together with O. Devik and T. Hesselberg, he published two volumes of a book, *Statistics* in 1910 and *Kinematics* in 1911. When he founded the Leipzig Institute in 1913, in Fig. 13.1, Bjerknes tried to finish the third part of the book, called *Dynamics*.

Due to the outbreak of World War I, research efforts were reduced, so the group in Leipzig never completed this third part of the book. Instead, Bjerknes and his collaborators, joined by Sverdrup, performed a three-dimensional analysis of atmospheric conditions based on data collected during the International Aerological Days from 1900 to 1913.

In 1913, Bjerknes set the following several goals and philosophies on which he based future research:

1. The belief that the problem of weather forecasting can be solved by physical methods. It is the so-called rational physical approach.
2. The need and possibility of applying physical laws to states at every moment of time, and not only to the intermediate state. This is the principle of "case studies".
3. The need to collect observational data in such intervals of time and space that differentials can be found. It is the principle of a denser measuring network;

Fig. 13.1 Members of the Leipzig Institute: (from left to right) Heinz Letau, Paul Mildner, Wilhelm Bjerknes, and Ludwig Weikmann

4. Recognizing the need to improve synoptic tools. For the first time since 1860, synoptic maps are significantly improved. The orography on the maps is presented. The wind field is shown by means of isogons and isotachs. Altitude data are displayed at constant pressure levels.
5. Each individual piece of data must be in agreement with the general picture of the selected situation.

These principles of the Leipzig school were a necessary and sufficient condition to build a new idea. But basic conditions were needed for work. Life and work in Germany during the war became more and more difficult. The culmination of difficulties was during the winter of 1916–1917. There were restrictions on food and other needs, so Bjerknes arranged to return to Norway with his family and assistants.

13.3 General Conditions for the Development of Meteorology in Norway

When V. Bjerknes returned from Leipzig with two assistants (son Jacques and Solberg) in 1917, there was lively activity in Bergen regarding the establishment of the Geophysical Institute within the Bergen Museum (PAGEOPH 1980). At the head of these activities was the oceanographer Björn Hellen-Hansen. On March 17, 1917, the Council of the Bergen Museum sent an invitation to V. Bjerknes to fill the second position of professor at the Geophysical Institute. After they satisfied him with some other conditions, which enabled him to continue his work from Leipzig, Bjerknes accepted the invitation.

Several factors encouraged the development of meteorology within the Bergen School, in a direction opposite to that planned in Leipzig (Crew and DeSalvio 1939; Bergeron 1959; Bundgaard 1979; Jewell 1981; Namias 1980–1981, 1983). Namely, in Leipzig, Bjerknes set himself the task of finding a theoretical solution to atmospheric dynamics. But in Bergen, the development went in a different, more practical direction. The first factor is that a less formal academic life was lived in Bergen. Conditions for receiving meteorological data were better. New colleagues, especially the oceanographer Hansen, brought new vitality to the work. One factor decisively contributed to them turning to practical work instead of theoretical work. Namely, in Norway, there was a problem with the food supply, because the harvest in 1917 was very bad. That is why the Government in the fall of 1917 and the first weeks of 1918 supported every idea that would improve the production of agricultural products. This was the reason why Bjerknes turned from theoretical to practical work.

On February 13, 1918, the newspaper *Tidens Tegn* published a report in Kristiania according to which meteorologists from the Meteorological Service of Sweden supply their farmers with meteorological forecasts by telephone. Journalists asked the director of the Norwegian Meteorological Service, Hesselberg, whether something similar could be organized in Norway. The answer was that there are some problems of a practical nature that prevent it.

The previous answer created a key historical document on the history of the Bergen School. Namely, on the occasion of that interview, V. Bjerknes wrote the following letter to Director Hesselberg on February 18: "Helan-Hansen and I are very disturbed to read your answers in 'Tidens Tegn' of February 13. The situation is such that we must do everything we can to preserve meteorology in the country because situations like this will never happen again, for meteorology to be an important resource of the country."

There were numerous reactions regarding the problem of food supply. Bjerknes concluded numerous contracts, military and civilian. He also had a short conversation with Prime Minister Gunnar Knudsen, after which the government approved the financial resources needed to offer a weather forecast that would improve agriculture.

After that, Bjerknes writes a letter to Arrhenius: "Life is fateful. I suddenly become a practicing meteorologist. We will do everything we can to provide weather forecasts for farmers. This kind of work has only been attempted in the eastern part of Norway. But now, at my suggestion, we have received 100,000 kroner to establish a forecast service over the whole of western Norway, from Kristiansand to Trondheim."

Jakob Jacques Bjerknes, Fig. 13.2, worked intensively on the preparation of daily weather forecasts based on the increased number of measuring points established under the contract with the government.

Fig. 13.2 Jakob Jacques Bjerknes draws a weather map

V. Bjerknes wrote to Arrhenius on September 18, 1918: "Jacques really enjoys working here in Bergen. The squall lines and steering lines pass so often that they give him the best material to work with that can be." In November 1918, Jacques Bjerknes completed a brilliant eight-page work. In it, he presents a revolutionary cyclone scheme. In that cyclone concept, there is a storm line and a vertical line. Later, these lines were called "cold and warm fronts," which constitute the two cornerstones of the new meteorology.

There Are No Nobel Laureates in Meteorology

The Nobel Prize is the most prestigious prize awarded for achievements in various fields of science. It was established in 1901. The prize is awarded by the Swedish Academy of Sciences on the proposal of the Nobel Committee. In the field of meteorology, no one has received this award so far, because meteorology, as a science, is not included in the list of scientific fields for which it is awarded. One of the scientists who experienced this was Wilhelm Bjerknes. He was nominated for the award in 1923, 1924, 1926, 1929, and 1930. In cases where all proposals were rejected, the physicist Dr. S.V. Autumn consistently served as the member of the Nobel committee. In the explanation, Osen pointed out that meteorology cannot be a field for the awarding of the award, and meteorology cannot even be considered a part of physics, and therefore is outside the field for awarding the award. He also pointed out that Bjerknes's circulation theorem had already been worked out by Helmholtz, and that the polar front and cyclone theory had not been confirmed as a meteorological phenomenon in Sweden. Bjerknes, however, received other distinguished awards. Thus, in 1932, he received the gold medal of the British Royal Meteorological Society. In 1940, his son, Jakob (Jak), received the same award.

13.4 Thor Bergeron Is Coming to Bergen

In October 1918, V. Bjerknes and his two assistants, Jakob Bjerknes and Halvor Solberg, completed the work *On the Structure of Moving Cyclones* (Namias 1991). Immediately after that, J. Bjerknes and H. Solberg went to Christiani, Stockholm, and Copenhagen, in order to get data from second and third-order stations, as well as precipitation stations. Of course, along the way they tried to find people who would help them to establish a more intensive cooperation, and to help in the research work. At that time, university-educated people easily found well-paid jobs in Norway. There were not enough educated people in meteorology in Norway. That is why four young Swedes came to Bergen in 1919. Two of them were Thor Bergeron and Carl Gustav Rosby. None of them, with the exception of Bergeron, had prior experience in synoptic meteorology.

V. Bjerknes had the sense to organize young people to work harmoniously in the small town of Bergen, Figs. 13.3 and 13.4.

Fig. 13.3 Workroom in Bergen. From left to right are: Thor Bergeron, Carl Gustav Rosby, Sven Roseland, Jacques Bjerknes (standing), Sver Gasland, and Johan Larsen. The young lady is Gunvor Ferstad who received the data by phone and plotted it on the maps

Fig. 13.4 Bergen School in 1934 (from left to right): H. Solberg, H. Sverdrup, S. Petersen; sitting: T. Bergeron, V. Bjerknes, J. Bjerknes, K. Godske

Fig. 13.5 The house where V. Bjerknes lived, on the ground floor with his family, and the Bergen School was in the attic

They did not have fixed working hours, but very often stayed to work late into the night. From time to time, V. Bjerknes would leave his mathematical analyses and ask them: "Does anyone have a new discovery tonight?"

In the environment of long snowy winters, working in a warm room, which was heated by a stove from one corner, must have been inspiring, Fig. 13.5.

V. Bjerknes, who lived on the ground floor of the house, was a constant source of inspiration and encouragement, especially for those who were separated from their families for the first time in a long time.

A further description of the contribution made in Bergen can best be seen by following the life path of one of the most influential members of that group, Thor Bergeron. He is most responsible for spreading the method developed in Bergen to other countries. In order to understand the main character's traits and many of his talents (he spoke six languages), one should come from the deeper roots of his family.

13.4.1 The Origins of Thor Bergeron

In 1974, the book *Wendella* by Brita Hebe was published in Stockholm. The book describes the life of Wendela Hebe (1808–1888), a woman of great possibilities and many talents. She married Clemens Hebe (1804–1894), a landowner from the province of Smeland (southern Sweden) and a great adventurer. He had to leave Sweden with his wife and three young daughters for economic reasons. He lived in England for several years and then moved to the USA. In America, he was engaged in various jobs, from journalist, lawyer, translator, and historian, to politician. He actively participated in the campaign for the election of the president of the USA, Franklin Pierce.

When Clemens left Wendela in 1839, she returned to Stockholm, where she worked as a journalist and writer. Her three daughters were also very gifted artists. The middle daughter, Thekla (1835–1861), was an actress but died early of tuberculosis. She is Thor Bergeron's grandmother. Thekla gave birth to her son, Armand, in France in 1861, in an extramarital relationship with the well-known publicist and politician August Solomon (1824–1874). She died immediately after giving birth. Baby Armand (father of Thor Bergeron) was adopted by the Swedish married couple Fredrik and Konstanz Bergren, owners of a hotel in Paris. Many Swedish travelers to Paris, poets, and other artists stopped by the hotel. The Bergrens changed their surname to Bergeron (Bergeron), to correspond to the French language.

Armand Bergeron (1861–1898) worked as a stamp dealer. He married a singer and singing teacher, Hilde Stau. Their only child, Thor Bergeron, was born in 1891 in Godstone, England. Armand died of rheumatic fever in 1898, at the tender age of 37. The son and wife were left to live in poverty. When great-grandmother Wendela died in 1899, she left 10,000 Swedish kroner intended for Thor Bergeron's education. Money is really well invested.

Thor Harold Percival Bergeron (1891–1977)

He wrote these many names for himself in the biography he wrote himself in 1966. However, he always used only Thor Bergeron. He was born on August 15, 1891, in Godstone, Shure, England. He completed secondary education on June 13, 1910, in Lunzberg, Sweden. He finished his studies on October 28, 1916, at the University of Stockholm. He received his doctorate on September 20, 1928, at the University of Oslo. He worked as a young meteorologist at the Swedish Meteorological and Hydrological Institute (Institute) from 1919 to 1922, and at the Norwegian Weather Service, at the Meteorological Institute in Bergen, from (1919) 1922–1929 (There is some overlap here data from the biography he wrote himself). Then, until 1935, he worked as a consultant at the Meteorological Institute in Oslo. From 1936 to 1947, he worked again in the Meteorological Service of Sweden. From 1947 to 1961, he worked as a professor at the Department of Synoptic Meteorology at the University of Uppsala. Since 1953, he managed the large, long-term project "Pluvius" at the University of Uppsala. He was an instructor in the application of the new Bergen weather forecasting model around the world. On that occasion, he stayed in September and October 1953 in Belgrade. He held a series of lectures at the Federal Hydrometeorological Institute as part of the UNESCO technical assistance program. He was a member of a large number of commissions of the World Meteorological Organization. He communicated easily in all those missions because he spoke six languages. He married Vera Romanovskaya in 1932 during his second stay in Moscow. In the fall of 1976, right after he turned 85, he fell ill. He had pancreatic cancer. He died on June 13, 1977, in Uppsala.

13.4.2 Schooling and Meteorological Observation

While attending high school, Bergeron stayed at a boarding school in Lundsberg, near the town of Filipstand, in central Sweden. In 1908, his professor tasked the 17-year-old Bergeron with reading the barometer every morning for a month. That measurement acted as a real trigger for him. From the very beginning, the boy became a keen observer of meteorological phenomena and remained so for the rest of his life. He was truly a true meteorological observer, or phenomenon in the sky, or on a synoptic map. Since then, he has been observing visibility, wind, clouds, etc. Lundsberg is an ideal place for observing visibility. There are quite a few remote objects that served as rappers. He soon noticed that the visibility was changing a lot.

From 1909, Bergeron spent almost every summer in Ahn, near the Norwegian border. From there, visibility was up to 45 km, and the clouds could be followed as they moved from the Atlantic toward the mainland. There, in the lee of the mountains, the young observer often observed clouds that followed mountain waves, and "cloud systems," although this term had not yet been introduced into meteorology.

These observations of the young student determined his occupation and the various scenarios he saw in the sky encouraged him to research. His mother saw his interests, so she met Nils Ekholm, a meteorologist, polar explorer, and from 1914 to 1918, the director of the Swedish Meteorological Institute. That acquaintance was of great benefit to Thor. He received good advice from an experienced scientist and could feel at home at the Institute. He got to know Ecolm's method of forecasting based on wind speed and isallobar. When Bergeron began his studies in 1910, Ecolm advised him to master mathematics and physics well if he wanted to be a good meteorologist.

Around 1915, he became interested in atmospheric electricity. He found a connection between the turbidity of the air and the electrical conductivity of the air.

In the summer of 1917, Bergeron and one assistant spent 6 weeks in a tourist lodge in the Ara Valley (63°N, 850 m above sea level). This place is not far from Ana. All the time he was engaged in observation work. The place was ideal for observing the visibility. Once he saw mountains at a distance of 70 km. He later included those observations in condensed form in his doctoral dissertation, which he defended in 1928 in Oslo.

After returning to Stockholm, he often visited the Meteorological Institute (institute), in order to, analyze weather maps, and come to the conclusion that a significant change in visibility is accompanied by a change in wind direction. Then Sandström, former assistant to V. Bjerknes, and later director of the Institute, trained him how to draw currents on weather maps. Then Bergeron observed that the change in visibility occurs when the convergence line (which is now called the cold front) passes. Thus, in the fall of 1917, he set out on the path of the concept of air masses.

The following summer (1918), Bergeron carried out the same measurement program, but in Abisko, on the shores of Lake Tornetrek, in northern Sweden. He made visibility observations both during the day and at night. Based on these measurements, he concluded that there is no relationship between visibility and brightness.

This opened the way for the optical theory of visibility. He also showed that there is no relationship between visibility and relative humidity. He also elaborated on these observations in his doctoral dissertation.

13.4.3 Bergeron's Stay in Bergen in 1919

When Bergeron came to Bergen in the spring of 1919, V. Bjerknes and his assistants had already established a denser network of meteorological stations. They had gained practical experience in weather forecasting the previous summer. In 1919, representatives of the fishing industry contacted V. Bjerknes to conclude a contract on storm forecasts. However, he and his son, Jacques, hesitated to take on such a responsibility because they did not have enough experience in it.

Before coming to Bergen, Bergeron already had experience in forecasting, because he worked in Stockholm for about 2.5 months. He brought the method of isallobaric analysis to Bergen, so already in July 1919, they were practically doing it. During 1919, other habits were gradually introduced in Bergen. Thus, until then, the cold front (disaster line) was drawn on maps mostly in red and the warm front (management line) in blue. In the summer of 1919, Rosby suggested that the two colors change their meaning because it is more logical for a cold front to be marked with a blue line and a warm one with a red line. It was accepted, and to this day, it is done in all countries of the world.

Later, Bergeron suggested how to mark fronts when maps are made in black and white. He did this on a letterhead that he sent to Jacques Bjerknes from Leipzig on January 8, 1924, Fig. 13.6.

The names "cold and warm front" have been used since 1919. So, 1 year after the end of World War I, these names were logical, given that the line between the two armies could be considered similar to the line separating the two air masses.

On November 18, 1919, Bergeron observed and defined the occlusion front. Namely, that day, until late at night, he analyzed the weather situation from November 12, when very deep snow had fallen. He wanted to find out why it happened. He observed that the cyclone model of Jacques Bjerknes with a cold front and a warm front was inappropriate for this situation. Bergeron drew a vertical section and established that the cold and warm waves at some distance from the center of the cyclone merge into one line and that at the height above that line, there is warm air displaced. He called that part of the line the occlusion front. This made the Bergen cyclone model more comprehensive, and it is today, as V. Bjerknes said, "the alphabet of every meteorologist." Jacques Bjerknes did not initially accept any modification of his cyclone model. Later, after long discussions, he accepted Bergeron's concept of the cyclone with occlusion.

The third significant contribution of the Bergen School was the discovery of Halvor Solberg in February and March 1920. He noticed that, in one situation, the line separating the two air masses extends continuously around the entire pole, so

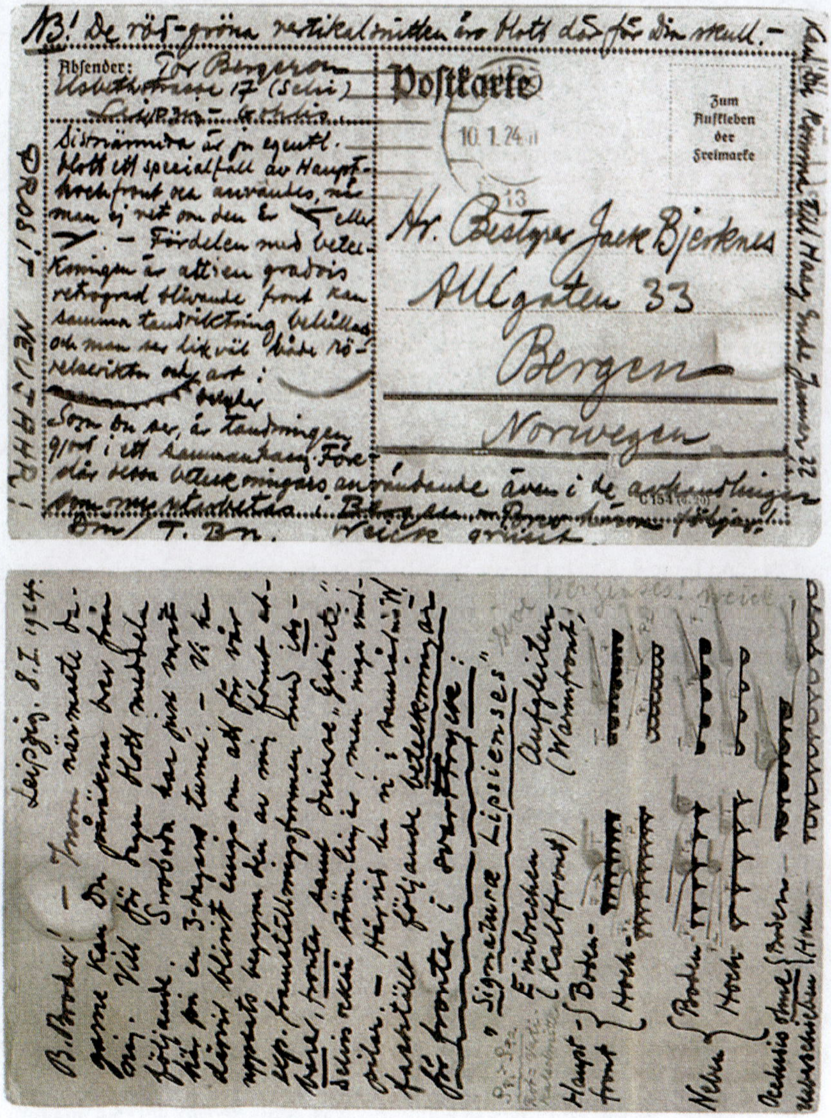

Fig. 13.6 Front and back of the Bergeron postcard with symbols for different types of fronts

there is a whole series (family) of cyclones. Thus, he observed a more complex organization of meteorological processes than a single cyclone.

In general, it is difficult to say strictly who discovered what or who proposed which name in the Bergen School. Because they worked as a group, they discussed among themselves. Both Bergeron and Jacques Bjerknes and Solberg observed a series of cyclones. Nevertheless, in 1922, Ž. Bjerknes and Solberg published the paper "Cyclone Development and Polar Front-Theory of Atmospheric Circulation" (Friedman 1987). There they also talked about the occlusion process, but Bergeron was not mentioned anywhere. Bergeron seems to have been angry that he was unfairly left out as a co-author of this work. That is why he does not mention in his autobiography that he worked in Bergen until 1922. Bergeron left Bergen shortly before Christmas 1919 and returned again in May 1922.

13.4.4 Work in Bergen from 1922 Until His Doctorate

Bergeron returned to Bergen again in May 1922, as he said: "for all time." He sought to find a practical method for analyzing air masses and fronts. For 1 year while he was away, J. Bjerknes was replaced by Bergeron as head of the Bergen Prognostic Department. In addition to other prognosticators, the German G. Schindl was also working at that time. He was very careful when analyzing the cards. In 1926, Bergeron married Schindl's sister Elfriede for the first time.

Many foreign meteorologists came to Bergen to learn their methods of weather analysis and forecasting. In the spring of 1923, a seminar was held for them. At that time, mostly meteorologists from small countries came. They came less often from bigger countries because it was a matter of prestige. Among the participants of the seminar was the young Sver Petersen. He studied in Oslo and became interested in meteorological work in Bergen. He asked Bergeron if it was worth it to study meteorology. Bergeron, although he was only 31 years old, answered him like an old man with great experience: "If you have a great interest in meteorology, if you will not give up because of hard work, and if you can bear disappointment, then you can study meteorology." Petersen not only did not disappoint but left an indelible mark in meteorology, working as a professor at the prestigious American Massachusetts Institute of Technology.

In the summer of 1923, Ludwig Weikmann visited Bergen to be informed about the results of the work. During that time, he stayed with the family of V. Bjer-knes, and they became close friends. Weikmann was supposed to take the position of professor of geophysics at the University of Leipzig from the fall of that year (a position held by W. Bjerknes from 1913 to 1917). They then agreed that Bergeron would travel to Leipzig to analyze weather situations together and that Bergeron would work on his doctoral thesis, which he was supposed to defend in Leipzig. Before going to Leipzig, Bergeron spent a few weeks in England. He traveled from London as the only passenger in a single-engine plane with two crew members to Rotterdam. It was his first airplane flight. The flight was at an altitude of 1000 m, below the base of the dense altostratus layer. When he arrived in Leipzig the next day, the same type of cloud was also there. When he took off from London, the London-Paris flight was canceled due to rain and low clouds. That weather situation, which Bergeron carefully observed during the summer, interested him and he asked Bergen to send him all the tickets for that day to Leipzig. He was told that the

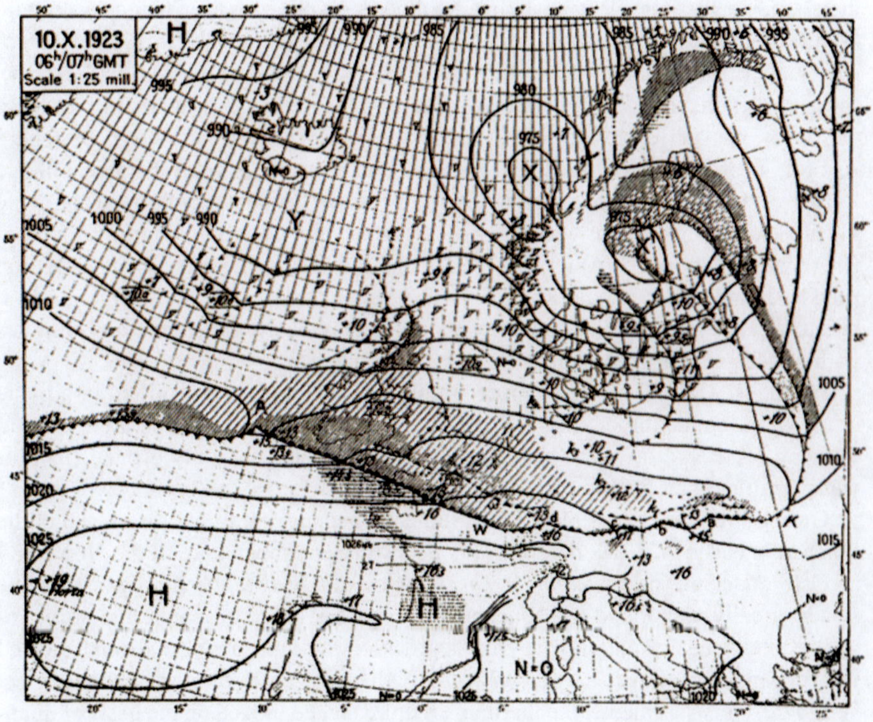

Fig. 13.7 Morning weather situation on October 10, 1923. Around noon, Bergeron flew from London to Rotterdam. The denser hatching represents precipitation, and the sparser the altostratus layer

Czech meteorologist Gustav Svoboda, who was staying in Bergen, also wanted to analyze the situation in more detail.

When Svoboda left for Prague, he stopped in Leipzig with a large roll of various weather maps. There, Bergeron and Svoboda agreed to write a paper related to the analysis of that situation. Over time, the work was expanded, so they published a monograph entitled *Waves and eddies on the quasi-stationary frontal surface over Europe*. A synoptic map on the day of Bergeron's flight from London is shown in Fig. 13.7.

That work was written in German, in 1924, and did not receive a significant response, among other things, because there was already more current literature written in English at that time. Bergeron could not even get a doctorate in that field, because according to the rules, the work from the doctoral dissertation had to be independent, without co-authors. So Bergeron had to start working again in another field for his doctoral thesis, but he still got a lot—Svoboda and Weikman as friends for life.

Bergeron returned to Bergen in February 1925. He chose a new topic for his thesis, but his income was not enough to live on, he had to be helped by his mother from Stockholm. V. Bjerknes gave him a salary supplement under the contract with the Carnegie Institute. Bergeron and Svoboda negotiated with Cherechevsky and Wehrle from the Central Meteorological Bureau in Paris to establish a small meteorological institute in Saarbrücken, which then belonged to the French (now German). In Saarbrücken, soundings would be made by plane every 6 h. At that time, there was a lot of experience with aerial sounding in the Netherlands. From Sesteberg Airport, near Utrecht, flights were made by the famous pilot Bekenes. Since Bergeron envisioned using aircraft measurements for his doctoral thesis to complete the theory, the negotiations with Saarbrücken were very important for him. However, Cherechevsky moved to America in 1926, and the plan failed. Rosby and Bergeron tried to organize altitude measurements by kite on an island west of Bergen. And it failed because Jacques Bjerknes was against it. Obviously, there were big stumbles within the small Bergen group.

In August 1925, Bergeron's close friend Ernest Calvagan was killed in a plane crash while collecting aerological data combined with cloud observations. From that day on, Bergeron never set foot on any plane. He used only ground (land or water) traffic on all his trips.

During the winter of 1926–1927, Bergeron began writing his thesis, which he finished in early 1928. He defended his thesis in September 1928 at the University of Oslo under the title *Combined Three-Dimensional Synoptic Analysis*. In it, he systematically elaborated the concept of air masses, tropical, polar, and arctic, as well as their transformation when moving from the source area. He also introduced the concept of "indirect aerology." Namely, he showed how, even without direct aerological measurements, the thermal structure can be judged on the basis of quantitative measurements of the type of precipitation, cloudiness, and visibility. He covered part of his doctoral thesis in the textbook *Fundamentals of Dynamic Climatology*.

13.4.5 Bergeron's Promotion of the Bergen School

Bergeron was the most important promoter ("apostle") of ideas and techniques of the Bergen School Jewell (1981, 1984). First, right after his doctorate, he spent the winter of 1928–1929 in Malta. But before that, an event took place that would take him to Moscow after Malta. Namely, in the spring of 1928, the Italian dirigible "Italia" made flights from Spitsbergen to the North Pole. On its return, on May 25, the airship hit an iceberg. A large international operation to rescue passengers has been launched. All means of transport, including icebreakers, airplanes, and ice transporters were used. Russia sent the powerful icebreaker "Krasin," which rescued the crew of the airship.

After the ship's return to Leningrad, they stopped in Bergen, so that Dr. Pishe, head of the Leningrad Meteorological Service, as part of the ship's expert team, would visit the Geophysical Institute in Bergen. There they were informed about new methods of synoptic meteorology. Since Pichet was impressed with the results, he invited Jacques Bjerknes or Bergeron to visit Russia as instructors of Russian meteorologists. Russia at that time was quite a closed country to the rest of the world, and Bjerknes did not want to accept the invitation, but Bergeron did.

Fig. 13.8 Thor and Vera Bergeron in Uppsala in 1969

When the ship Krasny returned from its voyage, the meteorological service was moved from Leningrad to a wonderful building in the center of Moscow. The synoptic maps were analyzed in a room where artistic paintings and icons were previously exhibited. On the first day of his first stay in Moscow (from September 23 to December 18, 1930), Bergeron met Vera Romanovska, Fig. 13.8, which drew synoptic maps.

She was a student of mathematics and physics. In February 1932 (during his second stay, from December 18, 1931, to October 31, 1932), Vera became Mrs. Bergeron. Since then, she has been his constant companion and helper in scientific work.

In Russia, he held courses that were attended mostly by older and more prominent meteorologists from all parts of the Soviet Union. They were very interested in Bergeron's lectures. Among the participants were forecasters Khromov and Djerdjeevski. The lectures were in German, and the notes taken by Vera were translated into Russian. The text was corrected by Bergeron, Khromov, and Djerdjeevski. During both stays, three textbooks in Russian were made from the notes, which were published in 1934. These three books were translated mainly unofficially and outside of Russia. Chromov expanded Bergeron's lectures and published the content under his own name in 1934 as a synoptic meteorology textbook. The second edition was published in 1937, which Svoboda translated into German and published in Vienna in 1940. Despite the fact that some believed that Chromov's textbook was largely Bergeron's, Bergeron remained good friends with Chromov after this.

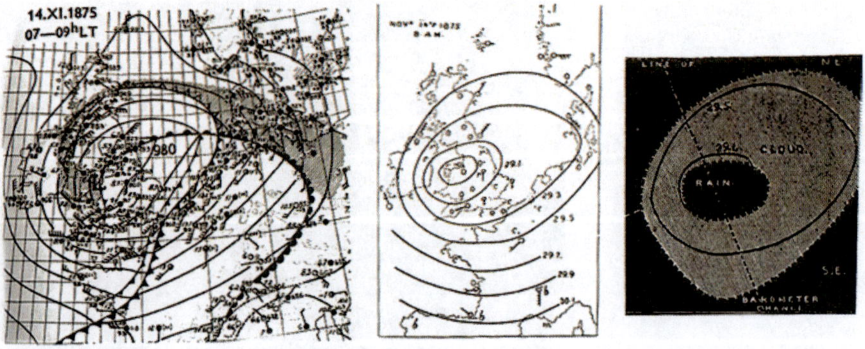

Fig. 13.9 Analysis of the same weather situation after (**a**) and before (**b**, **c**) accepting the Bergen School model

In September and October 1953, Bergeron held a course in synoptical meteorology in Belgrade. The lecture was organized by the Federal Hydrometeorological Institute, as part of the UNESCO technical assistance program. The course was attended by participants from all republics of SFR Yugoslavia.

How much the Bergen school cyclone development model improved the interpretation of the measured data can be seen in Fig. 13.9.

It shows an analysis of the same weather situation, based on the same data from November 14, 1875, after the acceptance of that model, Fig. 13.9a, and much earlier, Fig. 13.9b, c.

13.5 Acceptance of the Bergen School in the USA

For decades, weather maps at the US Weather Service have been produced by drawing a separate map for each meteorological element. For example. Maps of pressure, pressure tendencies, clouds, cloud movements, precipitation, etc., were created (Fig. 13.10).

This represented a pile of maps that the chief forecaster had to look at and formulate a weather forecast based on all of them. This was very difficult to do because of the lack of a single map on which all meteorological elements are located. In order to facilitate the work of prognosticators, the book *Weather Forecast in the USA* was published in 1916. Many prognostic rules were collected in this book, which was difficult to apply because there was no clear concept of the development of meteorological processes. One can imagine how a meteorologist felt, who should gain insight into the weather situation from a multitude of maps.

Despite this state of affairs, there were individuals in the US Weather Service in the 1920s who tried to improve the existing bad situation. Among them, two pioneers in this field should definitely be mentioned, Dr. Leroy Meisinger and Dr. Francis Reichelderfer. Meisinger was a great enthusiast who, as a good student of

Fig. 13.10 Forecasters at the US Weather Service in 1926, made maps with a single meteorological element

meteorology, went up with a balloon to study the upper layers of the atmosphere. He performed one such series of 10 climbs in 1924. Before the last, tenth lift, he was advised not to lift due to the possible development of a strong cumulonimbus. He decided to complete a series of 10 ascents and, with Lt. Neely of the Reporting Service, ascended in a balloon that was filled with hydrogen. Unfortunately, the electric discharge ignited the hydrogen balloon and both lost their lives. Meisinger was only 29 years old at the time of his death.

Despite his short life, Meisinger was considered the most prominent person in American meteorology. The American Meteorological Society awards regular awards in his name. He is one of the first to analyze height maps and who knew and accepted the theory of the polar front, developed in Bergen. He worked on his doctoral thesis with Professor Humphrey, who looked at the atmosphere like a physicist, who only understands what happens in a small volume of fluid, and did not have a broad view of the atmosphere, as is done today in dynamic meteorology. In his book *Physics of Air*, the first edition printed in 1920, Humphrey devoted only one page to air masses and the polar front. And in such circumstances, Meisinger paid attention to the new concept of the front and the air masses.

And Reichelderfer was a pioneer in the application of the theory of air masses and fronts in America, Fig. 13.11. He became acquainted with this theory in 1920, while as a naval officer, he encountered work in that field. He was the first in America to start analyzing the front on maps at the Navy Center in Virginia. Immediately after that, he was sent to Europe to learn the technique of map analysis. When he arrived at Berlin's Templehof Airport, Dr. Noth, who worked as a meteorologist, said to him, after learning that he wanted to travel to Bergen: "Can you

Fig. 13.11 Francis Reichelderfer (right) as head of the US Weather Service in 1962 with Joe Smagarinsky, head of the General Circulation Research Laboratory in Washington. In front of them is a computer map of isolines for the Northern Hemisphere calculated on the STRETI computer (in the background)

expect anything of scientific importance to be found in tiny Norway?" in the minds of certain people was a significant division into "big and small countries." Reichelderfer did not listen to him and continued to work on the acceptance of this theory in America.

Another important event that contributed to the spread of the Bergen model in America is related to the arrival of Carl Gustav Roseby in 1926 in America, as a scholarship holder of the Guggenheim Foundation. Reichelderfer helped Rosby a lot to stay in America, and to develop the famous school of meteorology at the Massachusetts Institute of Technology (MIT).

At first, Rosby was not welcome in America. The Meteorological Service did not accept Rosby's interest in the ideas of the Bergen School. The servicemen were conservative and told him that Norwegian ideas could not be accepted on the "large North American stage." Rosby did not give up on introducing a more scientific approach to weather forecasting. He analyzed weather maps but also simulated the behavior of the atmosphere in the "revolving tank" experiment, Fig. 13.12.

With the financial help of Harry Guggenheim, Rosby established the first proper department of meteorology at MIT in 1928. His first task was to expand the frontal theory and the theory of air masses so that a complete three-dimensional picture is obtained. While Rosby was still in the weather service, he noticed young H. C. Willett. He studied mathematics and physics at Princeton University and opted for weather forecasting. Rosby provided him with a scholarship for a one-year stay in Bergen in

Fig. 13.12 Rosby conducts an experiment simulating the rotation of the atmosphere in 1927, in the American Weather Service

1929. After his stay in Bergen, Rosby hired him at MIT. He brought together other capable members, O. Lang from Darmstadt as a specialist in instruments, G. Emmons for map analysis, and C. Pekeris for atmospheric dynamics. Having secured the teaching staff, Rosby made efforts to attract students to MIT's Department of Meteorology. Through Reichelderfer, who was then commander of the Navy, he ensured that the Navy sent students to graduate studies for many years. Studies included analysis of North American fronts and air masses, atmospheric thermodynamics, tephigrams, fog, stratus clouds, etc. Later, he managed to send postgraduates from all branches of the army to Rosby. The education they received there was extremely useful during World War II.

In 1934, President Roosevelt established the Oversight Committee on Meteorology, which consisted of: Carl Compton of MIT, Isaac Bowman of John Hopkins University, and Carl Milliken of the California Institute of Technology (KIT). The Committee's job was to advise and supervise the implementation of reform in the Meteorological Service. When the members of the Committee asked why they were not applying the Bergen Method, Chief Marvin replied, "Because the man who was sent to learn about the Bergen Method left our service."

The supervisory committee ordered that the method of analysis of fronts and air masses be applied. Horace Byers was the key person in charge of implementing the methods. Byers worked with Roseby in 1920, after completing his studies at the

University of California, Berkeley. It was there that Rosby noticed that Byers was a very capable meteorologist.

Byers encountered a number of problems in trying to introduce air mass analysis into the weather service. First of all, they placed his group in a corner of the weather forecasting department, at a sufficient distance from the classic forecasters. The group had daily discussions of polar front analyses, in which Byers was most assisted by Harry Wechsler and Jerome Neims. In 1931 Neims was employed in Washington as a receiver of meteorological data. He organized a small group of young people who were interested in attending evening seminars on the polar front. He himself was a lecturer at those seminars, where they quickly expanded the horizon of meteorological knowledge. Neims came to MIT to study in 1931. He began by writing a series of articles in the Bulletin of the American Meteorological Society, entitled "Analysis of Air Masses and Fronts." Later, from those works, he wrote the book *Air Masses and Isentropic Analysis* with his collaborators. One of the co-authors, along with Willett and Robert Stone, was Bernard Hauwitz, who visited Bergen and wrote a significant paper on the Norwegian wave theory. The book was very popular.

Neims worked at MIT under the direction of Rosby and Willett on the three-dimensional analysis of fronts. Vertical sections were made based on "direct aerology" (direct height measurements), Figs. 13.13 and 13.14, unlike indirect aerology, developed out of necessity (lack of data) in Bergen.

Because of the better analysis of vertical sections, Rosby then formulated the well-known diagram called the "Rosbygram." The purpose was to identify air masses using conservative properties of potential temperature and specific humidity. One axis of that diagram represented the potential temperature, and the other the water vapor mixture ratio, with slanted lines representing the equivalent potential

Fig. 13.13 A plane with a meteorograph hanging on the wing flies over Washington in 1934

Fig. 13.14 Pilot balloon measurements into the US Weather Service around 1930

temperature. Given the great expanse of America, then, for the first time, large-scale weather phenomena were realistically observed.

In the early 1930s, Dr. Irwin Crick founded the Department of Meteorology at the California Institute of Technology (KIT), chaired by Nobel laureate Milliken. Crick had no formal training in meteorology. He was tied to meteorology by Byers because Crick married Byers' sister-in-law (his wife's sister). Crick quickly learned meteorology, and as an excellent lecturer, he attracted a sufficient number of students to meteorology. He analyzed ground maps well and issued short-term forecasts for air traffic, as well as for the famous California film industry.

In the mid-1930s, Jörgen Holmboe, fig., also worked with Rosby at MIT (Fig. 13.15). Although he excelled in analyzing weather maps, he was primarily interested in dynamic meteorology. In mid-1940, he transferred to the University of California, Los Angeles (UCLA), where he was an associate of Jacques Bjerknes (who was then working there). From 1935 to 1940, a large number of papers and monographs were written in the field of isentropic analysis. That period is called the "isentropic era" by some. It was an effective method of monitoring the behavior of the "tongue" of moist and dry air, which was used in precipitation and cloud forecasting. Later satellite images of cloud cover confirmed the validity of the isentropic analyses. However, that method was replaced by weather analysis at selected heights of the atmosphere during World War II, due to its cumbersome nature.

Fig. 13.15 Jürgen Holmboe (right) with Karl Ludwig Godske

A major impetus to the development of knowledge about cyclone series and long waves before World War II was the contract signed under the auspices of the Government between the Meteorological Service, MIT, and the Bureau of Agricultural Economics. The goal was to study the conditions under which long-term droughts occur in agricultural regions of the United States, with an emphasis on the possibility of predicting their occurrence. This project accelerated the research of "centers of action" (cyclones and anticyclones), and cyclone series. The group at MIT was mainly concerned with long waves and related weather phenomena. A model of long waves, which represent a significant part of the general circulation of the atmosphere, was formulated.

Thanks to Rosby and his collaborators, a barotropic model of the atmosphere was formulated at MIT. Based on it, a method of maintaining absolute vorticity along the trajectory of air particles was formulated. The calculation of the vorticity flux at that time was done using the graphical method developed by Fjortoft, Fig. 13.16.

Later (in 1948), this method was used to "calculate" the weather forecast using the first computers. This was done by Jules Charney with Norwegian collaborators Fjortoft and Eliasen, Fig. 13.17.

With his efforts, Rosby managed to break the conservatism that reigned in the meteorological service of America in the 1930s. He trained a research group at MIT, and after World War II, he moved to the University of Chicago. He chose Sver Petersen as his successor at MIT, who sought to introduce objective forecasting methods using kinematic techniques. Rosby established a meteorological

Fig. 13.16 Rajnar Fjortoft, 1960

Fig. 13.17 A short break in work by the lake in Chicago, 1948. From right to left: Rosby, Jules Charney, Erwin Beal

department at the University of Chicago, with Horace Byers, who was still with the weather service, working there. And that's where Rozby gathered the meteorologists he had seen. Among others, the permanent visiting professor was Eric Palmen, Fig. 13.18, who later, in 1969, wrote the well-known textbook *General Circulation of the Atmosphere* together with C. Newton.

Some giants of meteorological science in America gathered in 1952 at MIT, at the Conference "On Large-Scale Processes in the Atmosphere", Fig. 13.19. From

Fig. 13.18 Eric Palmen

Fig. 13.19 Giants of meteorological science in America at the Conference from March 25–28, 1952 at MIT (see text)

left to right, front row: Henry Samella, Hard Willett, Hiao-Lan Kuo, Sam Solot, Ralph Shapiro, Jacques Bjerknes, Edward Lorenz, Victor Starr. Second Row: Bill Widger, Dave Fultz, Sherman Lowell, Major Don Martin, Seymour Hess, George

Benton, Harry Wexler, Jules Charney. Third Row: Lt. Edward Dolezer, Dan Cooley, Robert White, Eberhard Wahl, Robert Long, Yale Mintz. Fourth Row: Gunter Lesser, Thomas Kegan, Major Philip Thomson, Heinz Letau, Julius London.

References

Bergeron, T. 1959. Methods in scientific weather analysis and forecasting: An outline in history of ideas and hints at a program. In *The atmosphere and the sea in motion*, ed. B. Bolin, 440–474. Rockefeller Institute Press.

Bundgaard, R.C. 1979. Sverre Petterssen, weather forecaster. *The Bulletin of the American Meteorological Society* 60: 182–195.

Friedman, R.M. 1987. Constituting the polar front, 1918–1919. *Isis* 73: 343–362.

Galileo, G., trans. H. Crew and A. DeSalvio. 1939. *Dialogues concerning two new sciences*, 64–68. Chicago: Northwestern University Press.

Jewell, R. 1981. The Bergen School of Meteorology: The cradle of modern weather-forecasting. *The Bulletin of the American Meteorological Society* 62: 824–830.

———. 1984. The meteorological judgment of Vilhelm Bjerknes. *Social Research* 51: 783–807.

Namias, J. 1980–1981. The early influence of the Bergen school on synoptic meteorology in the United States. *Pure and Applied Geophysics* 119: 491–500.

———. 1983. The history of the polar front and airmass concepts in the United States: An eyewitness account. *The Bulletin of the American Meteorological Society* 64: 734–755.

———. 1991. Francis W. Reichelderfer, August 6, 1895-January 26, 1983. *National Academy of Sciences Biographical Memoirs* 60: 272–291.

PAGEOPH. 1980. *Bergeron memorial volume*. Basel: Birkhauser Verlog.

Chapter 14
Clouds and Precipitation

14.1 Introduction

There have been many theories about the origin of clouds and precipitation since ancient times (Howard, 1804, 1812, 1932). However, in those early periods, it could be said that these were speculations rather than theories. If the term "theory of clouds and precipitation" means attempts to explain the transition from water vapor suspended in the air to the formation of raindrops or snowflakes, then it can be said that these real theories begin in the seventeenth and eighteenth centuries (Aristophanes, 1973). Of course, even then there were incorrect understandings.

Obviously, this problem consists of two parts: how the tiny droplets form and how they form raindrops. Until the middle of the eighteenth century, in the absence of knowledge, these two steps were generally seen as one. Only in a few papers are there fundamentally correct views. Thus, in 1637, Descartes claimed that raindrops are formed by the merging of smaller droplets. For that to happen, there should be a lot of water vapor, and it should be "quite cold." As far as the coalescence of droplets is concerned, it was most commonly thought that it occurs due to air compression during the confrontation of winds from opposite directions. Such views were represented, for example, by Hadge Hamilton (1729–1805) from Dublin in 1765, Peter von Muschenbrock (1692–1761), or Jean Andre de Luc (1727–1817) in his work "Modifications." Another group of scientists had the opposite view, that is, in order for clouds and raindrops to form, air-spirit thinning should occur. Edmond Halley (1656–1742) was among the first to advocate such views in his work published in 1693. Dicarla–Bonifas Mareolin (1738–1816) published a remarkable work in Geneva in 1780, entitled *Des meteors locaux*. In it, he elaborated the theory of air thinning in updrafts, which leads to condensation. The original idea of dilution and condensation undoubtedly comes from Otto von Guerick (1602–1686) and his experiment in a "cloud chamber" from which he extracted air with a newly

M. Ćurić, V. Spiridonov, *History of Meteorology*,
https://doi.org/10.1007/978-3-031-45032-7_14

invented pump. However, until 1755, it was not known that when the air becomes rarefied, the temperature also decreases.

In his work in 1693, Halley pointed out that "there must be some kind of salts or particles raised from the ground that enable rapid condensation into the rain." Later, in 1715, Edward Barlow (1639–1719) claimed that the formation of rain required the presence of minerals and sulfur fumes. The Cartesian philosopher Antoine le Grand (1629−1699), who lived in Oxfordshire, wrote a paper published in London in 1680, in which he pointed out the following: "In summer, the raindrops are larger because they fall from greater heights, and traveling a long way through the air increases in size by collecting water (small droplets)." This opinion was expressed completely independently by Edley Mariot (1620–1684) in his work published in Paris in 1686. Marriott noted that the winter clouds are lower. In 1739, in "Physical treatises," published in Leiden, Muschenbrock pointed out that drops of different dimensions fall at different speeds, which are proportional to the volume of the drop. He described that raindrops quickly reach a maximum speed corresponding to their dimensions. He also explained that, due to the force of air resistance, raindrops cannot be larger than about 6 mm in diameter. Those who dealt with electricity used this knowledge to explain the formation of raindrops. So Benjamin Franklin (1706–1790) wrote a letter in 1771 to Dr. Thomas Percival (1740–1804) in which he says: "Every raindrop grows continuously as it falls. This happens because of two mechanisms. The first is the same as when a bottle filled with cold water collects water vapor (humidity) from the apparently dry surrounding air. The second is the same as when an electrified body attracts dirt. The raindrops coming from above are cold and electrified." Franklin obviously did not have enough knowledge about this to develop these claims into a meaningful theory.

In the works from the end of the eighteenth century, we find wonderful ideas that turned out to be completely correct. They were only formulated after about 150 years as a fundamental theory about the origin of rain. One is again related to Franklin. On December 22, 1784, he sent the paper "Meteorological imaginations and conjectures" to the philosophical and literal society in Manchester. One of the ideas is the following: "It is possible that the rain comes to the ground in the summer as such, and that it originated from the snow at a height, which melts falling through the warm air near the ground. Therefore, it changed from snow to rain." A similar opinion was expressed by Johann Henrich Lambert (1728–1777) in his work published in 1765, at the Munich Academy. He expressed such an opinion based on the observation that rainwater is often colder than ground air.

From that period, one interpretation of the origin of rain can in the true sense be called the theory of the origin of rain. It originates from the Scottish geologist James Hutton (1726–1797), whose work *The Theory of the Earth* is considered the cornerstone of his profession. In 1784, he published the work *The Theory of Rain*, which he presented before the Royal Society in Edinburgh. The basis of his theory is that he assumes that air at a certain temperature can hold a certain amount of water vapor. If there is more, it condenses. This is a completely correct position, which was expressed much later by Clausius and Claiperon. He further claims that rain and snow are caused by the mixing of two different heated airs. When both are sufficiently saturated with water vapor, condensation occurs. He explained it with a scene that is familiar to everyone in life. Namely, on cold winter days, when the door of the warm room is opened, condensation occurs due to air mixing. His theory could explain the cloud that appears in front of our noses on cold days due to breathing or the cloud above the pot in which something is being cooked. A significant shortcoming of his theory is that it is qualitative rather than quantitative. If he had tried to calculate anything, he would have noticed that mixing can explain the formation of some types of fog, but not heavy rain.

And the grandfather of the famous scientist on evolution, Charles Darwin (1809–1882), and Erasmus Darwin (1731–1802) dealt with the problem of the formation of rain. He published his views in the Philosophical Review in 1788. He connected the occurrence of rain with the barometric condition and air temperature. His views were not widely accepted mainly, because he expressed himself too poetically, and not strictly exactly, as it must be in science.

The eighteenth century was characterized by two important breakthroughs in science that are significant for meteorology. The first period, somewhere between 1730 and 1760, is characterized by the rapid expansion of knowledge about electricity, and the second, falling in the last three decades, is characterized by the "chemical revolution." The influence of the acquired knowledge about electricity on meteorology was not so great. Franklin, Watson, and Knolip mostly tried to apply knowledge from that field to meteorology through speculation. Franklin first tried to explain the formation of rain by means of electricity on April 29, 1749. He then sent a letter to John Mitchell, a Fellow of the Royal Society, in which he outlined possible mechanisms for the effect of "electric fire" on rain. The work was not well received and was published only in 1751. The chemical composition of the formation of precipitation was significantly more pronounced, and this development will continue in the nineteenth century.

As time went on, the views on the origin of precipitation came closer and closer to today's understandings. Thus, in 1876, Osborn Reynolds (Fig. 14.1) presented his

Fig. 14.1 Osborne Reynolds

work before the Philosophical and Literary Society in Manchester, in which he rejected the theory according to condensation can lead to the formation of large drops. He concluded this, because he believed that the released heat of condensation could not be effectively removed from the droplet. He considered that the removal of heat is done by radiation, and he did not take into account the transfer of heat through molecules. However, it should be borne in mind that the kinetic theory of gases had just begun to develop at that time. As far as the growth of the drops is concerned, according to Reynolds, it happens like this: "It is clear that the only way of the growth of the falling drops is to collect moisture particles from the air… If we take into account the total water content in the cylinder through which the falling drop passes, it is not difficult to explain the attainment of large drop sizes… Raindrops do not exceed a critical size for two reasons. Firstly, only a part of the water from the path of the drop is attached, and secondly, when the drop reaches a critical dimension, it is not maintained due to surface friction." Reynolds had an answer to the question of why all clouds do not rain. According to him, this happens because "under normal conditions, these cloud particles have equal dimensions, and do not have inertial movements. Dimensional changes occur due to some irregularity or disturbance in the cloud." Reynolds was clearly ahead of his time in terms of attitudes. It was only in 1935 that Bergeron elaborated "disruption," that is, "irregularity," as a trigger for precipitation.

14.2 New Technique and New Data

In the last quarter of the nineteenth century and the first quarter of the twentieth century, significant efforts were made to learn the characteristics of clouds and precipitation (Burkhardt 1977; Kikuchi 1979; Hamblyn 2001). Thus, in 1921, Keller caught drops of fog on a wire coated with a thin layer of grease. He determined their dimensions using a microscope. The second, indirect method of determining the dimensions of the drops was based on the measurement of the diameter of the corona ring produced by the cloud drops when they were illuminated by light from the Sun or the Moon. Keller made tens of thousands of such measurements. He found that the drops of such clouds usually have a diameter of 17 μm. In 1936, Hagemann used oil-coated film to study captured fog or cloud droplets. In 1904, Lenard used colored paper filters to determine the dimensions of the droplets, so that the drop in the soluble dye would leave a better mark. In 1953, John and Dean used a camera to photograph raindrops on the ground floor. In recent times, electro-optical drop meters are used. The first devices of this kind were developed by Nollenberg in 1970. Measurement of the dimensions of raindrops was carried out very precisely by Gan and Kinzer in 1948. Then they formulated the well-known law of distribution of drops by dimensions.

When height measurements were started, using balloons, airplanes, or observation along the mountainsides, one important fact was established. There can be liquid drops in the clouds even at temperatures well below 0 °C. It was first observed

Fig. 14.2 International Aerological Commission in 1921, in Bergen, Norway

by Fahrenheit in 1724. Such, so-called supercooled clouds were carefully studied by the Swiss naturalist Saussure in 1783. The Aerological Commission played a significant role in the establishment of height measurements, in general, even for the need for better knowledge about clouds. It was composed of eminent scientists. The composition of the International Aerological Commission in 1921 can be seen in Fig. 14.2.

Knowledge about the formation of clouds was also obtained from laboratory experiments. That is how the French scientist Paul-Jean Coulier (1824–1890) investigated the formation of clouds in a closed chamber. He varied the pressure, temperature, and humidity, seeing how it reflected on the cloud. After a series of successful experiments, he was suddenly faced with the fact that a cloud would not form despite sufficient moisture content at a given temperature and pressure. When he added some outside air to the chamber, a cloud formed immediately. Coulier correctly concluded that condensation requires dirt, which we now call condensation nuclei—tiny particles on which water vapor collects. Those particles were washed away by Coulee with many conducted experiments so that the air in the chamber remained clean. There are always a lot of such particles in the air, which is why a cloud appeared when it introduced some outside air.

The more detailed nature of these particles was studied by the Scottish scientist John Aitken (1839–1911). These particles are invisible to the human eye. He determined their size, composition, and origin. The smallest particles have dimensions between 10^{-7} and 10^{-5} cm, (known as Aitken particles), the large ones are between 10^{-5} and 10^{-4} cm, while the giant ones are those larger than 10^{-4} cm. And the American meteorologist James Espy (1785–1860) made a simple "chamber" in 1849, which he called "nephelescope." In it, he investigated the conditions necessary for the formation of clouds.

14.3 Wegener as the Originator of a New Idea About Clouds

A significant step in the development of cloud physics was the knowledge that a cloud with ice particles in a multitude of supercooled droplets represents a thermodynamically unbalanced environment, which manifests itself in the growth of crystals and in the evaporation of droplets. It is a consequence of nature's tendency to establish thermodynamic equilibrium. The first to clearly point out this fact was Alfred Lothar Wegener, a German meteorologist. In order to understand the physics of clouds, it is necessary to draw attention to the contribution of this energetic man to meteorology, and to science as a whole.

Alfred Lothar Wegener (1880–1930)

He was born in Berlin on November 1, 1880. He studied meteorology, geology, and astronomy at universities in Heidelberg, Innsbruck, and Berlin. He received his doctorate in Berlin in 1905. With his older brother Kurt, also a meteorologist, Wegener made several balloon flights. On one occasion, they achieved a world record in flight length of 52 hours. Wegener participated in four expeditions that studied the ice in Greenland.

Wegener married Elsa Köppen in 1913, daughter of the famous climatologist Vladimir Köppen. During the second expedition to Greenland, Elsa lived in Bergen, in the house of Wilhelm Bjerknes. Thus, the Bergen group and Thor Bergeron, in particular, were familiar with his thermodynamics, which Bergeron cited in 1933 when he published a famous paper on cloud physics. In 1924, Wegener and Köppen published the book *Climate of the Earth's Past*. They entrusted this voluminous work in the writing phase to our scientist Milutin Milanković for editing, whose astronomical climate theory they included in their work. They were friends with him and regularly visited each other in Belgrade and Austria. Kepen always addressed letters to Milanković in Cyrillic.

Wegener spent from the spring to the fall of 1929 in Greenland, in the third preparatory expedition for the fourth, which began in the spring of 1930. The fourth expedition surpassed all previous ones. They were supposed to spend the whole year (during the winter) in three base stations (one on the western side, the second on the eastern side, and the third in the middle of Greenland, at an altitude of 3 km and 400 km away from the other two).

(continued)

(continued)

Wegener left on November 1, 1930, (on his birthday) from the central station to the western station, with his companion Ekim Rasmus. Wegener's heart did not last (he suffered from heart problems). Rasmus buried him in an ice grave. During the summer of the following year, an auxiliary expedition found him in the grave, completely preserved, with his eyes open, with a calm and gentle expression on his face. Later, his wife Elsa wrote a comprehensive book about Wegener's travels based on his surviving diary.

During the first expedition, Wegener had the opportunity to observe and study the growth of frost. Then he thoroughly studied the existence of three phases (ice, water vapor, and supercooled water). In 1910, he published a paper on this topic in the Meteorological Zeitschrift. Immediately, the following year in 1911, he published the textbook *Thermodynamik der Atmosphera*. This textbook was one of the basic meteorology textbooks of that era. The chapters "General Thermodynamics of real gases" and "Physics of Clouds" are especially important for cloud physics. Interpreting the phase diagram for water, he writes: "When the three phases are in contact, the water vapor pressure will adjust between the maximum value relative to ice and water. As a consequence, continuous condensation must occur on the ice, while liquid water will evaporate at the same time. This process will continue until the liquid phase is completely consumed."

The application of this Wegenerian view of crystal growth at the expense of supercooled water did not occur until long after the invention. This seems to have happened, because Wegener was giving a lecture in a classical physics course that emphasized equilibrium thermodynamics. Several years later, the importance of this process was clearly observed by Bergeron.

14.4 Bergeron Mechanism

When Bergeron was returning from Stockholm to Bergen in 1922, he spent several weeks in February recuperating on Voksenkollen hill, 470 m above sea level. It is located north of Oslo, near the better-known place of Holmenkollen. That hill was often shrouded in stratus clouds, which manifested itself as fog on the ground. Bergeron often walked along a path through the fir forest that covered the hill. He noticed that the fog did not descend into the air layer to the height of the fir tree when the air temperature was between −5 and −10 °C. When the temperature was above freezing, the fog descended to the ground, both between the trees and along the path he walked on. Bergeron sketched those situations in his autobiographical diary (Fig. 14.3).

He interpreted the absence of fog at temperatures below the freezing point by the fact that supercooled droplets from a layer of air as high as the height of conifer

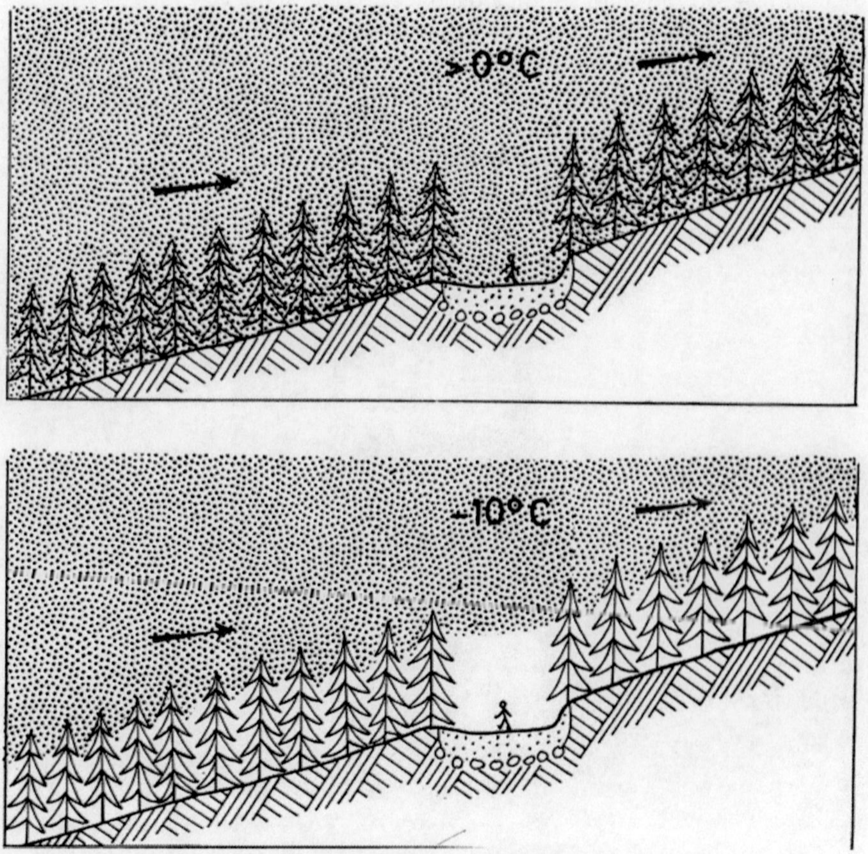

Fig. 14.3 Bergeron's drawing, from February 1922, with which he showed that fog does not descend between painted trees at temperatures below freezing

trees evaporated (sublimated, according to newer terminology) and the released water vapor was deposited on frost crystals that had formed on fir needles. He did not publish his conclusion, because he wanted to publish it in his doctoral dissertation, which had to contain original contributions that he had not published before. So this remained unknown until 1928 when Bergeron received his Ph.D. And since then, that work was not widely known, because the thesis was of "limited" scope, until 1933, when Bergeron presented it in Lisbon at the International Conference of the Union for Geodesy and Geophysics on September 19, 1933. The papers from that conference were published in 1935.

In that paper, Bergeron showed that a relatively small number of ice particles in a cloud of supercooled droplets can individually grow to such dimensions that they begin to fall, thereby disrupting the "colloidal" stability of the cloud. In that paper, he cited Wegener's textbook from 1911, and his doctoral dissertation from 1928. He concluded in his paper: "I assume that almost all raindrops (diameter larger than

0.5 mm) and all snowflakes form around ice crystals in the way I have explained." The results of this work are widely accepted as a new mechanism of precipitation.

Bergeron's hypothesis about the formation of rain (not taken into account here torrential rain, drizzle) represented the cornerstone of cloud physics. Theodore Robert Walter Findeisen (1909–1945), a German experimental meteorologist, greatly contributed to its acceptance and dissemination among the scientific public.

Theodore Robert Walter Findeisen (1909–1945)

He was born in Hamburg, on July 23, 1909. He studied in Karlsruhe and Hamburg with A. Wegener, A. Pepler, and W. Pepler. He earned his doctorate in 1931. His dissertation was on the dimensions, concentration, and coagulation of fog droplets. He was the head of the Meteorological Service at the Air Force in Munich. Later (during the Second World War), he was head of the cloud research department at the German Meteorological Service in Prague. He carried out many measurements of the characteristics of clouds with airplane flights. He was a specialist meteorologist for laboratory experiments on clouds. That is why he is considered the first real cloud scientist. After the end of the Second World War, he was shot around May 9, 1945, in Prague, as a member of the German army.

In 1938, he wrote a comprehensive overview of the known knowledge of cloud physics. There he developed Bergeron's hypothesis into a complete conceptual model of precipitation formation. This is why this mechanism is often called the Bergeron-Findeisen or Wegener-Bergeron-Findeisen process in the literature. Findeisen distinguished two types of cloud nuclei, condensation, and sublimation. He claimed that the collision and coalescence occur only between small droplets and that this is how drizzle is formed, while he believed that rain is formed through

Bergeron's process by the melting of snowflakes when falling through the warm part of the cloud. We can see that he has made a mistake in merging the larger drops. Today, it is known that the collection and coalescence of larger drops are an important mechanism of precipitation. Gravitational and acoustic gathering occurs in clouds.

14.5 Drop Growth by Coalescence

During the Second World War, the number of airplane flights over almost all parts of the Earth increased sharply. Pilots and meteorologists, as frequent companions of those summers, recorded that rain also appeared from warm clouds. It referred to tropical areas. In 1953, using radar and other measurements in his doctoral dissertation, Batan showed that in summer, precipitation often occurs in moderate latitudes due to the gravitational coalescence of drops. Ćurić and Vuković showed in 1988 and 1991 that in cumulonimbus clouds, rain often occurs due to the acoustic collection of droplets, which occurs due to sound waves caused by electrical discharges (Ćurić and Vuković, 1991). This put beyond doubt the question of whether the precipitation is produced by a mechanism called "warm rain."

The next important step in understanding the mechanism of rain formation is the knowledge about the accidental collision and coalescence of drops of similar size. The basic theory of the accidental joining of particles was elaborated by Smoluchovski in 1918. The application of that theory to the growth of fog drops was done by Schumann in 1940, while Hitchfield, a famous Canadian cloud researcher, explained the growth of drops in clouds with that theory. Random collisions and coalescence of drops expand the range of droplet sizes in the cloud. Edward Berry dealt with this in detail in his doctoral dissertation, which he defended at the University of Nevada in 1967.

A significant step forward in the field of cloud physics was made by the English meteorologist John Mason (1918–1998), a longtime professor of cloud physics at Imperial College in London and director of the English Meteorological Service. He dealt with virtually all areas of cloud physics. He wrote the first real textbook on cloud physics in 1957, "The Physics of Clouds."

14.6 Study of Ice Crystal Growth

Ice crystals and snow crystals have attracted people's attention since ancient times. However, the scientific understanding of the nature of crystals and their role in the occurrence of precipitation is only found after 1800. Thus, in 1820, **Scorby** interpreted the different shapes of snow crystals. Nevertheless, according to the depth of understanding of the formation of crystals, it is considered that Dobrovski was the first to take a serious step forward in explaining the connection between the shape of crystals and the temperature of the air where they form. Dobrovski was a scientist who was on the Russian ship Belgica, during the winter of 1888/1889, sailed between the Antarctic icebergs. He published his scientific observations on ice crystals in 1903.

An intensive study of the conditions under which ice crystals formed took place between 1930 and 1940. In addition to the detailed collection and analysis of natural ice crystals, laboratory crystals are created under controlled conditions in the laboratories of Hokkaido University, Japan. The research was carried out by Professor Ukichiro Nakaya (1900–1962) with his associates and postgraduate students. In 1936, Nakaya was the first in the world to produce an artificial snow crystal. He won an award from the Japanese Academy of Sciences for the discovery of the mechanism of the formation of snow crystals. In front of the University building where Nakaya conducted his experiments, a unique marble monument to the "snow crystal" was built (Fig. 14.4). In the building, a study room and a laboratory with Professor Nakaya's equipment are carefully kept.

Fig. 14.4 Monument to the first laboratory obtained snow crystal. In 1936, Ukichiro Nakaya successfully produced an artificial snow crystal for the first time in the world. The monument is located in front of the building of the museum dedicated to Nakaya in the university complex of the University of Sapporo

Ukichiro Nakaya (1900–1962)

He was born on July 4 in the city of Komacho, Ishikawa region, where he completed elementary and high school. He graduated from Tokyo Imperial University. After graduation, he devoted himself to the study of spark ignition, under the guidance of Professor Torahiko Terada. There he gained significant experimental experiences. After returning from his study stay in England, Nakaya took a job at the faculty of natural sciences at Hokkaido University in Sapporo. He chose snow crystals as his area of research. With this, he introduced experimental methods into meteorology. When he managed to produce an artificial snow crystal in a low-temperature laboratory (he could lower the temperature in the laboratory to −30 °C, which was rare at the time), his laboratory was visited in 1936 by His Majesty the Emperor of Japan. Then the documentary film "Snow Crystals" was shot. In 1940, Nakaya received an award from the Academy of Sciences for his work on snow crystals.

During the Second World War, he worked on many scientific programs but also engaged in literary work. He founded the Institute of Low-Temperature Science in Sapporo. He applied the experimental results to solve the practical problems of road icing, aircraft icing, and the appearance of fog over airports. At the end of the Second World War, in 1945, Nakaya had prepared for printing a book about snow crystals. When the printing was being done, the Allies bombed the printing house. Everything is destroyed. Fortunately, 1 copy of the manuscript and 3000 original photographs have been preserved. From this material, 10 years later, in 1954, Nakaina's unprecedented book in the English language "Snow Crystals" was printed.

(continued)

(continued)

After 1950, he continued his research in America. He was a participant in expeditions in the Arctic Sea and Greenland four times. He was elected vice president of the International Snow and Ice Commission in 1956. He wrote over ten works on film, painting, Japanese dances, etc. As much as he gave in the field of literature, he could be considered a very prolific author. Its greatness is even greater when it is known that he created it only in moments of rest, outside the cruelly cold conditions of his laboratories. Only in such uncomfortable conditions could he study crystals. Unfortunately, that also shortened his life. He died on April 11, 1962. When he was tired, his closest collaborator, Prof. Magono, noticed on Nakaya's lips the words "ko-o-ri, ko-o-ri," which means ice in Japanese.

Frozen Sailors in a Cumulonimbus Cloud

Five sailors found themselves with sailboats under a cumulonimbus cloud in 1930, in the area of the Rhine Mountains in Germany. They were caught in a strong updraft. Having lost control of the gliders, all five ejected. They too were lifted high by a strong updraft into an area with temperatures well below freezing. Then the descending current brought them down, and the ascending current lifted them up again. They experienced descent and ascent several times. At the same time, they were completely covered with a layer of ice. When they finally fell to the ground, they were completely pale, and only one of them survived.

A similar thing happened in 1982, in Australia. There, a paratrooper spent half an hour in a cumulonimbus cloud, caught in an updraft. When he opened his parachute, the current lifted him from 2 to 4 km high. To avoid further lift, he cut the parachute cords. Then he began to fall through the cloud. At 500 m above the ground, he opened an auxiliary parachute and thus happily landed alive.

14.7 Cloud Charging

The Italian scientist Giovanni Battista Becceria (1716–1791) studied electricity in the laboratory and in the atmosphere. He found that the water coming out of the faucet splits into smaller droplets that are charged. Based on this, he misinterpreted the formation of rain due to the charging of droplets. Today, it is known that when large drops are dispersed, the larger fragments are positively charged, and the smaller ones are negatively charged.

Separation of Charges by Droplets
When we are near a waterfall, or when we shower in the bathroom, we are surrounded by a large number of droplets. Then a large number of negative ions appear. Negative ions have been found to give a good feeling and contribute to a good mood. The negative ions of the air are partially composed of oxygen, while the positive ions are partially composed of carbon dioxide. Fresh air usually contains 700–1500 oxygen ions in 1 cm^3. The air in the forest contains much more about 15,000. The content is especially high after the arrival of thunder clouds with rain. There are areas that contain more negative than positive ions. Air with more negative ions is better. In urban areas, the air contains five to 10 times fewer negative ions than the air in unpopulated areas. Because of this, people lose one of the vital elements that give them strength and mood. The negative influence of positive ions is reflected in the increase of serotonin, a substance in the body that makes a person feel nervous, depressed, and irritable.

A large amount of electric charge occurs in cumulonimbus clouds. Electric discharges that occur due to a strong electric field in a cloud were first studied by Benjamin Franklin (1706–1790), an American scientist and politician who participated in drafting the Declaration of Independence.

Benjamin Franklin (1706–1790)

He was born in Boston, Massachusetts, the 15th of 17 siblings. He is one of the founders of the American nation. In his life, he was engaged in various jobs, from printer, writer, and diplomat to scientist. He invented the lightning rod and improved bifocal lenses. Franklin advanced the science of electricity. He was the first to study the Gulf Stream. He interpreted the great cold that gripped Paris in 1783 as the result of a volcanic eruption. He represented the USA during the Revolutionary War in the Paris court. He is one of the select members of the famous Great Americans.

In 1750, Franklin sent a letter to the Royal Society of London in which he suggested how to protect a building from lightning strikes. He describes that a long steel spike insulated from the underside can be placed at the highest point of a building and thus collect electricity from the clouds. To demonstrate this possibility practically, he tied a spike to a kite he was holding with a string. A combed silk ribbon was tied at the end of the string. He held the rope with insulator gloves. When he encountered a cumulonimbus cloud, he noticed that the silk fibers straightened. He touched the end of the string that held the key. Then a powerful spark appeared. He could have died from that if he didn't have protective gloves.

On April 12, 1753, Franklin used the aforementioned spike to collect electricity in the Leyden bottle (a container first used in the Dutch city of Leiden for storing electricity). Based on the comparison of that charge with the positive obtained by rubbing the glass rod, he concluded that "the storm cloud is most often negatively charged, sometimes positively, with the latter, I believe, rarely occurring."

After that, a large number of people devoted themselves to the study of the mechanisms by which the charge is separated in the Cb cloud and the nature of electrical discharges (Schonland 1956; Frisinger 1965; Pierce 1985; Vuković and Ćurić 1998). It was found that in one cloud, there are several areas with separated positive and negative charges. The nature of lightning was studied by researchers from South Africa, given that discharges are extremely frequent and intense there. In 1933, Schönland and his collaborators wrote significant works in that field.

Choi Magono (1916–1985)

He was born on September 22 in Komachu City, Ishikawa Prefecture, Japan. He is the oldest child of his parents. His father died in 1923. He completed primary and secondary school in his hometown and completed his studies in natural sciences in Kanazawa. As an exceptionally good student, as soon as he graduated he was invited to work at the newly founded Faculty of Natural Sciences of Hokkaido Imperial University. He accepted it, because Professor Nakaya, his compatriot, worked there. Under Nakaji's guidance, he worked on the creation of needle-shaped ice crystals in the area of small supercooling. During the Second World War, he spent 4 years in the army. During that period, Professor Nakaye led a military research team dedicated to the study of sea fog. After the war, food production was significant in Japan, so Magono founded the "Agronomical-Physical Research Institute."

(continued)

(continued)
Then he constructed an "integral thermometer" and an "integral solar radiation meter." He was engaged in analyzing the temperature of liquid water to prevent heat shock in rice fields.

Since 1955, Magono has been lecturing at Hokkaido University's regular and postgraduate courses in cloud physics and atmospheric electricity. He also held seminars on how to handle laboratory equipment (camera, working in a dark room, etc.). With his work, Magono contributed a lot to cloud physics and atmospheric electricity becoming integral fields of meteorology. He wrote an outstanding textbook in 1982, *The Thunderstorms*.

In contrast to the study of the nature of electrical discharges, Nakaya's coworkers from Sapporo University were concerned with the mechanisms of charge separation. Choji Magono (1916–1985) was particularly successful in this. He studied the electric field of the Earth's surface during rain and snow, and the charging of large raindrops, but he devoted most of his research to the generation of charges by various mechanisms related to ice crystals. Thus, he concluded that snow crystals become positively charged when the melting begins. He measured the electric charge created during the formation of frost in the natural environment, as well as the electrical discharges that occur due to the blowing of granular snow by the wind. During various snowfalls, he measured the charge of precipitation elements and the disturbance of the electric field at the ground caused by those precipitations. He also carried out experimental measurements of the charging mechanisms of crushed stone and granular snow in laboratory institutions. Magono, together with Kikuchi (the third generation of scientists: Nakaya–Magono–Kikuchi), carried out experiments on the artificial clearing of fog by a downward current created by dispersed falling water drops.

14.8 Weather Modification

Many times throughout history, people have tried to change the natural course of meteorological processes, to increase precipitation, to prevent the formation of hailstones, and to prevent strong electrical discharges. All this is connected to modifying the life cycle of the cloud, and that is why we talk about cloud modification or artificial cloud modification (Hess 1974). Rainfall stimulation was the most widespread. The word "Pluviculture" was also coined, originating from David Starr Jordan in 1925. By that word was understood the description of the formation of precipitation in an artificial way and the marketing related to it. In the early period of these jobs, there were good successes, but also scientifically unfounded jobs.

A turning point in cloud modification was made in July 1946 by three American scientists: Nobel Laureate Irving Langmuir (1881–1957), Vincent Schaefer (1906–1994), and Bernard Vonnegut (1888–1888) (Fig. 14.5). They worked at the

Fig. 14.5 Irving Langmuir (left) and Bernard Vonnegut watch Vincent Schaeffer which blows into the cooling chamber to form a cloud of supercooled droplets which they would sow with particles of dry ice (around 1947)

General Electric Company on the problem of aircraft icing at higher altitudes. They conducted the experiments in an open refrigerator, which was just then being used. During the hot summer of 1946 in Massachusetts, it was difficult to achieve low-freezing temperatures. Therefore, Schaefer used a piece of dry ice (solid carbon dioxide) to lower the temperature. He noticed that millions of tiny ice crystals immediately formed in the chamber.

Based on that accidental discovery, the formation of artificial ice crystals was tested in a real environment. The first such experiment was conducted on November 13, 1946. From an airplane flying through a supercooled stratus cloud over Massachusetts, Schaefer dropped 3 kg of dry ice granules. Then the first shower of artificial snow appeared from the clouds. Since then, the modern era of cloud modification begins (Fletcher, 1962; Dennis, 1980; De Mott, 1990). For this purpose, a wide variety of chemical agents and different methods of seeding clouds and fog are used. Today, weather modification is a modern field of meteorology.

14.9 Classification of Clouds

In addition to knowledge about the internal structure of clouds, it was necessary to make a simple classification of clouds, which would be used in the practical recording of cloud characteristics at meteorological stations (Nikitin 1947; Kington 1969; Khrgian 1955; Ludlam 1972; Hildebrandsson 1880). The first classification of clouds was published at the beginning of the nineteenth century. It was the work of

Lamarck from 1802. This famous naturalist did not intend to classify all possible clouds. He decided only on certain cloud forms which he assumed were formed under normal conditions and which, in his opinion, would be useful to distinguish from each other (Corsi, 1988). This classification, despite its real value, did not make any particular impression, even in France, so it seems that no one used it. The reason for this may have been his choice of somewhat specific names that were not easily accepted in other countries, or perhaps his work was discredited by the very fact that it was published together with forecasts based on astrological data (in the publication Annuaire Météorologique).

Jean Baptiste Pierre Antoine de Monde de Lamarck (1744–1829)

As a young man, he was wounded during the Seven Years' War. Lamarck came to Paris in 1760 and immediately devoted himself to scientific work. He was most interested in botany and meteorology. He was a curator in Paris when in 1799 he printed the first publication *Annual Meteorological Review* (Annuaires Météorologiques). In the third issue, he presented the first classification of clouds "Sur la forme des nuages, Annuaire pour Météorologique pour l'an X de la Republique Francaise, № 3"—On the forms of clouds, Meteorological yearbook for the tenth year of the Republic of France. It can be seen that he marked the years in accordance with the new revolutionary calendar (the first year was 1893).

In his yearbooks, he published astrological weather forecasts for the entire year. Despite the poor prognosis, Lamarck sold them well but was increasingly hated by his colleagues. He was never an easy person to work with. Louis Côté accused him "that the press is a hindrance to the progress of science." Emperor Napoleon also noticed Lamarck's alienation from science, so he wanted to dissuade him from astrometeorological work. This happened in 1809 when Lamarck wanted to publicly present Napoleon with a beautifully bound copy of his last work *Philosophie Zoologique*. The emperor refused to accept the gift and advised Lamarck "to leave the Moon and the stars and the unsuccessful pursuit of the atmosphere, but to concentrate on evolutionary biology, the science of the future in France." This brought Lamarck to a sad and sudden end to meteorology. He never wrote anything again. He quickly lost popularity among the people. From then on, his life was full of misery. He died in 1829, blind, penniless, and completely unknown. Lamarck can be included at the top of the list of unfortunate people who fill the history of science.

A year later, in 1803, Luke Howard (1772–1864) published a classification of clouds in England, which, in contrast, was a great success and which represents the basis of today's cloud classification. While Lamarck limited himself to defining and assigning names to a certain number of interesting cloud forms, Howard intended to establish a complete classification that would cover all possible cases. Howard singled out three basic classes—Cirrus, Cumulus, and Stratus—from which (by transformation or association) all the others arise. This conception is in some sense inconsistent. Cirrus and cumulus are privileged in this classification, where the first of them represents the purest type of cloud, composed of ice crystals in the upper layers of the atmosphere, and the second, a cloud composed primarily of droplets in the lower layers of the atmosphere. However, the cloud that Howard calls Stratus is not defined by the physical state of its constituent particles, and according to this classification can be found at all altitudes. Nevertheless, from a practical point of view, Howard achieved almost the same results as Lamarck. Four of Lamarck's five main cloud types also appear in Howard's division, just under different names. It is interesting that these two scientists, from different scientific circles, who never came into contact, achieved, independently of each other, similar results.

The famous German poet Johann Wolfgang Goethe (1749–1832) dedicated four of his poems to Howard, as a sign of gratitude for his classification of clouds, which he published in 1803, under the title *On the Modification of Clouds*. By this term "cloud modification," Howard did not mean what we understand today, artificially induced changes in the characteristics of clouds, but changes in the form of clouds that occur in a natural way.

German meteorologist Kamz added Stratocumulus to Howard's cloud forms in 1840, with a precise definition of this cloud that is in line with today's.

Renaud, the director of the observatory of Saint Maur and Monsoury, gave in his "Meteorological instructions" in 1855 the classification of clouds from which the names of several clouds of today's classification derive: cirrocumulus, cirrostratus, altocumulus, and altostratus. He introduced the last two mentioned types of clouds in the "Bulletin of the Monsuri Observatory," and his example was soon followed by the observatory in Uppsala. Thus, by introducing clouds of medium height, between low clouds and those of the Cirrus family, he was the originator of the idea of adopting cloud height as a criterion for classification, which was later founded by Hildebrandson.

In 1863, Roeu observed clouds in Havana. He put forward some original ideas, which may not have been as well received as they deserved. He gave a very clear description of the sky in the center of the depression, distinguishing two cloud layers one above the other: altostratus (called pallio-cirrus) and a layer of fractostratus or fractocumulus (called pallio-cumulus).

Hugo Hildebrandson (1838–1920), director of the observatory in Uppsala, was the first to use photography in 1879 to study and classify clouds. In his paper entitled "On the Classification of Clouds Applied to the Uppsala Meteorological Observatory," he included an atlas with 16 photographs. He adopted Howard's classification, with a few modifications. These changes related in particular to nimbus, which was not the name of every complex of rain clouds (especially not cumulonimbus), but only the dark, lower layer of the rainy sky, then to Stratus, which was associated with the fog that rises from the ground and remains at a certain distance from the earth, as well as the cumulostratuses, which, following the idea of Camus, called the heavy, heaped masses of cumulus. Hildebrandson took over the stratocumulus from KCam-z. In his first work, Hildebrandson tried to follow Howard's procedure but at the same time took into account later works.

A little later, Welbach and Ritter proposed classifications that were very different from Howard's (which was mostly already accepted), which is why they had almost no chance of success—which was later the case with the classifications of Clayton and Clement Ley. These authors are still to be credited for interesting definitions of species (a subdivision of large genera) or subspecies (specific characteristics of clouds that can be observed at different heights), and Welbach especially for the introduction of cumulonimbus or thundercloud, which is clearly different from cumulus.

Finally, in 1887, Hildebrandson and Ralph Abercrombie (1842–1897) published a classification of clouds in which they tried to harmonize the previous achievements; adhering to Howard's system, they also included newer ideas, especially those due to Reno (introduction of altocumulus and altostratus, distinguishing separated and continuous clouds at each level) and Welbach (introduction of cumulonimbus, classification of cumulus, and shrubs-wine clouds into different families). Abercrombie first went around the world twice (giving a good example of scientific conscientiousness) to check whether cloud shapes are the same in all parts of the world—a fact that is only true to a first approximation. One of the main characteristics of this approximation is the importance given to height as a criterion, since, according to these authors, cloud observations were made primarily to determine the direction of the wind at different heights. They grouped the clouds into four levels, the mean height of which they determined orientationally, based on measurements made in Sweden. Today's international classification of clouds arose directly from Hildebrands and Abercrombie's classification, without any major changes.

At the International Meteorological Conference, held in Munich in 1891, an unequivocal recommendation was made to use the cloud classification of these authors, and the establishment of a special committee was initiated, which was entrusted with the task of finalizing and publishing it in the form of a picture atlas.

This committee met in Uppsala in August 1894 and began the selection of images for the atlas. For this purpose, an exhibition of more than 300 photographs or sketches of clouds was organized. The committee in charge of publishing, consisting of Hildebrandson, Riggenbach, and Tisserand de Bohr, faced great technical and especially financial difficulties. In the end, Tisserand de Bor himself took over the creation of the atlas, which was published in 1896. That atlas had 28 color images, with accompanying text in 3 languages (French, German, English) with definitions and descriptions of clouds and instructions for observing them.

The cloud classification based on this International Atlas was soon formalized and came into almost general use in all countries. Almost all meteorologists who subsequently published studies on clouds adopted this classification, but it turned out to be insufficiently detailed. This prompted some meteorologists, notably Clayton and Vincent, to introduce new species or subspecies, which fitted the primary cloud forms.

Observations from airplanes made it possible to get to know the previously unknown details of the clouds, which completed and expanded knowledge about them. Finally, new theories based on hydrodynamic and thermodynamic interpretation of soundings gave a physical interpretation of clouds as well as their role in atmospheric disturbances. These new and very interesting points of view had to be definitely acknowledged.

The Commission for the Study of Clouds met in Copenhagen in September 1929. It was suggested that an extract from the Complete Atlas be published as soon as possible, intended for observers, in order to facilitate the application of the new International Key, in which observations of the state of the sky were widely represented. The opportunity to publish the new atlas was extremely favorable, thanks to the generous gift of the Catalan patron Rafael Pathot, who had already entrusted the science of clouds with the interesting work of the "Concepcio Ra-bel" Foundation. This made it possible to print 1000 copies of the Complete Atlas free of charge and to sell it, together with an abridged version, at a very favorable price. A subcommittee was also appointed, with Professor Siring as president, to prepare a program for the Year of Clouds (1932–1933) and to study the physical processes of the formation and evolution of clouds, with the idea of eventually creating a suitable addition to the General Atlas. It was proposed to prepare two more appendices, one on tropical clouds and the other on special local cloud formations, for the preparation of which Dr. Braque and Dr. Bergeron were appointed.

References

Aristophanes. 1973. The clouds. In *Lysistrata and Other Plays*. Trans. Alan H. Sommerstein. Harmondsworth: Penguin.

Burkhardt, R.W., Jr. 1977. *The spirit of system: Lamark and evolutionary biology*. Cambridge, MA: Harvard University Press.

Corsi, P. 1988. *The age of Lamarck: Evolutionary theories in France 1790–1830*. Berkeley: The University of California Press.

Ćurić, M., and Z. Vuković. 1991. The influence of thunderstorm generated acoustic wave on coagulation. Part one: Mathematical formulation. *Z. Meteor.* 41: 164–169.

De Mott, P.J. 1990. *Quantifying ice nucleation by silver iodide aerosols*. Ph.D. dissertation, Department of Atmospheric Science, Colorado State University, Fort Collins, 259.

Dennis, A.S. 1980. *Weather modification by cloud seeding*, 267. New York: Academic.

Fletcher, N.H. 1962. *The physics of rainclouds*, 309. London: Cambridge University Press.

Frisinger, H.H. 1965. Early theories on the cause of thunder and lightning. *Bulletin of the American Meteorological Society* 46 (12): 785–787.

Hamblyn, R. 2001. *The invention of clouds*. 1st ed. New York.

Hess, W.N. 1974. *Weather and climate modification*. New York: Willey. 842 str.

Hildebrandsson, H.H. 1880. *Sur la classification des nuages employée à l'observatorie météorologique d'Upsala*. 2d ed. Uppsala: C. J. Lundström.

Howard, L. 1804. *On the modifications of clouds*. London: J. Taylor & C.

———. 1812. The natural history of clouds. *Journal of Natural Philosophy, Chemistry and The Arts* 30: 35–62.

Howard, E.F. 1932. *Goethe and Luke Howard*, 66. F. R. S. Friends Quarterly Examiner.

Khrgian, A.Kh. 1955. *I. I. Kasatkin i ego raboty o grozakh i grozovykh oblakakh* (I. I. Kasatkin and His works on storms and storm clouds). *Meteorologiya i Gidrologiya*, No. 3.

Kikuchi, K. 1979. Profile of professor Choi Magano. *Journal of Faculty of Science, Hokkaido University* 6: 1–19.

Kington, S.J.A. 1969. A century of cloud classification. *Weather* 24: 84–89.

Ludlam, F.H. 1972. History of cloud classifications. In *Clouds of the world*, A Complete Colour Encyclopedia, ed. Richard Scorer. Newton Abbot: David & Charles.

Nikitin, D. 1947. *Ocherk razvitiya klassifikatsii oblakov* (An outline of the development of cloud classifications). *Meteorologiya i Gidrologiya*, No. 3.

Pierce, E.T. 1985. Storm electricity and lightning. In *Thunderstorm morphology and dynamics*, ed. E. Kessler, 277–288. Norman: University of Oklahoma Press.

Schonland, B.F.J. 1956. *The lightning discharge*, Encyclopedia of Physica. Berlin: Springer.

Vuković, Z., and M. Ćurić. 1998. The acoustic-electric coalescence and the intensification of precipitation radar echoes in clouds. *Atmospheric Research* 47–48: 691–693.

Chapter 15
Auxiliary Tools in Meteorology

15.1 Numerical Tables and Mechanical Calculators

Meteorological instruments provide a large amount of data that needs to be processed in order to be usable by meteorologists. Various types of calculations were used (Mul'tanovskii 1933; Boyer 1959; Haurwitz 1986). Some conversions had to be used due to the use of different scales in different countries. For temperature alone, the Celsius, Rheomir, and Fahrenheit scales were used for a long time. Corrections had to be made to the read state. Some quantities are calculated based on other measured quantities. For example, based on the dry and wet temperatures, the water vapor pressure is calculated. Some meteorological quantities had to be converted to standard conditions.

To solve the accumulated computing tasks, meteorologists used computing devices that had long been used by astronomers. In the second half of the nineteenth century, more tables appeared for the reduction of meteorological data. These tables represented, in fact, computing resources. Thus, in 1852, Arnold Jiat compiled 331 pages of *Meteorological and Physical Tables*. The fourth edition, from 1884, consisted of 700 pages of tables (Fig. 15.1). Those plates were also used in the twentieth century, and many are still used in the twenty-first century in many countries.

Blaise Pascal built the first adding machine in 1642. The first commercial mechanical calculating machine, the arithmometer, was patented in 1820 by Frenchman Charles Thomas. Until 1920, such machines were used for commercial and scientific purposes. There were well-known Odner machines (patented in 1891), Mercedes-Euclid machines (produced in 1910), etc. Many of these machines were used by meteorologists. Calculating climate changes based on astronomical theory was started by Milutin Milanković in 1912 on such a machine. Another important use of these machines was made by the English meteorologist Lewis Free Richardson. He used Odner's and Mercedes-Euclid's machines and calculated the first numerical weather forecast, which was published in 1922 (Richardson 1922).

M. Ćurić, V. Spiridonov, *History of Meteorology*,
https://doi.org/10.1007/978-3-031-45032-7_15

Reading of Thermometer, Fahr.		Temp of Dew-Point, Fahr.	Force of Vapor in English Inches.	Weight of Vapor.		Humidity, Saturation = 1.000.	Weight in Grains of a Cubic Foot of Air. Height of the Barometer in English Inches.						
Dry.	Wet.			In a Cubic Foot of Air.	Reqd. for Sat'n. of a Cubic Ft. of Air.		in. 28.0	in. 28.5	in. 29.0	in. 29.5	in. 30.0	in. 30.5	in 31.0
°	°	°	in.	gr.	gr.		gr.	gr.	gr.	gr.	gr.	gr.	gr.
62	62	62.0	0.559	6.25	0.00	1.000	491.2	499.9	508.7	517.5	526.3	535.1	543.9
	61	60.3	0.528	5.91	0.34	0.946	491.4	500.1	508.9	517.7	526.5	535.3	544.1
	60	58.6	0.499	5.58	0.67	0.893	491.5	500.2	509.0	517.8	526.6	535.4	544.2
	59	56.9	0.472	5.27	0.98	0.843	491.7	500.4	509.2	518.0	526.8	535.6	544.4
	58	55.2	0.445	4.99	1.26	0.798	491.9	500.6	509.4	518.2	527.0	535.8	544.6
	57	53.5	0.421	4.70	1.55	0.752	492.0	500.7	509.5	518.3	527.1	535.9	544.7
	56	51.8	0.397	4.44	1.81	0.710	492.1	500.7	509.5	518.4	527.3	536.1	544.9
	55	50.1	0.375	4.19	2.06	0.670	492.2	500.9	509.7	518.5	527.4	536.2	545.0
	54	48.4	0.354	3.95	2.30	0.632	492.4	501.1	509.9	518.7	527.6	536.4	545.2
	53	46.7	0.333	3.72	2.53	0.595	492.5	501.3	510.1	518.9	527.7	536.5	545.3
	52	45.0	0.315	3.52	2.73	0.563	492.7	501.5	510.3	519.1	527.9	536.7	545.5
	51	43.3	0.297	3.31	2.94	0.530	492.8	501.6	510.4	519.2	528.0	536.8	545.6
	50	41.6	0.280	3.13	3.12	0.501	492.9	501.7	510.5	519.3	528.1	536.9	545.7
	49	39.9	0.263	2.95	3.30	0.472	493.0	501.8	510.6	519.4	528.2	537.0	545.8
	48	38.2	0.248	2.77	3.48	0.443	493.1	501.9	510.7	519.5	528.3	537.1	545.9
	47	36.5	0.234	2.61	3.64	0.418	493.2	502.0	510.8	519.6	528.4	537.2	546.0

Fig. 15.1 Part of a table from Jiat Tables

15.2 Graphic Technique

A number of graphic techniques were developed in the first half of the twentieth century. Thus, in the first decade of the twentieth century, the Austrian meteorologist Felix Exner showed that the pressure tendency is inversely proportional to the area between successive isobars and isotherms. In 1923, Japanese meteorologist S. Fujiwara discovered how forecasters could calculate vorticity from weather charts using a celluloid wind gradient scale. In 1930, Jerome Neims of the Massachusetts Institute of Technology developed graphical methods for solving equations. One type of graphic technique is the tephigram, a diagram developed by William Naper Shaw in 1925 to forecast the development of clouds.

A number of graphical methods, called "nomograms," were developed. The name "nomogram" comes from the French engineer Maurice Ossagne who used a graphic technique in 1891. This technique is widely used by engineers of various professions, including meteorologists. An example of a nomogram for calculating the gradient wind based on the pressure gradient, latitude, and isobar curve radius is shown in Fig. 15.2. This nomogram was made by Herbert Bell of the University of Chicago and is shown in Humphrey's book *Physics of the Air*, which was printed in 1920.

Various logarithms (slide rulers) were also widely used. In 1948, John Bellamy found how to calculate absolute vorticity advection by the geostrophic wind from

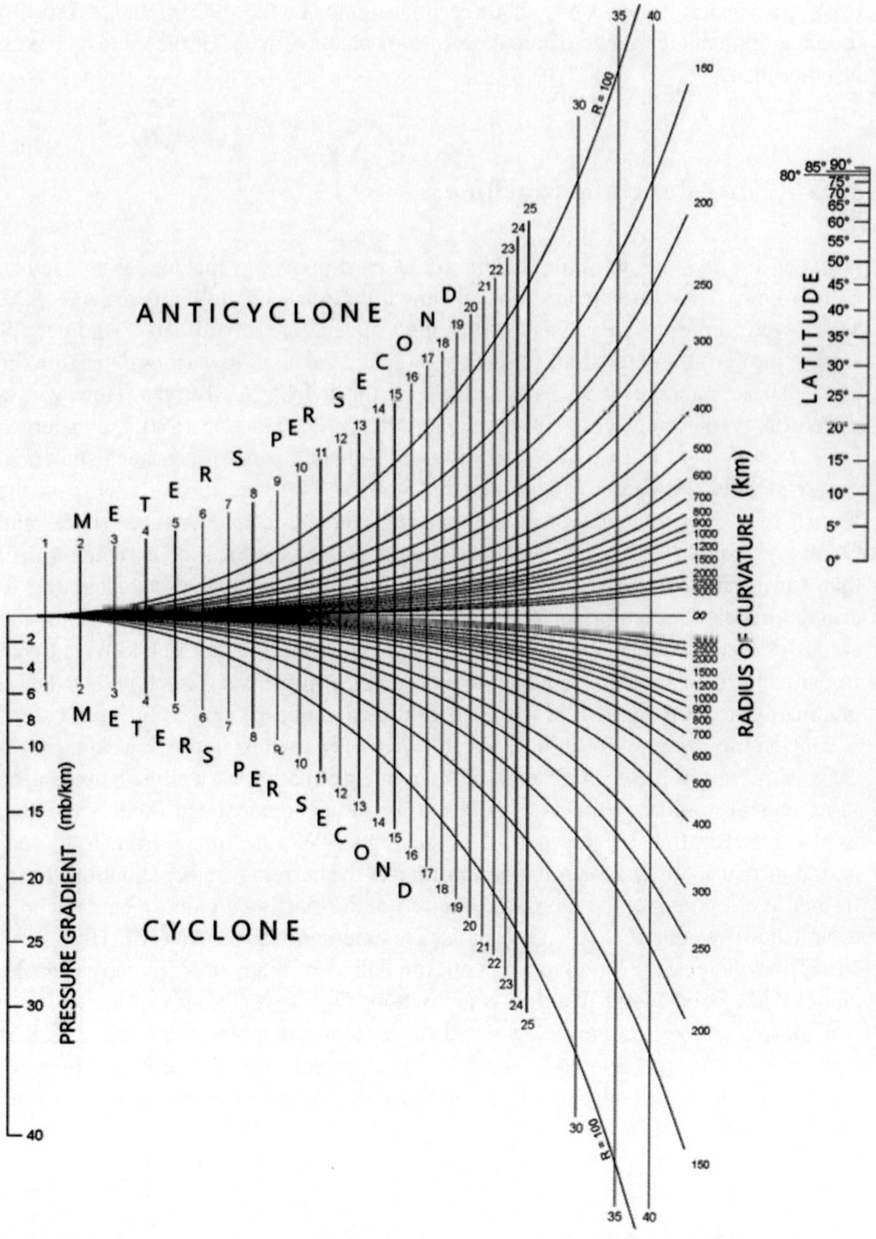

Fig. 15.2 Nomogram of gradient wind calculation using pressure gradient, isobar curvature, and latitude

Roseby's vorticity equation with three logarithms. In the 1950s, Rajnar Fjortoft used a graphical technique to calculate various characteristics of the atmosphere on weather maps.

15.3 Card-Punching Machine

Between the two world wars, an important card-punching machine was used in meteorology. These are the machines invented by Hermann Hollerath around 1880. There were three devices: one for punching special cardboard cards, another for sorting the cards, and the third for processing, and calculating various data from the cards. These machines were used in 1890 by the US Census Bureau. However, in meteorology, these devices were used only after 1920. Thus, in 1920, the meteorological service at the British Admiralty used punched cards for various statistical purposes. And the Dutch Meteorological Service borrowed those cards from the British in 1922. Immediately after that, they were used in Norway, France, and Germany. Around 1925, Czechoslovak meteorologist L. Polak constructed a drill that was installed at every meteorological station. In 1927, Polak used calculators to arrange tables of fractions of certain values of air pressure. Almost all European countries used machines for drilling and data analysis before the Second World War. In our country, this technique began to be used only at the end of the 1960s, when it practically stopped being used in other developed countries.

During the Second World War, punch cards were used to improve weather forecasts. Correlation between ground and height information as well as between the same meteorological elements at different locations is calculated. They were also used for forecasting by the method of analogies. Weather maps from a 40-year period were used to find the most similar map to the current weather situation. Then, further development of the time was found from the existing archive of cards, whose time features were punched on the cards. This method was used by American forecasters when deciding when to carry out the Allied landings in Normandy near the end of the Second World War. It was the famous D-day weather forecast.

Card-punching machines were also used for scientific purposes. Thus, V. Eckert showed how they can be used to numerically solve some types of ordinary differential equations. Eckert also wrote the book *Punched Card Methods in Scientific Computations* in 1940.

15.4 Analog and Digital Computers

During 1939 and 1940, John Mockley (1907–1980) constructed an electrical harmonic analyzer. Using it, harmonic coefficients with two significant digits could be calculated five to ten times faster than the fastest method up to that time. With this device, Mockley found the periodicity of precipitation, with a period of 13.5 and

27 days. In 1941, he constructed a harmonic analyzer or analog computer. Such devices were able to solve ten equations with ten unknowns. Meteorologists also used them to solve systems of nonlinear equations, which were set up by Edward Lorenz, as well as to represent functions in spectral form, according to the suggestions of George Platzman.

These devices provided low accuracy, usually one to two significant digits. That is why John William Mockley, together with John Presper Eckert (1919–1995), devoted himself to the construction of a digital computer. They succeeded in this and were the main designers of four famous computers: ENIAC, EDVAC, BINAC, and UNIVAC.

ENIAC (Electronic Numerical Integrator and Computer) is the first large digital computer. It consisted of 18,000 vacuum tubes. It weighed 30 tons, and a room of 150 m^2 was needed for its accommodation. EDVAC was the first computer with a stored program, while BINAC is the first such operating computer in the USA, and UNIVAC is the first commercial digital computer.

ENIAC was completed at the very end of the Second World War. At the time, John von Neumann (1903–1957) planned to build a more capable machine at the Institute for Advanced Study (IAS) in Princeton with which he could advance mathematical science. In order to demonstrate the revolutionary ability of computers in solving specific scientific problems, a meteorological project was formulated at the Institute. The basis of the project was a weather forecast model formulated by Jules Charney (1917–1981) with a group of collaborators (Fig. 15.3). The model solved

Fig. 15.3 The Princeton team in 1949 working on the first numerical weather forecast: John Fritman (seated), then from left to right Gilbert Hunt, Jules Charney, Joe Smagarinsky, Margaret Smagarinsky, Ellen Eliasen, and Arn Eliasen

the air vorticity equation. They got the first successful numerical weather forecast in March 1950, after 24 hours of calculation on the ENIAC.

Similar numerical forecasts were obtained in 1952 on the IAS computer. After that, the computers are used in several university centers and meteorological services. Since 1957, numerical weather forecasts have been made every day, and computers have been used to process meteorological data for various research purposes.

Computers became more and more capable, performed more calculations per second, and had more and more memory space. Meteorologists simply encouraged the improvement of computers for their own needs. So, in 1990, the fastest computer was the CRAY Y–MC C90. It performed 16 milli-yard operations per second. Today, in 2005, it already has very poor capabilities compared to the current best computer, the Japanese Earth Simulator (Fig. 15.4).

Despite the fact that since the appearance of the first ENIAC computer, the dimensions of the computer have been drastically reduced, to house the Earth Simulator, a special building had to be built, the size of a large football stadium, 65 m × 50 m × 17 m. Considering the grandiose unique capabilities of the computer, the building in which it is housed is completely protected from seismic, electromagnetic, and other influences. The computer is an ultrafast parallel system with a speed of 40 Tera FLOPS and a main disk that can store 10 terabytes.

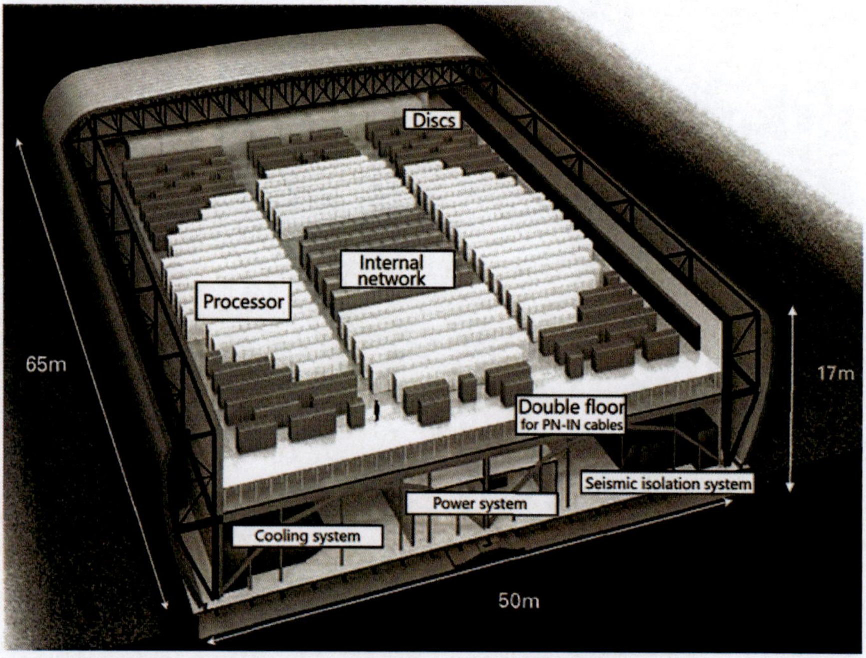

Fig. 15.4 The most capable computer of 2005, Japan's Earth Simulator. It performs 40 tera operations per second (i.e., 40,000 billion operations per second), and the disk has a capacity of 10,000 billion bytes

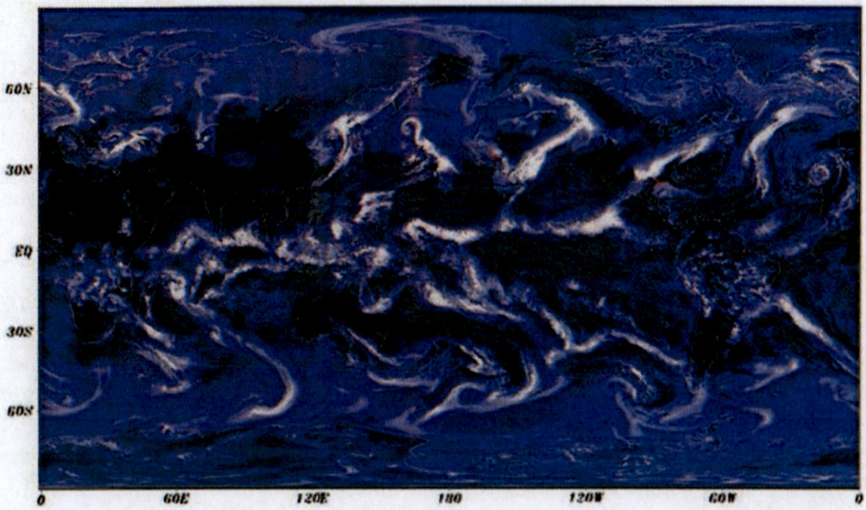

Fig. 15.5 Calculated amounts of precipitation over the whole country (mm/h) using the Earth Simulator

Such a capable computer is used to calculate forecasts of atmospheric and oceanic changes. It is possible to unify all detailed microprocesses (such as cloud elements) and global processes (those occurring over the entire Earth). One example of calculated precipitation amounts in the whole world is shown in Fig. 15.5. Precipitation amounts are expressed in shades of white. It is really fascinating how the image of calculated precipitation matches the measured image of clouds by satellite.

15.5 Modern Meteorological Technical Means

Modern meteorology uses many very expensive and complicated technical means every day. Just as at the beginning of the era of quantitative meteorology (seventeenth century), the thermometer and barometer were basic measuring tools, so today meteorological radars and satellites can, without exaggeration, be classified as basic tools. The first meteorological radar was produced in 1942, and the first meteorological satellite was launched in 1960.

A radar is an electronic instrument that sends out a narrow beam of very short pulses of radio waves (electromagnetic waves). Objects in the atmosphere (raindrops, snow crystals, hailstones, dust particles, etc.), over which the waves pass, partially reflect (reflect) them toward the radar antenna. The returned signals are electronically converted into an image containing information about the position, type, and "intensity" of clouds and precipitation. Some radars have devices that can measure the difference in the frequency of the sent and returned waves. That difference contains information about the speed of movement of objects in the direction of wave propagation. It is based on the well-known Doppler effect, a principle discovered in 1842 by the Austrian scientist Christian Doppler (1803–1853). Radars that have such a device are called Doppler radars. They give us, in addition to the information already mentioned, useful information about wind speed. Nowadays, Doppler radars are mostly used instead of ordinary ones. In 1990, 160 such radars were installed in the USA. Since then, the network of Doppler radars continues to be completed.

Weather satellites provide useful information about the location and type of weather systems over the entire Earth. The first non-meteorological satellite was launched in Russia in 1957. By analyzing the information obtained from it, scientists concluded that satellites can be a significant source of data on weather characteristics. Immediately after, in April 1960, Tiros, the first meteorological satellite, was launched in America. The first satellites had cameras for visible and infrared radiation, which allowed us to have a continuous picture of cloud cover (both day and night). Today, clouds and other meteorological features are monitored using geostationary and rotating satellites.

Geostationary satellites are located in the equatorial plane and are 36,000 km away from the Earth. At that height, even though they are orbiting, they are constantly over the same area of the Earth. Today, the entire Earth is covered with seven such satellites (Fig. 15.6).

The satellites shown are equipped with the latest generation of microwave scanning radiometers (Fig. 15.7) that operate at 36 wavelengths. In this way, after every 15 min, a lot of measured data from the atmosphere and from the ground is sent from them. Recently, special attention has been paid to measuring the condition of the ice sheets of the poles and Greenland. So efforts are being made to figure out what the Earth's climate future will be like.

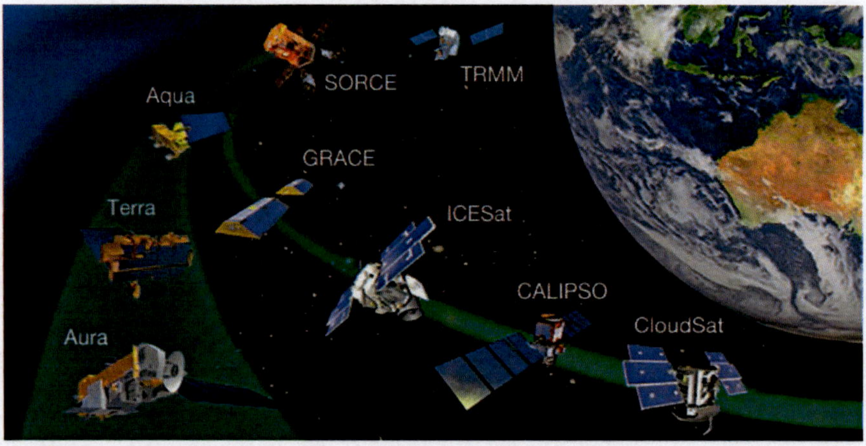

Fig. 15.6 Geostationary and orbital meteorological satellites

Fig. 15.7 Thirty-sixth channel radiometer installed on satellites AQUA and TERA

Orbiting satellites orbit the Earth in planes that are slightly tilted in relation to the poles, which is why they are called polar orbiting satellites. They orbit at a height of 870 km from the ground, so they circle the Earth 14 times in 24 h. Since the Earth rotates, each such satellite passes over any small area of the Earth twice in 24 h.

Meteorology as a science is experiencing rapid progress. Among other things, it contributes to the rapid development of electronics, applied mathematics, and the introduction of new measuring techniques. The most modern measurements are made using sodar, lidar, radar, and meteorological satellites in aviation laboratories. Achievements in the past, present, and future challenges make meteorology the most interesting branch of all modern sciences.

References

Boyer, C.B. 1959. *The history of the calculus and its conceptual development*, 187–223. New York: Dover Publications.

Haurwitz, B. 1986. Meteorology in the 20th century: A participant's view. *Bulletin of the American Meteorological Society* 66: 282–291, 424–431, 498–504, 628–633.

Mul'tanovskii, B.P. 1933. *Sovremennoe sostoyanie metoda dolgosrochnykh predskazanii* (The present state of the method of long-term predictions) – Meteor. Vestn., No. 5–7.

Richardson, L.F. 1922. *Weather prediction by numerical process*. Cambridge: Cambridge University Press.

Chapter 16
Development of Modern Meteorology

16.1 Development of Modern Meteorological Measurements

Modern meteorology uses many very expensive and complicated technical means every day. Just as at the beginning of the era of quantitative meteorology (seventeenth century), the thermometer and barometer were basic measuring tools, so today meteorological radars and satellites can, without exaggeration, be classified as basic tools. The first meteorological radar was produced in 1942, and the first meteorological satellite was launched in 1960.

16.1.1 Meteorological Radars and Lidars

A radar is an electronic instrument that sends out a narrow beam of very short pulses of radio waves (electromagnetic waves). Objects in the atmosphere (raindrops, snow crystals, hailstones, dust particles, etc.), over which the waves pass, partially reflect (reflect) them toward the radar antenna. The returned signals are electronically converted into an image containing information about the position, type, and "intensity" of clouds and precipitation. Some radars have devices that can measure the difference in the frequency of the sent and returned waves. That difference contains information about the speed of movement of objects in the direction of wave propagation. It is based on the well-known Doppler effect, a principle discovered in 1842 by the Austrian scientist Christian Doppler (1803–1853). Radars that have such a device are called Doppler radars. They give us, in addition to the information already mentioned, useful information about wind speed. Nowadays, Doppler radars are mostly used instead of ordinary ones. In 1990, 160 such radars were installed in the USA. Since then, the network of Doppler radars continues to be completed. More capable, so-called dual-polarized radars have also been developed.

M. Ćurić, V. Spiridonov, *History of Meteorology*,
https://doi.org/10.1007/978-3-031-45032-7_16

The development of radar technology is closely related to the development of physics. A significant contribution was made by the German physicist Heinrich Rudolf Hertz (1857–1894) with the experimental discovery of electromagnetic waves in 1888. They were previously predicted theoretically by Maxwell's equations. In 1904, German engineer Christian Hülsmeyer (1881–1957) patented a device that could detect radio waves reflected from ships. He called the device a telemobilscope. Sometimes he is called the inventor of radar, even if his device could not determine the distance of a distant object. The inventor of radar is Robert Watson-Watt (1892–1973), who in 1935 conducted experiments with radio waves for the needs of the Royal Air Force using a device he constructed called radio direction finding.

A breakthrough in the development of radar technology was made at the beginning of the Second World War by the British physicists John Randall (1905–1984), Harry Boot (1917–1983), and James Sayers (1912–1993). They constructed a hollow magnetron that would serve as an emitter (transmitter) of powerful electromagnetic waves of high frequencies. The first radar was constructed immediately afterward in the research laboratory of the Massachusetts Institute of Technology in Cambridge near Boston. The development of similar devices took place simultaneously in other countries during the Second World War for military purposes. They very quickly found applications in meteorology as well. The first observation of precipitation by radar was in Great Britain in 1941 (Raghavan 2003). The first monitoring of tropical cyclones was done shortly thereafter (Maynard 1945). Very soon radars began to be used in thundercloud research projects. They were used by Buers and Braham (1949) in a large research called the Thunderstorm Project.

It is believed that the first radar designed exclusively for meteorological applications was the AN/CPS-9 which operated at a wavelength of 3 cm. It was constructed in the USA in 1949. In Canada, the scientific project Stormy Weather started in the army in 1943, and then it was transferred to the University of Montreal at McGill University. The project was managed by John Stewart Marshall (1911–1992), who is rightly called the father of radar meteorology (Raghavan 2003). In addition to Marshall, the pioneers in the development of radar meteorology were David Atlas (1924–2015) and Louis Joseph Battan (1923–1986).

Great progress in the development of radar technology has been made with the use of electromagnetic waves that have different polarizations (horizontal, vertical, circular). Such radars are dually called polarization radars. They entered operational use after 2000. Polarization radars could transmit and receive two different polarized radar signals, which provide more information about the shape, size, and phase of the hydrometeor (ice or water).

LIDAR (light detection and ranging) is a device for remote measurements in meteorology that could emit laser light of wavelengths from the ultraviolet, visible, or near-infrared part of the spectrum. Thus, the wavelengths used by lidars range from about 250 nm to about 10 μm.

Lidars, like radars, can be CW or pulse lidars, monostatic or bistatic, can use the Doppler technique, etc. Also, they can be mounted on the ground, on airplanes, or on satellites. The lidar transmitter is a laser that emits directional light in a desired direction, while the receiver is a telescope that detects the backscattered laser light from the target. The distribution of atmospheric constituents (water vapor and aerosols), as well as temperature and clouds, was determined from the intensity of the returned laser energy. The vertical wind profile is obtained from the Doppler frequency shift, as with Doppler radars.

Doppler lidars operating at wavelengths from 0.3 μm to 10 μm receive scattered laser radiation from various particles suspended in the atmosphere, such as sea salt particles, haze, smoke, and water droplets (Beran 1997). These particles can be suspended in the atmosphere for days and can serve as tracers for lidar measurements of vertical wind profiles (Eberhard et al. 1989; Zuev et al. 1992).

Lidar measures and calculates various parameters at heights from 100 m to 105 km from the ground at different wavelengths using different types of lasers. They are useful for examining highly variable atmospheric parameters. Lidars can probe a wide range of atmospheric processes, from the smallest scales to synoptic scales (De Jong et al. 2007). Meteorological phenomena such as fronts, tropical cyclones, or leeward mountain waves can be successfully monitored with lidars. They can be used to identify polar stratospheric clouds based on their light-scattering characteristics. They can tell whether they are droplets or ice crystals. Also, they can contribute to our knowledge about the effects of aerosols on the climate of planet Earth. Therefore, it can be said that the typical application of lidar is for the research of the middle atmosphere, the planetary boundary layer, and urban meteorology, i.e., examination of the so-called urban heat islands.

16.1.2 Meteorological Satellites

Weather satellites provide useful information about the location and type of weather systems over the entire Earth. The first non-meteorological satellite was launched in Russia in 1957. By analyzing the information obtained from it, scientists concluded that satellites can be a significant source of data on weather characteristics. Immediately after, in April 1960, Tiros, the first meteorological satellite, was launched in America. The first satellites had cameras for visible and infrared radiation, which allowed us to have a continuous image of obscurity (both day and night). Today, clouds and other meteorological features are monitored using geostationary and rotating satellites.

Geostationary satellites are located in the equatorial plane and are 36,000 km away from the Earth. At that height, even though they are orbiting, they are constantly over the same area of the Earth. Today, the entire Earth is covered with seven such satellites (Fig. 16.1).

The latest generations of microwave scanning radiometers (Fig. 16.2) operating at 36 wavelengths are located on the satellites shown. In this way, after every 15 min, a lot of measured data from the atmosphere and from the ground is sent from them. Recently, special attention has been paid to measuring the condition of the ice sheets

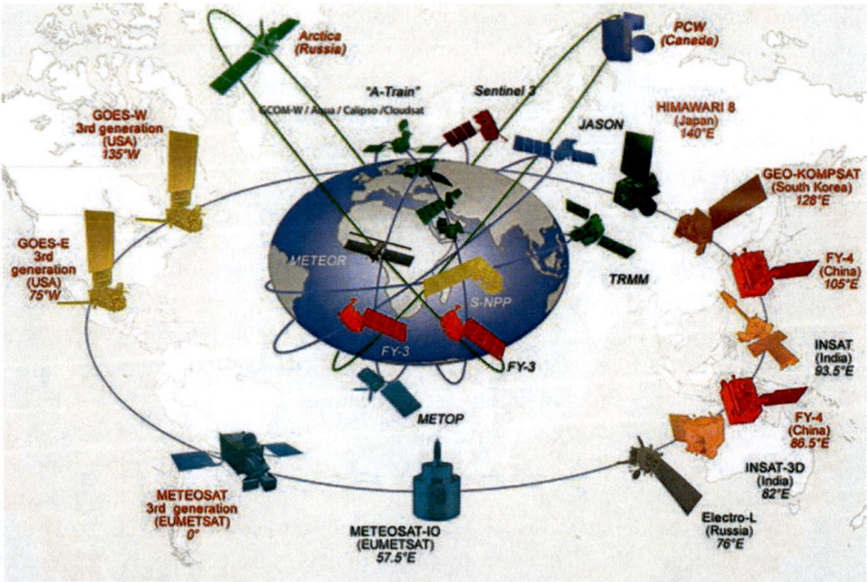

Fig. 16.1 Geostationary and polar-orbital meteorological satellites

Fig. 16.2 Thirty-six-channel radiometer installed on satellites AQUA (left) and TERA (right)

of the poles and Greenland. So efforts are being made to figure out what the Earth's climate future will be like.

Polar-orbiting satellites orbit the Earth in planes that are slightly tilted in relation to the poles, which is why they are called polar-orbiting satellites. They orbit at a height of 870 km from the ground, so they circle the Earth 14 times in 24 h. Since the Earth rotates, each such satellite passes over any small area of the Earth twice in 24 h.

The sequence of development of satellite technology is as follows: The launch of the first artificial satellite, Sputnik 1, on October 4, 1957, by the Soviet Union, began the space era. Four months later, in January 1958, the USA launched its own artificial satellite, Explorer 1.

The first satellite carrying a meteorological instrument, Vanguard 2, was launched on February 17, 1959. It was the first successfully launched Suomi radiometer satellite, named in honor of Werner Suomi (1915–1995), who is considered the father of satellite meteorology.

Soon, nine more satellites from this series were launched. TIROS 10 was launched on July 2, 1965, and was the last satellite of this series. The TIROS series of satellites introduced several technical improvements. For example, TIROS 2 introduced the scanning radiometer and the MRIR (Medium Resolution Infrared Radiometer). TIROS 3, 4, and 7 carried improved Suomi radiometers. TIROS 8, from 1963, introduced automatic picture transmission—APT (Automatic Picture Transmission). The problem with the first eight satellites in this series was that they rotated around their own axis, which meant that the cameras only saw the Earth a quarter of the time. The new generation of satellites had sensors constantly pointed at Earth.

Five polar-orbital meteorological satellites are currently active as a joint project of the American NOAA agency (English National Oceanic and Atmospheric Administration) and the EUMETSAT agency (English European Organization for the Exploitation of Meteorological Satellites). Russia sent Meteor-M N2 satellites into orbit, launched in 2014, or Meteor-M N2-2, launched in 2019. They belong to the Meteor series, which started in 1969.

All satellites until the mid-1960s of the last century were polar orbital. The first semi-operational geostationary satellite, a prototype of geostationary satellites, intended for synoptic meteorology was the SMS (Synchronous Meteorological Satellite), which was launched in 1974. It was the beginning of the first generation of geostationary meteorological satellites. The fourth-generation geostationary satellites GOES–R and GOES–S are currently in operation, launched in 2016 and 2018, respectively. They cover the Western Hemisphere, located above the equator at 75.2° and 137.2° west longitude. The plan includes new operational GOES meteorological satellites of the fourth generation, which are planned to be launched by 2024.

The European Space Agency (ESA) launched its first geostationary meteorological satellite, Meteosat 1, in 1978. Today, four satellites of this series (Meteosat 8–11) are in operational use and cover Europe, Africa, and the Indian Ocean.

16.1.3 Measurement of Air Pollution

Air pollution is a state of the atmosphere in which pollutants are present under certain weather conditions in concentrations harmful to human health and the human environment. Pollutants can be of natural or anthropogenic origin. They can be present in all three aggregate states. Atmospheric conditions affect the transport, distribution, and distribution of pollutant deposits. Air pollution affects both weather and climate. The air is never completely clean.

In relation to the chemical composition, air pollution can appear as a result of the presence and activity of the following chemical components: sulfur, nitrogen, carbon, halogen, toxic, and radioactive. A review of satellite measurements of aerosol particles is given by Kaufman et al. (2002).

Based on the phase state, atmospheric pollutants are divided into gaseous, liquid, and solid. According to the way they reach the atmosphere, they are divided into primary and secondary. Primary pollutants are emitted directly from the source, while secondary pollutants are formed in chemical reactions between primary pollutants under normal atmospheric conditions.

The primary pollutants are particles (in English particular matter or abbreviated PM), sulfur dioxide, nitrogen oxides, volatile organic compounds (VOCs), carbon monoxide, and lead. Secondary pollutants are sulfur smog and photochemical smog, described in more detail in the chapter on fogs. As already mentioned, photochemical smog is produced by the action of solar radiation on nitrogen oxide in the presence of hydrocarbons. A series of chemical reactions create a wide range of secondary pollutants, such as aldehydes, nitrogen oxides (nitrogen dioxide), peroxyacetyl nitrates (PANs), tropospheric ozone, and volatile organic compounds (VOCs). Based on the spatial scale, polluting substances are divided into local, regional, and global.

The smallest meteorological element that can be observed without the use of special equipment is probably a raindrop or a snowflake. From the point of view of cloud microphysics, the elements that makeup precipitation are recognizable because of their size. As is known, cloud droplets have to increase in mass a million times to become raindrops. They form on aerosol particles that can be as small as 0.01 μm in diameter or smaller. An aerosol is defined as a suspension of solid and liquid matter (with a low deposition rate) in the atmosphere (Kahn et al. 2004; Kaufman et al. 2002).

It can be said that the first and most significant contribution to the characterization of these particles was made by John Aitken (1839–1919). Scottish physicist and meteorologist who, through a series of experiments and observations in which he used his own design apparatus, elucidated the crucial role that microscopic particles, now called Aitken nuclei, play in the condensation of atmospheric water vapor in clouds and fogs. He worked instead in the laboratory in his home in Falkirk. Much of his work was published in the journals of the Royal Society of Edinburgh, of which he was a member.

One of the oldest and most suitable devices for measuring the concentration of atmospheric aerosols is the Eitken counter. In this instrument, air is expanded with saturated water vapor, so that the water vapor becomes several hundred percent oversaturated with respect to water. At such high supersaturations, water condenses on practically all aerosol particles, and a cloud of small droplets is formed. The concentration of droplets in the cloud (close to the aerosol concentration) can be found in such a way that the droplets will settle on the substrate. Then they can be observed with a microscope or other optical technique. The concentration of aerosol particles measured with an Eitken counter primarily refers to the number of Eitken aerosol particles. The concentration of Aitken aerosols observed at the ground depends on the location. Likewise, the change in the concentration of Aitken aerosols with time at one location can be by more than one order of magnitude. Overall, the average concentrations of aerosol particles are 109 m^{-3} over the ocean, 1010 m^{-3} over rural areas, and up to 1011 m^{-3} over polluted areas of cities. Measurements also show that particle concentrations of Eitken aerosols decrease with height. This is the result of the fact that soil is an important source of aerosols, especially in human and industrial activities.

Aerosol particles are present in large concentrations in the atmosphere. Therefore, it is important to know the average characteristics of the entire aerosol population, and not each particle individually (Kahn et al. 2004). The particle sizes of atmospheric aerosols vary from 10^{-4} to tens of micrometers in diameter. Depending on the size of the particles, as well as the measurement location, aerosol concentrations vary from 1013 to 1 m^{-3}. Since the sizes of aerosol particles are very different, different measurement techniques must be used to measure their characteristics. Thus, aerosols with diameters between 0.003 and 1 μm can be measured with an electrical analyzer. First, the given particles are charged with a known charge, and then they are collected in a certain way by the action of an electric field. The variation in the size of the collected charge is interpreted through the corresponding distributions of aerosol particles by diameter size. Aerosol sizes from 0.3 to 30 μm can be determined by measuring the amount of light energy they scatter. Aerosol can also be collected on different surfaces and observed with optical or electron microscopes (de Jong et al. 2007). Larger aerosols (>l μm) can be detected by different analytical techniques (dispersive X-ray energy analysis). In order to determine the concentration of collected aerosols, the efficiency of the aerosol collection must be known. Larger aerosols are collected more efficiently than smaller ones. Therefore, only the concentration of aerosols with a diameter greater than 0.1 μm can be measured by direct collection.

16.2 Modern Developments in Theoretical Meteorology

Without fear of exaggeration, unfortunately, it must be said that there was no significant progress in theoretical, or dynamic, meteorology until the middle of the twentieth century. Basic physical laws were applied to the atmosphere. At the very beginning of the twentieth century, Wilhelm Bjerknes formulated a system of five equations, the solution of which should be used to obtain the basic meteorological quantities (pressure, air speed, and temperature) in space and time. From the formulation of the problem to the solution, nothing was done for the next 50 years. It was only in the middle of the twentieth century that Rossby, Charney, Eady, and others began to deal more seriously with this system of equations.

16.2.1 Cyclones as Baroclinic Instability

Baroclinic instability is understood to be the dynamic cause of synoptic-scale mid-latitude storms. It is the result of vertical shear in the basic-state zonal wind. Its elucidation by J. Charney in 1947 (Jule Charney, 1917–1981) and Eric Eady in 1949 provided the springboard for much of modern dynamic meteorology. Formally, baroclinic instability results can be derived by generalizing the derivation for barotropic instability. Eric Eady (1915–1966) was a British meteorology researcher and author of the Eady Model of baroclinic instability, which gave rise to weather systems (Byers 1959; Mintz 1975). Eady received a BSc in mathematics, and in 1937, he became a weather forecaster in the UK Meteorological Office. In 1946, he started a Ph.D. in mathematics at Imperial College. His 1948 thesis was: The Theory of Development in Dynamical Meteorology, which was an early work on atmospheric instability and the development of weather systems.

Physically, the key difference between the two instabilities is the poleward advection of basic-state temperature by the perturbation meridional wind. The formal relationship is a consequence of the significance of the potential vorticity gradient for both instabilities. In baroclinic instability, the horizontal temperature gradient, which is proportional to the vertical shear of the wind through the thermal wind relation, is the key source of a change in the sign of the potential vorticity gradient. In fact, the conditions for instability can be satisfied even if the vorticity gradient is identically zero in the interior but a horizontal temperature gradient exists at the lower boundary. This is a generalization of the Eady problem in baroclinic instability theory. Because of the formal similarity to the barotropic instability problem, a semicircle theorem giving bounds for growth rate and phase speed is available for baroclinic instability as well. In baroclinic instability, the Rossby radius of deformation is the relevant horizontal scale of motion (Carl-Gustaf Arvid Rossby, 1898–1957, Swedish-American meteorologist).

The classic mechanism, as first developed in the 1940s and 1950s, plays a primary role in the development of synoptic-scale extratropical cyclones.

Baroclinic instability mechanisms are significantly modified by the presence of moisture: Latent heat is released through condensation of water vapor during ascent, and this additional heating reinforces the baroclinicity of the system and thus enhances growth. The most intense polar mesoscale cyclones tend to form as polar lows in maritime environments, where there is a ready supply of moisture, which suggests that any baroclinic dynamics are likely to be enhanced by latent heating. Indeed a number of numerical modeling case studies have found that without moist processes, the modeled polar low is unable to grow to the strength observed.

16.2.2 Quasi-geostrophic Equations

Atmospheric flow takes place over horizontal length scales which are very large compared to their vertical length scale, and so they can be described using the shallow water equations. The Rossby number is a dimensionless number that characterizes the strength of inertia compared to the strength of the Coriolis force. The quasi-geostrophic equations are approximations to the shallow water equations in the limit of a small Rossby number so that inertial forces are an order of magnitude smaller than the Coriolis and pressure forces. If the Rossby number is equal to zero, then we recover geostrophic flow.

The quasi-geostrophic equations were first formulated by Jule Charney. Jule Gregory Charney (1917–1981) was an American meteorologist who played an important role in developing numerical weather prediction and increasing understanding of the general circulation of the atmosphere by devising a series of increasingly sophisticated mathematical models of the atmosphere. His work was the driving force behind many national and international weather initiatives and programs. Considered the father of modern dynamical meteorology, Charney is credited with having guided the postwar evolution of modern meteorology more than any other living figure. Charney's work had a significant influence on his close colleague Edward Lorenz, a pioneer in the field of chaos theory who explored predictability limitations. Lorenz succinctly summarized this theory as "Chaos: When the present determines the future, but the approximate present does not approximately determine the future." Chaotic behavior is observed in various natural systems, such as fluid flow, heartbeat irregularities, weather, and climate. Edward Lorenz, born Edward Norton Lorenz (1917–2008), was an American meteorologist who discovered the underlying mechanism of deterministic chaos, a fundamental concept in the study of complexity.

The three scalar equations of motion supplemented by the thermodynamic equation and the continuity equation allow us, in principle, to find five dependently variable quantities (three components of velocity, pressure, and temperature). However, that system of equations is not suitable for solving, so it is necessary to find a simpler system of equations that will conveniently describe some basic characteristics of weather that are of synoptic scale. From this system of five equations, a system of only two equations is derived, by means of which we can unambiguously find the vertical velocity and tendency of the geopotential.

Those two equations contain only two dependent variables, geopotential and vertical velocity. Therefore, they form a closed system of equations for those two quantities and are collectively called the quasi-geostrophic system of equations. With suitable mathematical procedures, the geopotential tendency and vertical velocity can be found from them, knowing only the geopotential field at some point in time. Thus, the development of synoptic-scale processes in moderate latitudes can be obtained approximately by measuring only the geopotential and not the airspeed. From these two equations, we can find the geopotential tendency and the vertical velocity. They are of fundamental importance for weather forecasting.

16.2.3 The Omega Equation

Omega equation is a term used to describe vertical motion in the atmosphere. The omega equation used in numerical weather models is composed of two terms, the differential vorticity advection term and the thickness advection term. Put more simply, omega is determined by the amount of spin (or large-scale rotation) and warm (or cold) advection present in the atmosphere. On a weather forecast chart, high values of omega (or a strong omega field) relate to upward vertical motion in the atmosphere. If this upward vertical motion is strong enough and in a sufficiently moist air mass, precipitation results (Holton 2004).

Omega is the vertical velocity in pressure coordinates. Omega has units of pressure per time and is normally expressed in negative units due to pressure decreasing from the surface. Because the units are expressed negatively, it is actually the positive omega values on the map that equate to positive (i.e., upward) vertical motion. But simply based upon the relationship, a greater negative value expressed for omega, the greater the upward vertical motion (and vice versa). It is important to check the legend to see if the units are expressed positively or negatively. NOTE: Because much of the operational meteorology uses pressure surfaces, omega is a more common quantity to see.

One term in the omega equation is thermal advection. This essay will be interested in the operational meteorology interpretation of thermal advection and the contribution it gives to vertical motion. Thermal advection can be divided into warm air advection and cold air advection. We can also divide the troposphere into the low levels which are between the surface and 550 millibars and the upper levels which are between 550 millibars and the tropopause (~150 millibars). This gives us four operational interpretations of thermal advection.

The vorticity advection term is also called the upper-level divergence term (Ćurić 2014). The upper levels generally extend from 550 millibars to the tropopause. Upper-level divergence occurs when a mass of air is pulled away from a region faster than that mass can be replaced. This most commonly occurs when the upper-level wind field is strong and meridional (high amplitude upper-level waves).

The omega equation contains only derivatives in spatial coordinates, so we see that it is a diagnostic equation for the vertical velocity as a function of the geopotential field, which should be known at some point in time. This equation is clearly more convenient than any other equation for vertical velocity in view of the small requirements regarding the knowledge of other quantities.

16.2.4 Q-Vector

Q-vectors are used in atmospheric dynamics to understand physical processes such as vertical motion and frontogenesis. Q-vectors are not physical quantities that can be measured in the atmosphere but are derived from the quasi-geostrophic equations and can be used in the previous diagnostic situations. Evaluation of atmospheric vertical motion is now possible on widely available and inexpensive microcomputers. Using the Q-vector form of the omega equation, formulated by Hoskins et al. (1978), allows a meteorologist to evaluate vertical motion throughout the atmosphere.

The conventional method of quasi-geostrophic diagnosis has the disadvantage that we have to look at two main forcing functions, namely, vorticity and temperature advection, which, of course, can have opposite signs and therefore tend to cancel each other out. There is an approximation with only one forcing term, the advection of vorticity by the thermal wind, but by this method, the effects of the deformation with regard to the release of vertical motions are neglected. Since these effects are important for frontogenesis and frontolysis, it is better to use the alternative form of diagnosis, which contains also only one forcing. This form uses the so-called Q-vector, introduced by Hoskins et al. (1978). According to its definition, the Q-vector describes temporal changes in the temperature distribution at a pressure surface resulting from horizontal variations of the geostrophic wind. The other factor for changing this distribution is horizontally differing vertical motions. For a parcel within a geostrophic flow, omega vanishes and the change of the temperature gradient is totally determined by Q. The parcel can experience the temporal change either locally or along its path through the temperature pattern. The Q-vector can easily be computed but also manually estimated. The Q-vector points perpendicular to the right of the vector change. Its amount is proportional to the magnitude of the rate of vector change and to the magnitude of the temperature gradient (Sanders and Hoskins 1990). Using the Q-vector, the modified quasi-geostrophic omega equation has a form containing only one forcing function on the right-hand side, namely, the convergence or divergence of the Q-vectors. A forcing of ascent results in areas with convergence, and a forcing of descent in areas with the divergence of Q. This term replaces the two adiabatic forcing functions of the conventional form of the omega equation and allows direct computation of the total forcing without any simplifications. However, the use of the Q-vector has also other advantages for diagnostic purposes. The temperature gradient can change either its direction or amount or both. In order to estimate these changes, the Q-vector should be split into its components along and perpendicular to the isotherms. It can be shown that the component of Q along the isotherms reflects changes in the direction, and the component perpendicular to the isotherms changes the amount of the temperature gradient.

16.3 Climate Variability

Climate change refers to long-term trends in climatic averages such as global warming over the past century and long-term changes in the frequency, intensity, and duration of extreme cases. Climate change is influenced by three factors: astronomical, planetary, and anthropogenic. The influence of astronomical factors has been expressed for a very long time, tens of thousands of years or more. So, it cannot be considered when thinking about climate change for more than a century.

Planetary factors for climate change include strong volcanic eruptions and meteorite falls. During these processes, large amounts of aerosol cells are released into the stratosphere. These particles stay in the upper layers of the atmosphere for a long time and weaken the solar radiation that heats the Earth's surface. Another important activity during volcanic eruptions is the introduction of large amounts of CO_2 into the atmosphere. Although global warming has increased in the last century, no planetary phenomena have been found to cause such global warming. Anthropogenic factors for climate change occur because of human influences and uncontrolled CO_2 emissions and other gases in the atmosphere (greenhouse gases) (combustion of fossil fuels, agricultural activities, deforestation, urbanization, industrialization, waste treatment, and others).

Climate variability is a variation around the average climate. There are different models of variability: repetitive temperature models or other climate variables. They are quantified by different indices. Characteristics of the indices are simplicity and completeness, and each index usually represents the status and time of the climatic factor that represents it. By their nature, the indices are simple and combine many details into a generalized, complete description of the atmosphere or oceans that can be used to characterize the factors that affect the global climate system. Climate oscillation or climate cycle is any repetition within a global or regional climate. These fluctuations in air temperature, sea surface temperature, precipitation, or other parameters can be quasiperiodic, often occurring in intermittent, perennial, centuries, millennia, or longer periods of time. A typical example is El Niño (Southern Oscillation), which includes temperatures on the surface of the equatorial part of the Pacific and the western tropical part of South America, which affects the climate of the entire planet.

16.3.1 Milanković Cycles of Climate Change

Here we will talk about a man who never worked in any faculty institution that dealt with education and research in meteorology, no matter what it was called, nor in the professional meteorological service, the Institute. Despite this, seemingly unrelated to meteorology, his name and work deserve to be written in golden letters in the history of meteorology. It is about Milutin Milanković (1879–1958) and his colossal scientific work.

His work covers several scientific fields, above all mathematics, mechanics, astronomy, meteorology, and geophysics. That is why it is referred to when the history of any of the mentioned sciences is presented. For this reason, only some elements of his work related to meteorology will be mentioned here. Because of that, it is still mostly the only one that is currently in the world of science.

After arriving at the University of Belgrade, Milanković carefully searched for the field of science in which he would continue to work. He talked with Pavle Vujević, whom he met in Vienna during his studies (the same year they received their doctorates in Vienna), about the problems of meteorology, with Jovan Cvijić and others. He wrote the following about his impressions from those conversations in the book *Through the Universe and Centuries*: "Those who deal with the earth's climate, meteorologists, do not worry about the climates of other planets. And as for

Milutin Milankovic (1879–1958)

He was born in Dalja, Slavonia (then it belonged to Austria-Hungary), on May 28, 1879. His ancestors immigrated from Kosovo during the Great Migration of Serbs. In his family, there were distinguished priests, state officials, and merchants, but also important creators. His grandfather Uroš Milanković (1800–1849) was a philosopher of nature, and his father, who died when Milanković was 7 years old, was a very wealthy merchant and landowner. Milanković was the eldest of six children (four brothers and two sisters). In the family, he received a good basic education in Serbian with the governess, while another governess taught him, German.

When he turned 10, he moved to Osijek and enrolled in a real high school. With very little study, he immediately proved to be the best student. In the higher grades, he got a new class teacher, a young doctor of mathematics, the Serb Vladimir Varićak (1865–1942), who later became a famous professor and scientist at the University of Zagreb. Varićak left the strongest influence on Milanković. He taught him how to reliably develop his abilities with books. At the age of 17, he graduated from a real school. As the eldest son, he was assigned to take care of the family's agricultural property. Since he fell in love with science, he agreed that his younger brother Ljubiša should manage the estate and he should continue his education in Vienna. But to enroll at the University of Vienna, according to the laws of the time, one had to know Latin and Greek. They convinced him not to waste a year learning languages but to enroll in construction at the Technical University in Vienna.

At the end of his studies in Vienna, the Austrian technical colleges were given the right to award doctoral degrees. Milanković decided to stay another year in Vienna working on his doctorate. At that time, independent scientific

(continued)

(continued)
work had to be defended for a technical doctorate. He defended his thesis "Theorie der Druckkurven" on December 3, 1904. He was the first Serb to be promoted to a doctor of technical sciences on December 18, 1904. As a civil engineer, he worked in Vienna in a concrete construction company. At that time, the application of reinforced concrete in construction gained momentum. Milanković excelled in that business and often traveled as an inspector of the works that his company performed. One such trip brought him to Belgrade, where his company built the main Sava sewage collector. He once passed by the building of Belgrade University and saw the Faculty of Philosophy written on the blackboard. Under the Department of "Applied Mathematics" were written subjects that include rational mechanics, theoretical physics, and celestial mechanics. That fascinated him because that was exactly what interested him. His desire to transfer to the University of Belgrade and become a professor of the mentioned subjects came true on October 1, 1909. At the suggestion of Mihailo Petrović and Jovan Cvijić, he accepted the chair of applied mathematics. It was a great honor for him, associated with a huge material sacrifice. In the new environment, he turns only to science and begins work on the theory of climate change under the influence of astronomical factors. As early as 1912, he published the paper "Contribution to the Theory of Mathematical Climate" in the Voice of the Serbian Royal Academy. Later, he supplemented the theory but did not bother to publish numerous works. However, what he published about the theory of climate outlived Milanković and, it turned out, will live for a long time, because it represents a fundamental contribution to the theory of climate change. The more time passes since the creation of that work, the more relevant it becomes. Today it can be said that this is certainly the greatest contribution to world science made by a Serbian scientist. Milutin Milanković died in Belgrade on December 12, 1958, at the age of 80. He was buried in Belgrade, and in 1966, according to his personal wish, he was transferred to Dalj, to the family tomb in the Serbian Orthodox cemetery.

the earth's climate itself, there are those pure empiricists who do not care about complicated theories, nor would they know how to apply them. They do not think of entering the church through the tower. Why hit the road across the Sun to find out what is happening on Earth, when we have several thousand weather stations on Earth that accurately inform us about all the temperature conditions on Earth, more accurately than the most perfect theory could do? Our great geographer (Cvijić) was surprised when I told him about my intention to calculate the temperature of the layers of the earth's atmosphere."

Milanković understood that no one had yet seriously attempted to come up with a mathematical theory of climate due to the fact that it required solving a whole series of complicated problems from various fields of exact sciences. This is exactly what he set as his goal, to find a complete mathematical theory that connects the thermal regime of the planets with their movement around the Sun.

The amount of solar energy received by a unit of the surface at the upper boundary of the atmosphere depends on the distance of the Earth from the Sun (i.e., on the shape and size of the Earth's path) and on the angle at which the rays fall on the unit of surface (i.e., on the tilt of the axis of rotation of the Earth and geographic latitude). Under the influence of other planets, the parameters of the Earth's movement change slowly over time. This changes insolation and sun exposure, that is, warming of the Earth. Milanković took this into account through three contributions.

The first contribution comes from the change in shape (eccentricity) of the path of the Earth around the Sun. It changes from an almost regular circle to a slightly elongated ellipse. Such a change occurs in time intervals of about 100,000 years. Eccentricity affects the differences between the seasons. When the Earth is closest to the Sun, it receives most of its heat. The hemisphere that is closest to the Sun during the winter should have a mild winter because of this, and the one where this happens in the summer would have warmer summers because of this. When the eccentricity is the largest (about 9%), then the seasonal difference in the received heat is about 20%. Now the Earth is in a period of low eccentricity (about 3%), so the seasonal difference in received heat is about 7%.

Another parameter that Milanković takes into account is the change in the tilt of the Earth's axis of rotation in relation to the plane of the orbit (path around the Sun). The slope varies from 22.1° to 24.5°. These kinds of changes happen in the time of 41,000 years. After that, it continues to change periodically. When the slope is higher, at higher latitudes, the difference in the character of the seasons is more pronounced compared to lower latitudes. This influence is negligible at the Equator and is greatest at the poles. With an increase in inclination by 1°, the total energy received by the hemisphere during the flight increases by about 1%.

The third parameter refers to the revolution of the Earth's rotation axis (or precession, as it is otherwise called). This change lasts 23,000 and 19,000 years. Precession is a complex phenomenon and originates from the rocking of the Earth's axis and the rotation of the Earth's elliptical orbit. Precession affects the orientation of the axis, but not its tilt. The direction of the swing of the axis is opposite to the direction of the Earth's movement around the Sun. Because of this, the north pole of the Earth will not be facing the North Star for 11,000 years but will have a deviation of about 47°. As a consequence of this complex movement of the Earth, the days of the equinoxes do not always occur on the same date but shift slightly in the calendar. The main consequence of precession is that the relative length of the seasons changes cyclically over time.

Milanković described all these very complex changes with equations, from which he calculated the amount of heat that reaches certain latitudes in different years. He calculated both backward and forward, so that with this method, he could interpret the Earth's climatic past and forecast its future. He had a lot of time for calculations, despite the weak computer resources at that time. He spent almost the entire period of 4 years, which he spent as a prisoner in Hungary during the First World War, on these calculations. At the beginning of 1923, he published a paper about this, which was accepted by the famous climatologist Vladimir Keppen and the geophysicist Alfred Wegener in the form of the Milanković curve of insolation in the last 600 million years.

They invited him to collaborate in the preparation of the work "Climates of the Earth's Past." The oscillatory nature of secular (long-term) changes in insolation is clearly visible on the Milanković curve (Fig. 2.1 in Chap. 2). Oscillations have different amplitudes and durations. Their irregularity is a consequence of the secular changes of the three previously mentioned parameters, caused by the influence of the gravity of the Sun, the Moon, and the planets.

Milanković did not bother to publish a lot of shorter works, but he consolidated the most important results of 30 years of research and published them in the main part "Kanon der erdbestruhlung und seine anwendung das eiszeiten-problem"—The Canon of insolation of the Earth and the ice age problem—published by the Serbian Royal Academy in 1941. The war years since then, and later, were not favorable for the continuation of his scientific work. Nevertheless, he was satisfied, because he considered that he had completed his theory. Since then, he has participated in discussions at scientific meetings and followed responses in world scientific circles.

Milanković's Canon of insolation experienced a far greater affirmation in world science after than during his lifetime. More attention was paid to his work by John and Catherine Imbree, in the book *Ice Age* published in 1979. Several international conferences were devoted to him and his method. Thus, on the occasion of the centenary of Milanković's birth, in 1979, the Serbian Academy of Sciences organized an international scientific gathering in Belgrade dedicated to his life and work. Then, in 1982, the symposium "Milankovic and the climate" was held at the Lamont Doherty Geological Observatory, Columbia University in the state of New York. In 2004, the Serbian Academy of Sciences again held an important international meeting "Paleoclimate and the Earth's Climate System" in Belgrade (Fig. 16.3).

Fig. 16.3 The badge was worn by participants at International conferences dedicated to Milanković

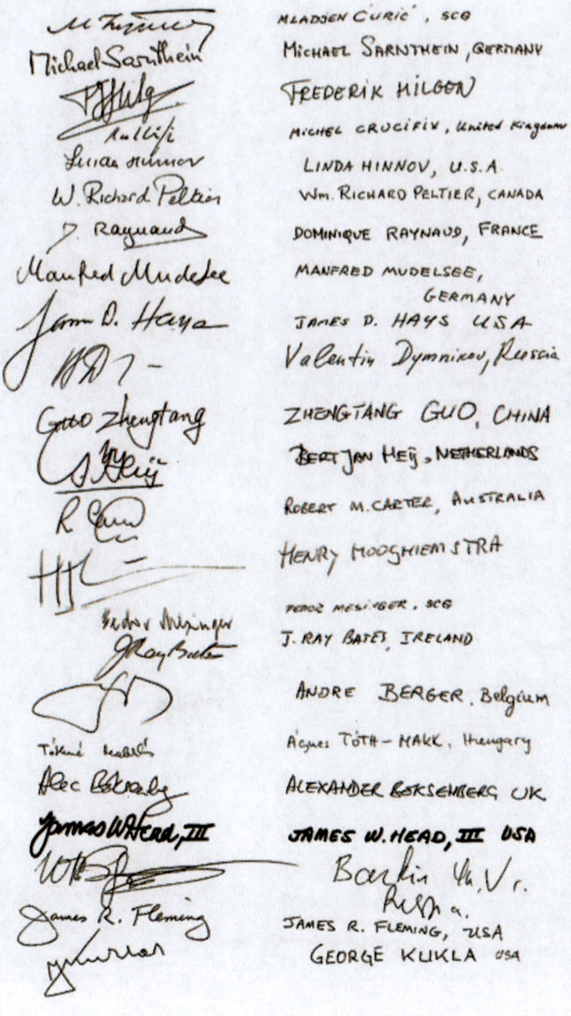

Fig. 16.4 Handwritten signatures of the participants of the symposium in Belgrade in 2004

A small number of prominent scientists from around the world participated in that symposium only at the personal invitation of the organizers. The author of this book wrote their full name and surname, the country they come from and signed, etc. (Fig. 16.4). The intention was to engrave their signatures on a stone monument dedicated to Milanković, which was prepared by our famous sculptor, the longtime head of the meteorological station in Nikšić, Mijo Mijušković (Fig. 16.5).

In addition to the large number of citations of Milanković's works dedicated to climate, he was awarded many honors of another kind. Thus, the European Geophysical Society established the Milutin Milanković medal in 1993. A crater each on the Moon and Mars bears his name, etc.

Fig. 16.5 Stone of Milutin Milanković, work of sculptor Mijo Mijušković

16.3.2 Coupled Ocean-Atmosphere Dynamic

Analyses of the height synoptic maps showed that the westerly current in extratropical latitudes is in the form of an irregular meandering circumpolar current directed from west to east. It usually consists of four to six long waves whose dimensions are typically around 5000 km in the zonal direction and several thousand kilometers in the meridional direction. These long waves, which are also called planetary waves, move slowly from west to east, and they can also be stationary. They are also called Rossby waves, after the Swedish meteorologist C. G. Rossby who discovered and explained them (Carl-Gustaf Arvid Rossby, 1898–1957). They transport polar air toward the Equator and tropical air toward the pole. They play a major role in the formation of cyclones and anticyclones at higher latitudes. The meandering path of the polar jet stream, whose position is linked to the polar front, is actually Rossby waves.

The oceans cover almost 70% of the Earth's surface and are a great source of heat, thanks to their heat capacity. Thus, a layer of ocean water only 3 m thick has a heat capacity equal to the heat capacity of the entire atmosphere. The oceans act on the atmosphere primarily through heating and fluxes of sensible and latent heat. A column of water about 1 m thick evaporates annually from the surface of the ocean in moderate and lower latitudes. This corresponds to vertical heat transport from the ocean to the atmosphere of about 80 Wm^{-2}. On the other hand, the atmosphere acts on the ocean through the surface wind, which creates ocean currents on the surface. They are of great importance for weather and climate. Thus, for example, currents transport about 40% of heat from the Equator toward the poles. All these facts indicate the great importance of oceanic circulation.

Ocean circulation consists of two components. One consists of surface ocean currents that are created under the influence of the wind on the ground floor. They occur in the upper layer of the ocean, which is several hundred meters thick. In the deeper layers of the ocean, there are slower currents that arise due to the different temperatures and salinity of the layers of ocean water. This circulation is called thermohaline circulation. The name is a coin of two Greek words: thermo—temperature and hal (hals)—salt.

Surface winds move ocean water to the surface. Water that is moved in this way gradually rises, creating pressure differences in the water. This is how movement appears in a water layer of several hundred meters. The general circulation in the ground layers of the atmosphere creates surface ocean currents. As the resistance to movement in water due to its density is greater than in the atmosphere, ocean currents move more slowly than the corresponding wind over the ocean surface. Ocean current speeds typically range from several kilometers per day to several kilometers per hour. In the North Atlantic and North Pacific, winds over the ocean surface are predominantly clockwise and from subtropical high-pressure centers. In addition to wind friction and force due to the exchange of movement with deeper layers of water, the water surface is also affected by the Coriolis force, which turns the water to the right in the northern hemisphere (or to the left in the southern hemisphere). This deviation causes the water on the ocean surface to move at an angle between 20° and 45° to the direction of the wind. Changing the direction of water movement with depth and the measure of air movement with height was studied by Ekman and is called the Ekman spiral effect. Because of this, the water tends to move in a circle, basically following the circulation in the subtropical high-pressure centers. The interaction between the atmosphere and the ocean can be seen by analyzing the large eddy in the North Atlantic. The warm Gulf Stream flows northward along the east coast of the USA. It transports huge amounts of warm tropical water to higher latitudes. To the north and west of the small subpolar gyre, cold water moves along the Atlantic coast of North America. This Labrador current brings cold water to the south. Along the coast of Newfoundland, the Gulf and Labrador currents flow in opposite directions next to each other, and there are large temperature contrasts. This is an area known for frequent and long-lasting sea fog. The Gulf Stream moves away from the coasts of North America and turns toward Europe under the influence of the prevailing westerly currents. Gradually, it widens and slows down and transitions into the broader North Atlantic Current. Approaching Europe, this current splits into two currents. One moves north along the coasts of Great Britain and Norway, causing much milder winters than would otherwise be the case at such high latitudes. Another current flows southward as the Canary Current, transporting cold water toward the Equator. The Atlantic vortex closes in such a way that the Canary Current changes into the North Equatorial Current, which moves westward and draws its energy from the northeast trade winds. The ocean circulation in the North Pacific is similar to that in the North Atlantic. In the western part of the Pacific, the warm Kuroshio Current flows northward and gradually changes into the North Pacific Current. One branch of this current flows along the west coast of the USA as the cold California current.

In the Southern Hemisphere, the surface ocean circulation is very similar to that in the Northern Hemisphere. But now the direction of motion in the eddies is counterclockwise. It is consistent with the flow direction in the subtropical high-pressure centers in the Southern Hemisphere. Ocean currents here move more in the west-east direction at higher latitudes than in the Northern Hemisphere. This is especially true for the Antarctic Circumpolar Current. This zonal flow somewhat limits the poleward inflow of warm tropical water. In the Southern Hemisphere, there are far smaller differences in temperature between the ocean surface and the atmosphere than in the Northern Hemisphere. This prevents more intense convective activity over the ocean. In the Indian Ocean, the monsoon circulation disturbs the ocean circulation.

The general characteristics of the ocean circulation in the surface layer of water are that next to the eastern coasts of the continents usually flow warm ocean currents that transport huge amounts of warm water from the Equator toward the poles. In contrast, along the western coasts of the continent, cold ocean currents flow from the pole toward the Equator. Atmospheric and oceanic circulations are closely related. Namely, the wind blows over the surface of the ocean and creates surface ocean currents, while both movements play a major role in the transport of heat from the Equator to the poles.

Ocean circulation significantly affects the climate as well as climate change. In this regard, the large oceanic conveyor belt within the thermohaline circulation is particularly significant. It transfers a huge amount of heat from the tropics to the polar regions. This energy is equal to the energy produced in over a million of the largest nuclear power plants. If the conveyor belt in the areas east of Greenland were to be interrupted due to the melting of the ice in the polar areas due to global warming, because the salinity of the ocean water would decrease and it would not sink as lighter, it would reduce the temperatures in the eastern parts of North America and parts of the western part of the country in 10 years and Northern Europe by 6 °C. The mean temperature of the Earth's surface was lower by that much compared to today at the peak of the last ice age.

16.3.3 Anthropogenic Climate Change

According to some scientists, the trend of global warming observed since the mid-twentieth century is due to human influence and the greenhouse effect as key elements for climate change on Earth (Harding 1982; Cools 2002 Imbrie and Imbrie 1994). The greenhouse effect is a natural process that causes additional heating of the Earth's surface due to the presence of greenhouse gases. When solar short-wave radiation reaches the Earth's atmosphere, some of that radiation is reflected into space, and the rest is absorbed and remodeled by glass-linen gases. In this way, the absorbed energy heats up the Earth's atmosphere and surface. Global warming is believed to be due to the increased greenhouse effect, which is due to the increased concentration of greenhouse gases in the atmosphere (carbon dioxide, water vapor, methane, nitric oxide, ground ozone, chlorofluorocarbons, etc.). They are the gases in the atmosphere that absorb and emit radiation in the infrared spectrum carbon dioxide as the main greenhouse gas (covering 75% of total greenhouse gas emissions) acts as a greenhouse gas, trapping heat in the atmosphere.

Statistics and scientific data show that in the next 100 years, the average global air temperature has risen by about 0.74 °C, and the global sea level has risen by 17 cm, in part because of melting snow and ice from the mountains and polar regions. In addition to these, other changes have been observed, including changes in Arctic temperatures, changes in the salinity of ice and oceans, changes in the movement of winds, droughts, precipitation, and warm waves, and changes in the frequency and intensity of the tropics cyclones. In the next two decades, the temperature is expected to rise by 0.2 °C. Continued greenhouse gas emissions will cause further increases in global temperatures. If the flooding continues for the next few centuries, it will lead to the melting of the icy cover, rising global sea levels by about 7 m, and changing ecosystems.

16.4 Contemporary Weather Forecast

16.4.1 Introduction

The English mathematician Louis Fry Richardson, a century ago, proposed a method by constructing a systematic process based on mathematics for forecasting the weather. In his 1922 book, *Weather Prediction by a Numerical Process*, Richardson attempted to write an equation that he could use to solve the dynamics of the atmosphere based on hand calculations.

It didn't work, because not enough was known about atmospheric science at the time. "Perhaps one day in the dark future it will be possible to advance computations faster than time advances and at a lower cost than the saving to mankind of the information gained. But it's a dream," concluded Richardson.

Fig. 16.6 A modern numerical concept for weather forecasting

A century later, modern weather forecasts are based on the kind of complex calculations Richardson envisioned, and they've become more accurate than anything he imagined. Especially in recent decades, steady advances in research, data, and computation have enabled a "quiet revolution in numerical weather prediction" (Fig. 16.6).

For example, a forecast of heavy rainfall 2 days in advance is now as good as a forecast for the same day in the mid-1990s. Errors in predicted hurricane tracks have been cut in half over the past 30 years.

There are still major challenges. Thunderstorms that produce tornadoes, large hail, or heavy rain remain difficult to predict. And then there's chaos, often described as the "butterfly effect"—the fact that small changes in complex processes make time less predictable. Chaos limits our ability to make accurate forecasts beyond about 10 days.

16.4.2 *Application of Modern Technologies for Weather Forecasting*

The modern development of weather forecasting is highly dependent on advanced technology—which is based on modern instruments and devices. Modern technology and instrumentation for weather equipment should be modernized in the future and include:

- Remote monitoring and measurement systems
- Advanced radars

- Instrumented balloons
- Meteorological satellites

In doing so, computers and communication systems are used, which would enable:

1. Interpreting data and images
2. Processing and modeling of atmospheric processes
3. Preparation of forecasts
4. Generation of maps and visualization of products and 3-D images
5. Dissemination of processed information to end users and the media

The development of technology and equipment is used to measure meteorological parameters that are essential in weather analysis: air temperature, atmospheric pressure, relative humidity, amount of precipitation, height of snow cover, wind direction and speed, visibility, UV -radiation, etc. In addition, we would improve the measurements of the ground layer of the atmosphere and the soil, as well as the height measurements (aerosonde measurements) of the vertical stratification of the atmosphere using balloons and airplanes, as well as the application of modern weather-monitoring radars and satellites to obtain a wider and more detailed picture of atmospheric phenomena and processes.

Upper air measurements and observations are of exceptional importance in the detailed analysis of the weather, weather modification, and air traffic—as well as for the purposes of preparation of the objective analysis and the initialization of the analysis in the numerical models for the weather forecast. Precise measurements of the vertical structure of the temperature field and the profile of water vapor in the troposphere are extremely important for all types of forecasting, especially for regional and local forecasting. The vertical structure of the temperature and water vapor field determines the stability of the atmosphere. Radiosonde measurements are of vital importance for studies related to environmental pollution as well as climate change at altitude. It is the task of every modern service to carry out such kinds of measurements. From there, one of the priorities would be the reintroduction of this neglected category of measurements, which is established in the observational practice of every meteorological service. Of course, for this purpose, it is necessary to purchase weather balloons that are equipped with measuring instruments (air probes). Aerosondes are instruments composed of sensors for measuring air pressure, temperature, and humidity. Furthermore, with the help of radio transmitters, a radio signal, they are transmitted to the representative station of the country. This information is essential for air traffic and weather forecasting. Also of fundamental importance is the modernization of the observation system with the acquisition and installation of modern radar(s) of the next generation of meteorological Doppler radars (NEXRAD), with dual polarization for high-resolution spatial scanning. These radars are modern tools for monitoring the movements of weather systems and detection of dangerous storms, tornadoes, and detailed display of snow storms as well as identification of precipitation areas, precipitation category, and intensity, convective cells, atmospheric movements, or wind. They can operate automatically and use certain algorithms for volume scanning in two basic modes that are selected by the operator:

(a) Atmospheric mode (slower scanning for analyzing the dynamics of the atmosphere during less activity in the area)
(b) Precipitation mode (faster scan to monitor active weather and dynamic state of the atmosphere)

In the context of the modern development and application of new technologies for diagnosing and forecasting the weather, satellite data, images, and weather and climate products from the European Organization for the Exploitation of Meteorological Satellites (EUMETSAT) are also used, which are obtained from geostationary (rotating at the same speed as the earth) or polar-orbiting satellites (used to get a closer picture of the atmosphere), as a top technology for monitoring the atmosphere, climate, and processes and phenomena occurring in the atmosphere of our planet as well as early detection and warning of weather disasters. The latest generation of MSG-2 (second generation) satellites provides weather and climate data for the entire planet. Meteosat-10 is a basic geostationary satellite that operates over Europe and Africa (placed at 0° latitude, at a height of 36,000 km above the Equator) and performs a full disk scan every 15 min and obtains detailed information, which is from a special value for weather forecasting-nowcasting and the early announcement of unfavorable weather phenomena.

Modernization in the more distant future can go in the sense of using the latest developments in weather technology, such as advanced interactive systems for processing and visualization of weather information, Doppler radars, automatic weather stations, supercomputers, aviation laboratories for research purposes and forecasting of certain weather systems and storms, etc.

16.4.3 From Short-Term to Long-Term Weather Forecast Information

Weather significantly affects everyone's health, safety, and economic well-being. Climate change and global warming contribute to the increasingly frequent occurrence of weather and hydrological disasters that have an increased intensity and negatively affect people and material goods, and hence sustainable development. Hence, continuous progress in terms of weather forecasting methods and tools, monitoring of new technologies in development, and application of numerical weather forecasting as a starting point for producing timely precise, and detailed weather information is of utmost importance. In this regard, the *European Center for Medium-Range Weather Forecasts* ECMWF stands out as the leading center in the world for the development of modern weather forecasts (Palmer 2018). ECMWF produces numerical weather forecasts, from deterministic to ensemble, from short- and medium-term forecasts to extended, global long-range, and seasonal forecasts up to a year ahead. The computer resources (the largest supercomputer facilities and meteorological data archives and data assimilation) at the disposal of this center enable the highest-quality preparation of initial data, reanalysis, as well as 3-D and 4-D variational analysis.

Considering the wave nature of atmospheric processes, their spatial and temporal variability (from seconds to seasons), from millimeter to planetary scales, and different weather scenarios, the following categories of forecasts are distinguished in modern meteorological and prognostic practice:

- Presentation—up to 2 h in advance (nowcasting)
- Very short-range from 0 to 12 h in advance
- Short-range from 12 to 72 h (3 days in advance)
- Medium-range from 3–10 days in advance
- Long-range (from 30 days to 2 years)
- Monthly outlook (up to 30 days in advance)
- Seasonal outlook
- Annual outlook
- Climate forecasts—climate scenarios.

Very short-term (0–12 h) forecasts cover two general areas: (1) forecasting the initiation, evolution, and movement of short-term, often intense weather phenomena such as tornadoes, thunderstorms, storms, and flooding, and (2) predictions of large-scale weather events such as which are rain or snow, cold, hot, or windy weather.

The ability and skill of forecasting rely on the skill of the forecaster. Very short-range (0–12 h) forecasts demonstrate considerable technique and skill, especially for predicting the formation and movement of large weather systems.

Short-range forecasts (12–72 h) of maximum and minimum temperatures and wind speeds are quite accurate. What's more, the forecast of precipitation amounts is much better than the forecasts made just two decades earlier. Medium-term forecasts (3–7 days in advance) show significant improvement over the last 20 years.

A medium-range weather forecast is a weather forecast for several days to 2 weeks. The ability to represent the location and timing of weather events decreases with increasing forecast length. In the current situation, medium-term forecasts are primarily based on global numerical forecast systems (models). On average, these systems produce skillful forecasts more than a week in advance, although their performance varies by season and region. Forecasters use general methods to increase the skill of medium-term forecasts by a day or more and get a picture of the potential skill of the forecast before it is verified. Over the past three decades, the range of medium-term forecasts has expanded by roughly 1 day per decade. Major winter storms can now be forecast a week or more in advance, allowing road maintenance personnel ample time for preparation by managers responsible for such weather emergencies.

Long-range weather forecasting is a category that is firmly based on statistical averages derived from past weather events, often known as climate data. The weekly, monthly, and seasonal weather outlooks prepared by the National Weather Service do not constitute weather forecasts in the general sense of the word. They only indicate whether or not the region will experience near-normal precipitation and temperatures. However, day-to-day weather predictability for periods beyond 7 days is usually less likely to be achieved or less reliable.

Seasonal climate forecasts are increasingly used in disaster management, health, agriculture, forestry, fisheries, tourism, transportation, and energy sectors.

Weather forecasting requires constant observation of the state of the atmosphere and the Earth's surface analysis of weather data. These analyses are the starting point for weather forecasting at scales (scales) from individual clouds or cloud systems to the forecast of the entire climate system. In order to develop useful prognostic products for the future state of the weather, integrated information obtained from observations, analyses, and computer models is necessary, as well as the transmission of this information with a reliable forecast to the end users. The biggest challenge in the future, for each meteorological service, will be the further development and improvement of the short-term weather forecast, for the next 1–2 h, known in meteorology under the term—*Convective Scale Numerical Weather Prediction* CS-NWP (Yano et al. 2019) and Nowcasting. Nowcasting represents an integral-complex system for forecasting adverse weather phenomena, using modern radar technology, satellite data, and products, as well as sophisticated non-hydrostatic numerical models. This system is used in daily operational practice, with special emphasis when forecasting complicated processes, such as precipitation, vertical development clouds, hail clouds, snow storms, storms, intense precipitation, and electrical discharges, i.e., those atmospheric processes that develop or change relatively quickly and can adversely affect people, safety, infrastructure, and the economy. In June 2019, the World Meteorological Organization WMO launched the Open Consultative Platform (OCP), a partnership and innovation for Next Generation Weather and Climate Intelligence, in recognition that progress in weather and climate services to society will be sought throughout the community access to stakeholder participation from the public and private sector, as well as the academic community and civil society. OCP is expected to serve as a means of sustainable and constructive dialogue between sectors, to help articulate a shared vision for the future weather and climate companies in the next decade and wider (WMO 2021).

16.4.4 Establishment of Modern Centers for Forecasting and Warning of Extreme Weather Events

Climate changes and the increased frequency and strength of extreme weather events require the establishment and functioning of independent centers for forecasting and announcement of multilevel hazards. Using the technical capabilities and advanced systems available as well as the experience and skill of experts, warnings of adverse weather impacts will be prepared and issued as far in advance as possible. In order to prevent adverse weather situations before they intensify and turn into a disaster, these modern centers operate in the capacity of a system for forecasting and announcing a multilevel danger from adverse weather impacts, which will require close cooperation with the competent institutions and structures for dealing with disasters from national to local level.

Multilevel hazards include floods, strong winds, torrential rains, storms, fires, air pollution, and other adverse weather, hydrological, and environmental impacts. In this way, an operational system for weather forecasts and warnings will be established with a special emphasis on the multilevel presentation of the weather and announcement. A multilevel hazard reporting system would contain three components: government management, institutional coordination, and community participation. The government would establish the necessary legal and regulatory frameworks for coordination and cooperation between government agencies, institutions, and local self-government. When it comes to a multilevel system for early warning, the national regulations defined in the law on hydrometeorological activity are also implemented, where the roles, responsibilities, and competencies of the departments are clearly defined and their operational functions, as well as the protocol for communication with those responsible in the state. It would be constituted of two components: organizational and technical

A failure in one of the components or a lack of mutual coordination between them can lead to the failure of the entire system. Issuing the warnings is a national responsibility, so the roles and responsibilities of the various public and private stakeholders for the implementation of this system should be clarified and reflected in state-to-local regulatory frameworks, planning, budgeting, coordination, and operational mechanisms. There are several European and international systems for forecasting, and early warning of adverse weather phenomena (e.g., European Meteorological Services (EUMETNET), European flood warning systems—"EFAS"). There are also links to global versions of these systems for early warning of extreme weather events. Here the global flood information system GloFAS (Alfieri et al. 2013) is linked with a state-of-the-art web platform system to enable current and future flood information in a global framework and impact assessment based on sophisticated deterministic, ensemble, and seasonal forecasts of the European Centre for Medium-Range Forecasting of Time. Regarding tropical cyclones, there are regional meteorological centers for forecasting and warning of tropical cyclones (we will not list them all) such as NOAA National Hurricane Centre, Joint Typhoon Warning Centre (JTWC), PAGASA Tropical Advisory Committee, and the Global Disaster Warning and Coordination System with online access (GDACS).

All existing and newly developed systems (and those not mentioned here) are aimed at monitoring, reliable forecasts, and announcements of extreme weather events for a given geographic area to provide timely public information to reduce risks. Novel Thunderstorm Alert System—"NOTHAS" (see Spiridonov et al. 2021, 2023) can be classified as a newly developed tool that could have a consultative role

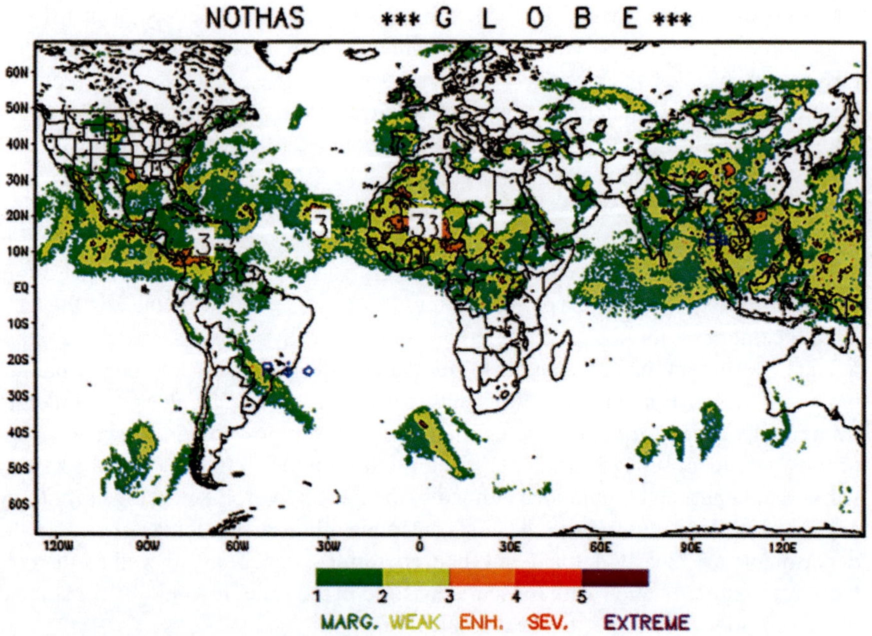

Fig. 16.7 NOTHAS Global-scale alert of severe convective weather events

in the process of disseminating information about the probability of occurrence of a certain category of risk from severe convective weather events (Fig. 16.7).

16.4.5 Application of Mobile Technology in the Modern Weather Display

For a longer period of time, observations have been a key component in weather forecasting. More recently, satellites and supercomputers are used for data acquisition. However, the next advancement in weather forecasting may come from smartphones and computers (Phan et al. 2018; Stewart and Bolton 2023).

One of the biggest revolutions in weather forecasting is literally in your hands: your cell phones or smartphones. Mobile phones—especially smartphones, due to their functionality, flexibility, faster communication—have become essential in our everyday life, from checking Facebook to using e-mail and many other applications.

Users have access to a lot of information that was previously restricted to using computers and the Internet. Now, researchers believe these phones could play an integral part in the collection of weather data—data that could one day improve weather forecasts. Researchers and developers of many smartphone applications have discovered a way to correlate smartphone battery temperature with the air

Fig. 16.8 Application of mobile technology in modern weather forecasting

temperature at a user's location. But temperature is not the only weather variable that many smartphones can sense.

A number of devices also contain pressure sensors. Smartphone pressure readings are believed to have even more potential for use in weather forecasting than temperature (Fig. 16.8). The rapid development of mobile technology will also enable the development of applications for the use of other weather variables, as a contribution to meteorology, and the improvement of short-term and localized weather forecasts. By using mobile weather services, you can get the latest forecasts directly on your mobile phone.

In order to use this advanced innovative mobile technology, in the context of weather information, the objective conditions and opportunities for developing this technology in our country should primarily be perceived. In that direction, communication with the management and design team of the mobile operators should be considered, regarding the definition of the joint platform for cooperation and creation of a business plan. Based on the real possibilities and resources available to the hydrometeorological service, the method and procedure for the realization of this project would be determined. This would use certain available resources from the service in terms of data, IT support, hardware, software, output data, and products from the numerical weather forecast model and current data from the network of automatic stations, radar, and even satellite data. Through the development of this application, numerical information and a symbolic representation of the weather would be obtained depending on the platform of mobile phones or smartphones. To begin with, weather information would be provided including a display of the weather forecast and announcement for all European cities up to 5 days in advance, as well as detailed weather information for each city up to 2 days in advance every hour. Among most of the products, users will have access to other information and forecasts about air temperature, relative humidity, precipitation, wind direction, and speed at any time, as well as warnings about adverse weather effects.

16.4.6 *Weather Forecasting and Artificial Intelligence (Machine Learning)*

As in many other scientific fields, the proliferation of tools like artificial intelligence and machine learning promises long-term weather forecasting (Bochenek and Ustrnul 2022). We've seen some of what's possible in our research on applying machine learning to high-impact weather forecasting. But we also believe that while these tools open new possibilities for better forecasts, many parts of the work are handled more skillfully by experienced people.

Today, forecasters' primary tools are numerical weather forecasting models. These models use observations of the current state of the atmosphere from sources such as weather stations, weather balloons, and satellites and solve equations that govern air movement. These models are excellent at predicting most weather systems, but the smaller the weather event, the more difficult it is to predict. As an example, think of a thunderstorm that dumps heavy rain on one side of a city and nothing on the other. Furthermore, experienced forecasters are remarkably good at synthesizing the vast amounts of weather information they must consider each day, but their memories and bandwidth are not infinite.

Artificial intelligence and *machine learning* (Fig. 16.9) can help with some of these challenges. Forecasters now use these tools in several ways, including predicting high-impact weather that models cannot provide. Researchers are also incorporating machine learning into numerical weather prediction models to speed up tasks that can be computationally intensive, such as predicting how water vapor turns into rain, snow, or hail.

It is possible that machine learning models will eventually replace traditional numerical weather forecasting models. Instead of solving a set of complex physical equations as models do, these systems will instead process thousands of past weather maps to learn how weather systems behave. Then, using the current weather data, they would make weather predictions based on what they learned from the

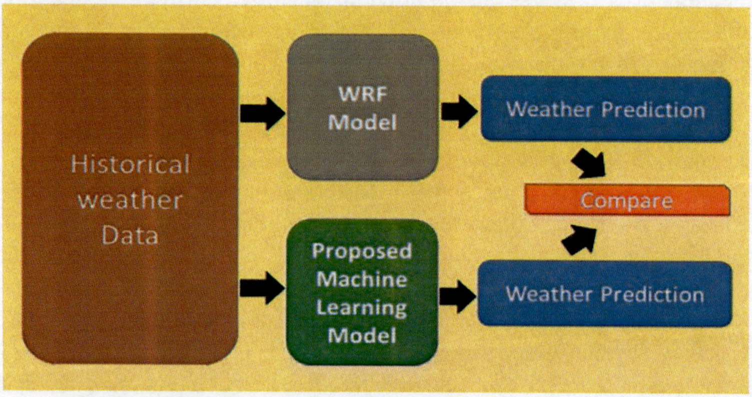

Fig. 16.9 Numerical weather forecast based on historical data and machine learning

past. Some studies have shown that machine learning-based forecast systems can predict general weather patterns as well as numerical weather prediction models while using only a fraction of the computing power required by the models. These new tools don't yet predict the local weather details that people care about, but as many researchers carefully test them and invent new methods, there is a promise for the future.

Unlike numerical weather prediction models, forecast systems using machine learning are not constrained by the physical laws that govern the atmosphere. So, it is possible that they will give unrealistic results—for example, predicting extreme temperatures outside the limits of nature. And it is unclear how they will perform in extremely unusual or unprecedented weather phenomena. Ideally, AI and machine learning will enable human forecasters to do their jobs more efficiently, spending less time generating routine forecasts and more time communicating the implications and impacts of forecasts to the public—or, for private forecasters, to their clients. We believe that careful collaboration between scientists, forecasters, and forecast users is the best way to achieve these goals and build confidence in machine-generated weather forecasts.

16.4.7 Quantum Weather Forecast

The preparation of a detailed weather forecast requires the analysis of a huge amount of data containing several dynamic variables, such as air temperature, pressure, and density. However, there are limitations to using conventional computers—even supercomputers—in developing numerical models for weather and climate prediction. Also, the process of analyzing weather data using traditional computers may not be fast enough to keep up with the ever-changing weather conditions, nonlinear processes in the atmosphere, and forecast uncertainty arising from the initial data.

Even local weather forecasts, which are constantly evolving rapidly, can benefit from improved forecasts. Take, for example, thunderstorms, where highly accurate and advanced forecasting through improved data analytics can minimize the damage caused, as there can be advance warning of potential power outages and increased preparedness, allowing the local community to respond more quickly to restore power. So how then can weather forecasting be improved both locally and globally?

Quantum machine learning *or quantum weather computing* (Fig. 16.10) has the potential to advance conventional numerical methods to improve weather monitoring and forecasting by efficiently and rapidly handling massive amounts of data containing many variables, harnessing the computing power of qubits, and using quantum-inspired optimization algorithms.

In fact, improving weather forecasting using quantum computing should become a reality in the not-so-distant future. The development of such technology may eventually be suitable for weather forecasting (Frolov 2017; Tennie and Palmer 2023).

Fig. 16.10 Quantum weather forecast opens the horizons of our future

Quantum computing will also benefit local long-term weather forecasts for more advanced and accurate warnings of extreme weather events, potentially saving lives and reducing property damage every year.

References

Alfieri, L., P. Burek, E. Dutra, B. Krzeminski, D. Muraro, J. Thielen, and F. Pappenberger. 2013. GloFAS – Global ensemble streamflow forecasting and flood forecasting. *Hydrology and Earth System Sciences* 17 (3): 1161–1175. https://doi.org/10.5194/hess-17-1161-2013.

Beran, D.W. 1997. Operational use of ground-based remote sensors: a review. WMO/TD-No. 860. https://library.wmo.int/idurl/4/41837

Bochenek, B., and Z. Ustrnul. 2022. Machine learning in weather prediction and climate analyses – Applications and perspectives. *Atmosphere* 13: 180. Bulletin of the American Meteorological Society, 2018, ff10.1175/BAMS-D-17-0125.1ff. final-02395934.

Byers, H.R. 1959. Carl-Gustaf Rossby, the organizer. In *The atmosphere and the sea in motion*, ed. B. Bolin, 51–59. Rockefeller Institute Press.

Cools, J. 2002. *Air pollution*. 2nd ed, 560. London: Clay's Library of Health and the Environment. Spon Press.

Ćurić, M. 2014. Dynamic Meteorology, AMG Book, Belgrade, 422 p.

De Jong, et al. 2007. The use of multi-band transmission data collecting at Scripps pier in November 2006. In *Proceedings of SPIE*, vol. 6708L.

Eberhard, W.L., R.E. Cupp, and K.R. Healy. 1989. Doppler lidar measurement of profiles of turbulence and momentum flux. *Journal of Atmospheric and Oceanic Technology* 6: 809–819. https://doi.org/10.1175/1520-0426(1989)006<0809: DLMOPO>2.0.CO;2.

Frolov, A.V. 2017. Can a quantum computer be applied for numerical weather prediction? *Russian Meteorology and Hydrology* 42: 545–553. https://doi.org/10.3103/S1068373917090011.

Harding, A.F. 1982. *Climate change in later pre-history*. Edinburgh University Press, Bintliff, J. L.

Holton, J.R. 2004. An introduction to dynamic meteorology", Elsevier Academic Press, Burlington, MA, pp. 535.

Hoskins, B.J., I. Draghici, and H.C. Davies. 1978. A new look at the ω-equation. *Quarterly Journal of the Royal Meteorological Society* 104(439): 31–38.

Imbrie, J., and K.P. Imbrie. 1994. *Ice ages, solving the mystery*. Harvard University Press.

Kahn, et al. 2004. Aerosol data sources and their roles within PARAGON. *Bulletin of the American Meteorological Society* 85: 1511–1522.

Kaufman, Y., D. Tanré, and O. Boucher. 2002. A satellite view of aerosols in the climate system. *Nature* 419: 215–223.

Maynard, R.H. 1945. Radar and weather, Journal of Metals 2: 214–226.

Mintz, Y. 1975. Jacob Bjerknes and our understanding of the atmosphere's general circulation. In *Western periodicals*, Select Papers of Jacob Aall Bonnevie Bjerknes, ed. M.G. Wurtele, 14–15.

Palmer, T. 2018. The ECMWF ensemble prediction system: Looking back (more than) 25 years and projecting forward 25 years. *Quarterly Journal of the Royal Meteorological Society* 145: S1.

Phan, M.D., B.E. Montz, S. Curtis, and T.M. Rickenbach. 2018. Weather on the go: An assessment of smartphone mobile weather application use among college students. *Bulletin of the American Meteorological Society* 99: 2245–2257.

Raghavan, S. 2003. Radar meteorology – atmospheric and oceanographic sciences library, Springer, 563 p. ISBN: 1402016042.

Sanders, F., and B.J. Hoskins. 1990. An easy method for estimation of Q-vectors from weather maps. *Weather and Forecasting* 5: 346–353.

Spiridonov, V., M. Curic, N. Sladic, et al. 2021. Novel thunderstorm alert system (NOTHAS). *Asia-Pacific Journal of Atmospheric Sciences* 57: 479–498. https://doi.org/10.1007/s13143-020-00210-5.

Spiridonov, V., M. Grcić, N. Sladić, et al. 2023. The capability of NOTHAS in the prediction of extreme weather events across different climatic areas. *Acta Geophysica*. https://doi.org/10.1007/s11600-023-01122-4.

Stewart, Alan E., and Matthew J. Bolton. 2023. Digital weather information in an embodied world. *Informatics* 10 (1): 13. https://doi.org/10.3390/informatics10010013.

Tennie, F., and T.N. Palmer. 2023. Quantum computers for weather and climate prediction: The good, the bad, and the noisy. *Bulletin of the American Meteorological Society* 104: E488–E500. https://doi.org/10.1175/BAMS-D-22-0031.1.

WMO. 2021. *Future of weather and climate forecasting*. Public-Private Engagement Publication No. 3. ISBN:978-92-63-11263-7.

Yano, J.-I., M.Z. Ziemiański, M. Cullen, P. Termonia, J. Onvlee, et al. 2019. Scientific challenges of convective-scale numerical weather prediction. *Bulletin of the American Meteorological Society* 99 (4): 699–710.

Zuev, V.E., V.V. Zuev, and B.S. Kostin. 1992. Optika Atmosfery i Okeana (ISSN 0869-5695). 5(10): 1076–1080.

Summary

The content consists of three main parts. The first part focuses on the general development of civilization, while the second part delves into the advancement of science. The third part provides a detailed account of the evolution of meteorology as a scientific discipline.

The sequence begins with an exploration of the earliest stages of planet Earth and its atmosphere. It then traces the early progress of meteorology, encompassing the earliest weather records, the teachings of ancient Greeks, Aristotle's contributions to meteorology, and the state of meteorology during the Dark Ages. A significant portion of the content is dedicated to the origins of quantitative meteorology, the invention of fundamental meteorological instruments, and the establishment of meteorological measurements and measurement networks. This section also covers the modernization of meteorological instruments, such as radar, satellites, and remote sensing technologies.

This book spans the development of meteorology from its early empirical applications for agriculture and navigation to the evolution of instruments, the establishment of the first meteorological networks and textbooks, and the emergence of semiempirical forecasting methods utilizing Bjerknes cyclone "model" and weather data. It further explores the advent of numerical weather forecasting following the invention of computers and continues into the era of weather satellites and beyond. Each of these milestones introduced new possibilities to the field and pushed its boundaries. This book also introduces the reader to the key figures who played pivotal roles in advancing meteorology.

M. Ćurić, V. Spiridonov, *History of Meteorology*,
https://doi.org/10.1007/978-3-031-45032-7

Index

M. Ćurić, V. Spiridonov, *History of Meteorology*,
https://doi.org/10.1007/978-3-031-45032-7